Approximation of Hilbert space operators

VOLUME I

Domingo A Herrero

Arizona State University

Approximation of Hilbert space operators

VOLUME I

Pitman Advanced Publishing Program

BOSTON · LONDON · MELBOURNE

PITMAN BOOKS LIMITED
128 Long Acre, London WC2E 9AN

PITMAN PUBLISHING INC
1020 Plain Street, Marshfield, Massachusetts

Associated Companies
Pitman Publishing Pty Ltd, Melbourne
Pitman Publishing New Zealand Ltd, Wellington
Copp Clark Pitman, Toronto

First published 1982

AMS Subject Classifications: Primary 47A55, 41A65, 47A60;
Secondary 47A15, 47A53, 81C12

British Library Cataloguing in Publication Data

Herrero, Domingo A.
Approximation of Hilbert space operators.
Vol. 1—(Research notes in mathematics; 72)
1. Hilbert space 2. Operator theory
I. Title II. Series
515.7'33 QA329

ISBN 0-273-08579-4

Library of Congress Cataloging in Publication Data

Herrero, Domingo A.
Approximation of Hilbert space operators.

(Research notes in mathematics; 72–)
Bibliography: v. 1, p.
Includes index.
1. Operator theory. 2. Hilbert space. I. Title.
II. Series: Research notes in mathematics; 72, etc.
QA329.H48 1982 515.7'24 82-10163
ISBN 0-273-08579-4 (v. 1)

Reproduced and printed by photolithography
in Great Britain by Biddles Ltd, Guildford

To Buenos Aires,
on her four-hundred first birthday

"A mí se me hace cuento que empezó Buenos Aires,
la juzgo tan eterna como el agua o el aire"

(Jorge Luis Borges)

Contents

Preface

The last decade has been fruithful in results on approximation of Hilbert space operators, due to a large extent to the impulse given by Paul R. Halmos in his famous survey article "Ten problems in Hilbert space".

The purpose of this monograph (and a second one, by C. Apostol, L. A. Fialkow, D. A. Herrero and D. Voiculescu that will follow and complete the results contained here) is to provide a set of general arguments to deal with approximation problems (in the norm-topology) related to those subsets of the algebra $L(H)$ of all operators acting on a complex separable infinite dimensional Hilbert space that are invariant under similarities.

Many interesting subsets of $L(H)$ have this property: nilpotent operators; algebraic operators (satisfying a fixed polynomial); polynomially compact operators; triangular, quasitriangular and biquasitriangular operators; cyclic and multicyclic operators; semi-Fredholm operators (with fixed given indices); operators whose spectrum is equal to a fixed compact subset of the complex plane $\mathbb{C}$, or whose spectra are contained in a fixed nonempty subset of $\mathbb{C}$; any bilateral ideal of compact operators, etc, etc.

The following list illustrates the kinds of problems to be considered here:

a) Given a subset R of $L(H)$ invariant under similarities, defined in algebraic, geometric or analytic terms (e.g., the set of all algebraic operators, the set of all operators T such that T^3 is compact, the set of all cyclic operators), characterize its norm-closure in "simple terms". Since the spectrum and its different parts are the most obvious similarity invariants of an operator, these "simple terms" will usually be expressed in terms of properties of the different subsets of the spectra of the operators in the closure of R.

b) More generally, obtain a formula for the distance from a given operator to R or, al least, upper and/or lower estimates for this distance.

c) In a surprisingly large number of interesting cases, either R is invariant under compact perturbations, or its closure is contained in (or equal to) the set of all compact perturbations of R. Compact perturbations will be used as a useful tool for approximation and will be also analyzed with respect to the above mentioned peculiar properties.

d) Analogues to the problems in a) and b) for subsets of the quotient Calkin algebra.

In Chapter I, we shall obtain all the necessary conditions for approximation that can be easily derived from the Riesz-Dunford functional calculus and the well-known stability properties of semi-Fredholm operators. This chapter is followed by "an apéritif": the solution of several approximation problems in finite dimensional spaces, which only depend on the results of Chapter I and some "handwork" with matrices.

It is interesting to observe that, for a large number of approximation problems (but, unfortunately, not for all of them) the "obvious" necessary conditions derived from the results of Chapter I are actually sufficient, but the proofs of their sufficiency are very hard. These proofs are constructive to a large extent and the "tools" for these constructions are developed in Chapters III and IV: Rosenblum's corollary, Rota's universal model and extensions, Apostol's triangular representation, results on compact perturbations and results "borrowed" from the theory of C*-algebras (the Brown-Douglas-Fillmore theorem, Voiculescu's theorem, results on closures of unitary orbits). Except for those results related to the theory of C*-algebras (Chapter IV), the monograph is essentially self-contained.

Results on approximation of Hilbert space operators really begin in Chapter V with the characterization (due to C. Apostol, C. Foiaş and D. Voiculescu) of the closure of the set of all nilpotent operators. The closure of the set of all algebraic operators is then obtained as a corollary and this result is used to characterize biquasitriangularity, the closure of the similarity orbit of a normal operator with perfect spectrum and to give several results about the closure of the similarity orbit of an arbitrary operator.

Combining the above results with Apostol-Morrel "simple models", we shall obtain the Apostol-Foiaş-Voiculescu theorem on the spectral characterization of quasitriangularity. Since algebraic operators and operators that "look like backward shifts" are the simplest examples of triangular operators, this approach to quasitriangularity is, perhaps, more natural than the original one.

Finally, the last two chapters are devoted to a deep analysis of the structure of a polynomially compact operator and to the closure of its similarity orbit, respectively.

A large part of the material contained in the last chapter consists of unpublished results of C. Apostol-D. Voiculescu, J. Barría-D. A. Herrero and D. A. Herrero on closures of similarity orbits of essentially nilpotent operators.

The author is deeply indebted to Professors Constantin Apostol, I. David Berg, Charles A. Berger, Ronald G. Douglas, Alain Etcheberry, Lawrence A. Fialkow, Ciprian Foiaş, Carl M. Pearcy, Allen L. Shields and Dan Voiculescu, and to his professors and ex-fellow students from the University of Buenos Aires, Alejandro de Acosta, Mischa Cotlar, Beatriz Margolis, Lázaro Recht, Norberto Salinas and, very especially, to his wife Marta B. Pecuch de Herrero for infinitely many informal discussions and suggestions and, most important, for their friendly support during all these hard years.

The contents of this monograph have been developed in a "Seminar on Approximation of Hilbert Space Operators" at the University of Georgia (Athens, Georgia, USA), during the academic year 1980-1981. The author wishes to thank the authorities of the University of Georgia for their support and to his colleagues Edward Azoff, Richard Bouldin, Kevin Clancey, Douglas N. Clark and Derming Wang for their assistance during the preparation of the manuscript. Many of their valuable observations have been included here.

Domingo A. Herrero

Tempe, Arizona
June, 1981

1 Stability and approximation

As explained in the Preface, in "most" approximation problems related to similarity-invariant sets of operators, the "obvious" necessary conditions for approximation that can be derived from the Riesz-Dunford functional calculus and the stability properties of semi-Fredholm operators turn out to be also sufficient. In a certain sense, this short first chapter contains all the necessary conditions and the remaining of the monograph (and the second one that will follow [16]) is devoted to explain why these necessary conditions are also sufficient.

Throughout this monograph, the word *operator* will always denote a bounded linear transformation mapping a complex Banach space into another. If X and Y are complex Banach spaces, the Banach space of all operators mapping X into Y will be denoted by $L(X,Y)$. We shall write $L(H)$ for the Banach algebra $L(H,H)$, where H is a complex separable Hilbert space. Unless otherwise stated, H (H_o, H_1, H_2,..., etc) will always denote an infinite dimensional Hilbert space.

The algebra $L(H)$ contains the open subset

$$G(H) = \{W \in L(H): \ W \text{ is invertible}\}$$

(the linear group of $L(H)$), which will play a very important role here.

A subset R of $L(H)$ is called *invariant under similarities* (or *similarity-invariant*) if it is invariant under conjugation by elements of the group $G(H)$, i.e.,

$$T \in R \Rightarrow S(T) \subset R,$$

where

$$S(T) = \{WTW^{-1}: \ W \in G(H)\}$$

is the *similarity orbit* of T.

If $K(H)$ denotes the ideal of all compact operators acting on H and $\pi: L(H) \to A(H) = L(H)/K(H)$ is the canonical projection of $L(H)$ onto the (quotient) Calkin algebra, then the image $\pi(T) = T+K(H)$ of $T \in L(H)$ in $A(H)$ will also be denoted by $\tilde{T}$.

The reader is referred to [77], [119] for the general theory of Hilbert space operators.

1.1 Lower estimates derived from the Riesz-Dunford functional calculus

A nonempty bounded open subset Ω of the complex plane $\mathbb{C}$ is a *Cauchy domain* if the following conditions are satisfied: (i) Ω has finitely many components, the closures of any two of which are disjoint, and (ii) the boundary $\partial\Omega$ of Ω is composed of a finite positive number of closed rectifiable Jordan curves, no two of which intersect. In this case, $\Gamma = \partial\Omega$ will be assumed to be positively oriented with respect to Ω in the sense of complex variable theory, i.e., so that

$$\frac{1}{2\pi i}\int_\Gamma \frac{d\lambda}{\lambda-\zeta} = \begin{cases} 1, \text{ if } \zeta \in \Omega \\ 0, \text{ if } \zeta \notin \Omega^- = \Omega\cup\Gamma \end{cases}$$

(the upper bar will always denote closure with respect to the metric topology of the underlying space). Clearly, Γ is uniquely determined by Ω (and conversely). We shall say that Γ is a *rectifiable contour*. If all the curves of Γ are regular analytic Jordan curves, we shall say that Γ is an *analytic contour* (or Ω is an *analytic Cauchy domain*).

If A is a Banach algebra with identity 1 and $a \in A$, the *spectrum* of a will be denoted by $\sigma(a)$. The complement $\rho(a) = \mathbb{C}\backslash\sigma(a)$ of $\sigma(a)$ in the complex plane is the *resolvent set* of a and the function $\lambda \to (\lambda-a)^{-1}$ (from $\rho(a)$ into A) is the *resolvent* of a. It is well-known that $(\lambda-a)^{-1}$ is an analytic function of λ in that domain that it satisfies the *first resolvent equation:*

$$(\lambda-a)^{-1} - (\mu-a)^{-1} = (\mu - \lambda)(\lambda-a)^{-1}(\mu-a)^{-1} \quad (\lambda,\mu \in \rho(a)).$$

Furthermore, if a, b $\in A$ and $\lambda \in \rho(a)\cap\rho(b)$, then

$$(\lambda-a)^{-1} - (\lambda-b)^{-1} = (\lambda-a)^{-1}(a - b)(\lambda-b)^{-1}$$

(*second resolvent equation*).

If σ is a nonempty clopen subset of $\sigma(a)$, then there exists an analytic Cauchy domain Ω such that $\sigma \subset \Omega$ and $[\sigma(a)\backslash\sigma]\cap\Omega^- = \emptyset$; in this case

$$E(\sigma;a) = \frac{1}{2\pi i}\int_\Gamma (\lambda-a)^{-1}\, d\lambda$$

is an idempotent of A commuting with every b in A such that ab = ba. ($E(\sigma;a)$ is the *Riesz idempotent* corresponding to σ.)

The following theorem is just a quantitative form of the classical result on upper semi-continuity of separate parts of the spectrum. The reader is referred to [76], [153], [172], or [173,Chapter XIV] for the basic properties of the Riesz-Dunford functional calculus.

THEOREM 1.1. *Let* a *and* b *be two elements of the Banach algebra* A

with identity 1. *Assume that the spectrum* $\sigma(a)$ *of* a *is the disjoint union of two compact subsets* σ_0 *and* σ_1 *such that* σ_1 *is nonempty* $(\sigma_1 \neq \emptyset)$ *and let* Ω *be a Cauchy* ***domain*** *such that* $\sigma_1 \subset \Omega$ ***and*** $\sigma_0 \cap \Omega^- = \emptyset$. *If*

$$\|a - b\| < \min\{\|(\lambda-a)^{-1}\|^{-1}: \lambda \in \partial\Omega\},$$

(where $\|.\|$ *denotes the norm of* A*)* *then* $\sigma(b) \cap \Omega \neq \emptyset$ *and* $\sigma(b) \cap \partial\Omega = \emptyset$.

PROOF. Assume that $\|a-b\| = \delta < m = \min\{\|(\lambda-a)^{-1}\|^{-1}: \lambda \in \partial\Omega\}$ and let $a_t = (1-t)a+tb$, $0 \le t \le 1$; then

$$\|(\lambda-a_t) - (\lambda-a)\| = \|a - a_t\| = t\delta < \|(\lambda-a)^{-1}\|^{-1}$$

for all $\lambda \in \partial\Omega$, so that $\sigma(a_t) \cap \partial\Omega = \emptyset$ for all $t \in [0,1]$.

Thus, the idempotent

$$e_t = \frac{1}{2\pi i}\int_{\partial\Omega} (\lambda-a_t)^{-1}\, d\lambda$$

is a well defined element of A for all $t \in [0,1]$. Furthermore, if $0 \le t < s \le 1$, the second resolvent equation implies that

$$\|(\lambda-a_t)^{-1}-(\lambda-a_s)^{-1}\| \le \|(\lambda-a_t)^{-1}\|.|t-s|.\|(\lambda-a_s)^{-1}\| \le (m-\delta)^{-2}|t-s|,$$

whence it readily follows that $t \to e_t$ is a continuous mapping from $[0,1]$ into A.

Since $\sigma_1 = \sigma(a) \cap \Omega \neq \emptyset$, it follows that

$$e_0 = \frac{1}{2\pi i}\int_{\partial\Omega} (\lambda-a)^{-1}\, d\lambda \neq 0,$$

so that $\|e_0\| \ge 1$ (Recall that e_t is idempotent, $0 \le t \le 1$). By continuity, $\|e_1\| \ge 1$ and therefore

$$e_1 = \frac{1}{2\pi i}\int_{\partial\Omega} (\lambda-b)^{-1}\, d\lambda \neq 0.$$

This is clearly impossible, unless $\sigma(b) \cap \Omega \neq \emptyset$. □

Recall that if (X,d) is a metric space and $\mathcal{B}(X)$ $(\mathcal{B}_c(X))$ is the family of all nonempty bounded (closed bounded, respectively) subsets of X, then

$$d_H(A,B) = \inf\{\varepsilon > 0: B \subset A_\varepsilon, A \subset B_\varepsilon\},$$

where $A_\varepsilon = \{x \in X: \operatorname{dist}[x,A] \le \varepsilon\}$, defines a pseudometric in $\mathcal{B}(X)$ (a metric in $\mathcal{B}_c(X)$, resp.); $d_H(A,B)$ is the *Hausdorff distance* between A and B, A, B $\in \mathcal{B}_c(X)$.

The qualitative form of Theorem 1.1 is the following

COROLLARY 1.2. *(i) Let* a *be an element of a Banach algebra* A *with identity. Assume that* $\sigma(a)$ *is the disjoint union of two compact subsets* σ_0 *and* σ_1 *such that* $\sigma_1 \neq \emptyset$, *and that* σ_1 *is contained in a bounded open set* Ω. *Then there exists a* ***constant*** $C = C(a,\sigma_1,\Omega) > 0$ *such that*

$\sigma(b) \cap \Omega \neq \emptyset$ *for all* b *in* A *satisfying* $\|a - b\| < C$.

(ii) Given a *in* A *and* $\varepsilon > 0$, *there exists* $\delta > 0$ *such that* $\sigma(b) \subset \sigma(a)_\varepsilon$, *provided* $\|a - b\| < \delta$, *i.e., the mapping* $a \to \sigma(a)$ *from* A *into* $\mathcal{B}_c(\mathbb{C})$ *(Hausdorff distance) is upper semicontinuous.*

PROOF. (i) follows immediately from Theorem 1.1: Given σ_1 and Ω (as above), there exists a Cauchy domain Ω_1 such that $\sigma_1 \subset \Omega_1 \subset \Omega_1^- \subset \Omega$ and $\sigma_0 \cap \Omega_1^- = \emptyset$. Take $C = \min\{\|(\lambda - a)^{-1}\|^{-1}: \lambda \in \partial\Omega_1\}$.

(ii) Apply (i) with $\sigma_0 = \emptyset$ and Ω = interior $\sigma(a)_\varepsilon$. □

COROLLARY 1.3. *(i) If* Ω *is an open subset of* $\mathbb{C}$, *then* $\{a \in A: \sigma(a) \subset \Omega\}$ *is an open subset of* A.

(ii) If Σ *is a* G_δ *subset of* $\mathbb{C}$, *then* $\{a \in A: \sigma(a) \subset \Sigma\}$ *is a* G_δ *subset of* A. *In particular, the set* $\{a \in A: \sigma(a) = \{0\}\}$ *of all quasinilpotent elements of* A *is a* G_δ *in* A.

PROOF. (i) follows from Corollary 1.2(ii) and (ii) is an immediate consequence of (i). □

The limit case of Theorem 1.1 yields the following

COROLLARY 1.4. *Let* a, σ_1 ($\sigma_1 \neq \emptyset$) *and* Ω *be as in Theorem 1.1 and assume that* $b \in A$ *satisfies the inequality*

$$\|a - b\| \leq \min\{\|(\lambda - a)^{-1}\|^{-1}: \lambda \in \partial\Omega\};$$

then $\sigma(b) \cap \Omega^- \neq \emptyset$.

PROOF. Define a_t as in the proof of Theorem 1.1; then Theorem 1.1 implies that $\sigma(a_t) \cap \partial\Omega = \emptyset$ and $\sigma(a_t) \cap \Omega \neq \emptyset$ for $0 \leq t < 1$.

Since $\lim(t \to 1) \|b - a_t\| = 0$, it follows from Corollary 1.2(i) that $\sigma(b) \cap \Omega^-$ cannot be empty. □

It is convenient to observe that the result of Corollary 1.4(and, a fortiori, the result of Theorem 1.1 too) is sharp. In fact, we have

EXAMPLE 1.5. Let $P \in L(H)$ be a non-zero orthogonal projection and let $\Omega = \{\lambda: |\lambda - 1| < \frac{1}{2}\}$; then $\|(\lambda - P)^{-1}\|^{-1} = \frac{1}{2}$ ($\lambda \in \partial\Omega$). Thus, by Corollary 1.4, $\sigma(B) \cap \Omega^- \neq \emptyset$ for all B in $L(H)$ such that $\|P - B\| \leq \frac{1}{2}$.

On the other hand, if $A = \frac{1}{2}$, then $\|P - A\| = \frac{1}{2}$ and $\sigma(A) = \{\frac{1}{2}\} \subset \partial\Omega$.

If $T \in L(H)$, σ is a clopen subset of $\sigma(T)$ and $E(\sigma;T)$ is the corresponding Riesz idempotent, then the *range* ran $E(\sigma;T)$ and the *kernel* ker $E(\sigma;T)$ of $E(\sigma;T)$ are subspaces of H invariant under every B in $L(H)$ commuting with T (i.e., *hyperinvariant* for T), and H can be written as the algebraic (not necessarily orthogonal!) direct sum $H = \operatorname{ran} E(\sigma;T) \dotplus \ker E(\sigma;T)$. (Here and in what follows, *subspace* will always denote a closed linear manifold of a Banach space.) Furthermore, the spectrum of the restriction $T|\operatorname{ran} E(\sigma;T)$ of T to $\operatorname{ran} E(\sigma;T)$ coincides with σ and the spectrum of the restriction $T|\ker E(\sigma;T)$ coincides with $\sigma(T)\setminus\sigma$ [173,Chapter XIV].

In what follows, $\operatorname{ran} E(\sigma;T)$ will be denoted by $H(\sigma;T)$. If $\sigma = \{\lambda\}$ is a singleton, we shall simply write $H(\lambda;T)$ $(E(\lambda;T))$ instead of $H(\{\lambda\};T)$ $(E(\{\lambda\};T)$, resp). If $\sigma = \{\lambda\}$ and $\dim H(\lambda;T)$ is finite, then λ is called a *normal eigenvalue* of T; in this case, $H(\lambda;T)$ coincides with $\ker(\lambda-T)^n$ for some $n \geq 1$. The set of all normal eigenvalues of T will be denoted by $\sigma_0(T)$. Clearly, $\sigma_0(T)$ is contained in the *point spectrum* $\sigma_p(T)$ of T (i.e., the set of all eigenvalues of T). The *essential spectrum* of T, i.e., the spectrum of $\tilde{T}$ in $A(H)$ will be denoted by $\sigma_e(T)$.

COROLLARY 1.6. *Let* A, B $\in L(H)$; *then*

(i) Assume that σ *is a nonempty (clopen) subset of* $\sigma(A)$ *and let* Ω *(a Cauchy domain) be a neighborhood of* σ *such that* $[\sigma(A)\setminus\sigma]\cap\Omega^- = \emptyset$. *If* $\|A-B\| < \min\{\|(\lambda-A)^{-1}\|^{-1}: \lambda\in\partial\Omega\}$, *then* $\sigma' = \sigma(B)\cap\Omega \neq \emptyset$;

(ii) furthermore, $\dim H(\sigma;A) = \dim H(\sigma';B)$ $(0 \leq \dim H(\sigma;A) \leq \infty)$.

(iii) If σ *is a nonempty clopen subset of* $\sigma_e(A)$ *and the Cauchy domain* Ω *is a neighborhood of* σ *such that* $[\sigma_e(A)\setminus\sigma]\cap\Omega^- = \emptyset$, *then* $\sigma_e(B)\cap\Omega \neq \emptyset$ *for all* B *in* $L(H)$ *such that* $\|\tilde{A}-\tilde{B}\| < \min\{\|(\lambda-\tilde{A})^{-1}\|^{-1}: \lambda\in\partial\Omega\}$.

PROOF. (i) and (iii) follow immediately from Theorem 1.1, applied to $A = L(H)$ and to $A = A(H)$, respectively.

(ii) This follows from the proof of Theorem 1.1. Observe that, if $A_t = (1-t)A+tB$, then the continuity of the mapping $t \to E(\sigma(A_t)\cap\Omega;A_t)$ $(0 \leq t \leq 1)$ implies that the idempotents $E(\sigma;A) = E(\sigma(A_0)\cap\Omega;A_0)$ and $E(\sigma';B) = E(\sigma(A_1)\cap\Omega;A_1)$ necessarily have the same (finite or infinite) rank.□

Until now, we have only applied the arguments of functional calculus to a very particular class of functions analytic in a neighborhood of the spectrum $\sigma(a)$ of an element of the Banach algebra A; namely, the characteristic function of a suitable neighborhood of a clopen subset of $\sigma(a)$. Analogous results hold in a much more general setting; namely,

PROPOSITION 1.7. *Let* a *be an element of the Banach algebra* A *with identity* 1 *and let* f *be an analytic function defined in a neighborhood* Ω *of* $\sigma(a)$. *Given* $\varepsilon > 0$, *there exists* $\delta > 0$ *such that* f(b) *is well-de-*

fined for all b *in* A *satisfying* $\|a - b\| < \delta$ *and, moreover,*

$$\|f(a) - f(b)\| < \varepsilon.$$

PROOF. Let Ω_1 be a Cauchy domain such that $\sigma(a) \subset \Omega_1 \subset \Omega_1^- \subset \Omega$. By Corollary 1.2(i), there exists $\delta_1 > 0$ such that $\sigma(b) \subset \Omega_1$ for all b in A satisfying $\|a - b\| < \delta_1$. Clearly, f(b) is well-defined for all these b, by means of the integral

$$f(b) = \frac{1}{2\pi i} \int_{\partial\Omega_1} f(\lambda)(\lambda-b)^{-1} d\lambda.$$

The second resolvent equation implies that

$$\max\{\|(\lambda-b)^{-1}\|: \lambda \in \partial\Omega_1\} < 1+\max\{\|(\lambda-a)^{-1}\|: \lambda \in \partial\Omega_1\},$$

and

$$\|f(a)-f(b)\| \le (1/2\pi).\mathrm{length}(\partial\Omega_1).\max\{|f(\lambda)|.\|(\lambda-a)^{-1}\| \times(1+\|(\lambda-a)^{-1}\|): \lambda \in \partial\Omega_1\}.\|a - b\|,$$

provided $\|a - b\| < \delta_2$ (for some δ_2, $0 < \delta_2 \le \delta_1$).

It follows that, if $\|a - b\| < \delta$ for a suitably chosen δ, $0 < \delta < \delta_2$, then f(b) is well-defined and $\|f(a) - f(b)\| < \varepsilon$. □

The following particular case is especially important for our purposes.

COROLLARY 1.8. *Let* $a \in A$, *let* μ *be an isolated point of* $\sigma(a)$ *and let*

$$f(\lambda) = \begin{cases} (\lambda-\mu)^k, & \text{in some neighborhood of } \mu \\ 0, & \text{in some neighborhood of } \sigma(a)\setminus\{\mu\}, \end{cases}$$

for some $k \ge 1$.

Let $\{a_n\}_{n=1}^{\infty}$ *be a sequence of elements in* A *such that* $\|a - a_n\| \to 0$ $(n \to \infty)$. *Then* $f(a_n)$ *is well-defined for all* n *large enough and*

$$f(a) = 0 \text{ if and only if } \lim(n \to \infty) \|f(a_n)\| = 0.$$

1.2 Lower estimates for the distance to $N_k(H)$

Let $N_k(A) = \{a \in A: a^k = 0\}$ denote the set of all nilpotent elements of order at most k (k = 1,2,...) of the Banach algebra A, and let $N(A) = \cup_{k=1}^{\infty} N_k(A)$ be the set of all nilpotents of A. In order to simplify the notation, the set of all nilpotents (of order at most k) in $L(H)$ will be denoted by $N(H)$ ($N_k(H)$, resp), or simply by N if H is understood.

It is a trivial consequence of Proposition 1.7 that, given a non-constant polynomial p, the set $\{a \in A: p(a) = 0\}$ is closed in A. In particular, if we choose $p_k(\lambda) = \lambda^k$ $(k = 1,2,\ldots)$, then we obtain

COROLLARY 1.9. *(i)* $N_k(A)$ *is closed in* A *for all* $k = 1,2,\ldots$.
(ii) $N(A)$ *is an* F_σ *subset of* A.

The following result provides a partial answer to the problem of estimating the distance from a given operator to $N_k(H)$.

PROPOSITION 1.10. *(i) Let* $A, B \in L(H)$. *If* $\|A^k x\| = \|x\| = 1$, *but* $B^k x = 0$, *for some* x *in* H *and some* $k \geq 1$, *then*

$$\|A - B\| \geq \max\{(\|A\|^k + 1)^{1/k} - \|A\|, (\|B\|^k + 1)^{1/k} - \|B\|\}.$$

(ii) If $\|A^k x\| = \|x\| = 1$, *but* $B^k x = 0$ *for some* x *in* H *and some* $k \geq 1$, *and* $\max\{\|A\|, \|B\|\} = M$, *then* $\|A - B\| \geq 1/(kM)$.

(iii) Let $a, b \in A$ *(a Banach algebra). If* $\|a^k\| = 1$, *but* $b^k = 0$, *for some* $k \geq 1$, *then*

$$\|a - b\| \geq \max\{(\|a\|^k + 1)^{1/k} - \|a\|, (\|b\|^k + 1)^{1/k} - \|b\|\}.$$

(iv) If $\|a^k\| = 1$, *but* $b^k = 0$, *for some* $k \geq 1$, *and* $\max\{\|a\|, \|b\|\} = M$, *then* $\|a - b\| \geq 1/(kM)$.

PROOF. (i) Let $\|A\| = M$ and $\|A - B\| = \delta$. It is completely apparent that $\|B\| \leq M+\delta$, and therefore

$$1 = \|A^k x\| = \|(A^k - B^k)x\| \leq \|A^k - B^k\| \leq \sum_{j=0}^{k-1} \|A^{k-j}B^j - A^{k-j-1}B^{j+1}\|$$

$$\leq \sum_{j=0}^{k-1} \|A\|^{k-j-1}\|A-B\| \cdot \|B\|^j \leq \delta \sum_{j=0}^{k-1} M^{k-j-1}(M+\delta)^j = \delta M^{k-1} \sum_{j=0}^{k-1} (1+\tfrac{\delta}{M})^j$$

$$= \delta M^{k-1}[(1+\tfrac{\delta}{M})^k - 1](\delta/M)^{-1} = (M+\delta)^k - M^k,$$

so that $(M+\delta)^k \geq M^k + 1$. Hence, $\delta \geq (M^k+1)^{1/k} - M$.

If we assume that $\|B\| = M$ and $\|A - B\| = \delta$, then $\|A\| \leq M+\delta$ and we arrive at the same inequality, whence the result follows.

The remaining statements follow by the same argument. □

REMARKS. 1.11. (i) Since, in Proposition 1.10(ii) and (iv), M cannot be smaller than 1 and (by the mean-value theorem)

$$1/(2kM) < (M^k+1)^{1/k} - M \leq 1/(kM),$$

for all $k \geq 1$, the estimates of (i) and (ii) ((iii) and (iv)) are of the same order.

(ii) The argument of the proof of Proposition 1.10(i) also applies to the case when $\|p(A)x\| = \|x\| = 1$, but $p(B)x = 0$, for some polynomial p, $p(\lambda) = \Pi_{j=1}^{m}(\lambda - \lambda_j)^{k_j}$. Since $\|[p(A) - p(B)]x\| = 1$, $\|A - B\|$ cannot be "too small", where the words "too small" have a concrete numerical expression in terms of p and A. The same applies, of course, to (ii), (iii) and (iv).

1.3 Lower semicontinuity of the rank

PROPOSITION 1.12. *Let* $A \in L(H)$ *and let* $\{A_n\}_{n=1}^{\infty}$ *be a sequence of operators such that* $\|A - A_n\| \to 0$ $(n \to \infty)$; *then*

(i) rank $A \leq \lim\inf(n \to \infty)$ rank A_n.

(ii) *If* $\|A_n x\| \geq \varepsilon\|x\|$ *for some* $\varepsilon > 0$ *and all* x *in a subspace* H_n *with* dim $H_n \geq d$, *then* $\|Ax\| \geq \varepsilon\|x\|$ *for all* x *in a subspace* H_o *with* dim $H_o \geq d$.

(iii) *If* $A_n \in K(H)$ *for all* $n \geq n_o$, *then* $A \in K(H)$.

PROOF. (i) If $\lim\inf(n \to \infty)$ rank $A_n = \infty$, then there is nothing to prove.

Assume that $\lim\inf(n \to \infty)$ rank $A_n = d < \infty$. Passing, if necessary, to a subsequence, we can directly assume that rank $A_n = d$ for all n = 1, 2,... . If rank $A \geq d+1$, then there exist d+1 linearly independent vectors $y_1, y_2, \ldots, y_{d+1} \in H$ such that $\{Ay_j\}_{j=1}^{d+1}$ is a linearly independent set. Clearly, Y = linear span$\{y_1, y_2, \ldots, y_{d+1}\}$ has dimension d+1 and therefore $Y \cap \ker A_n \neq \{0\}$ for each n = 1,2,... . It is easily seen that $\{A_n\}_{n=1}^{\infty}$ is a bounded sequence and there exists $y \in Y$ and a subsequence $\{A_{n_j}\}_{j=1}^{\infty}$ such that $\|Ay\| = 1$ but $\lim(j \to \infty)\|A_{n_j} y\| = 0$, so that $\{A_{n_j}\}_{j=1}^{\infty}$ cannot converge to A, even in the strong operator topology, a contradiction. Therefore, rank $A \leq d$.

(ii) Observe that our hypothesis implies that $\|A^* - A_n^*\| \to 0$ and $\|A^*A - A_n^*A_n\| \to 0$ $(n \to \infty)$. Now the result follows from an elementary analysis of the spectral decompositions of the hermitian operators A^*A and $A_n^*A_n$ (see, e.g., [117]).

The third statement is trivial. □

It is convenient to recall that $A \in L(H)$ is compact if and only if ran A does not contain an infinite dimensional subspace [71],[96]. We shall use the following non-standard notation:

If $A \in K(H)$ is not a finite rank operator, then rank A will be defined as $\infty-$. Thus, rank $T = \infty$ will be an equivalent way to say that T is not compact. The different possible ranks will be linearly ordered by

$$0 \lneqq 1 \lneqq 2 \lneqq \ldots \lneqq n \lneqq n+1 \lneqq \ldots \lneqq \infty- \lneqq \infty.$$

1.4 Stability properties of semi-Fredholm operators

Recall that $T \in L(H)$ is a *semi-Fredholm operator* if ran T is closed and either nul T = dim ker T or nul T^* = dim ker T^* = dim H/ran T is finite (where T^* denotes the adjoint of T in $L(H)$). In this case, the *index* of T is defined by

$$\text{ind } T = \text{nul } T - \text{nul } T^*.$$

The following theorem resumes the main properties of the semi-Fredholm operators. The reader is referred to [106] and [153, Chapter IV] for details.

THEOREM 1.13. *Let* $T \in L(H)$ *be a semi-Fredholm operator; then*

(i) T^* *is also a semi-Fredholm operator and* ind $T^* = -$ind T;

(ii) There exists a constant $\delta > 0$ *such that* $\|Tx\| \geq \delta\|x\|$ *for all* $x \perp \ker T$ *and* $\|T^*y\| \geq \delta\|y\|$ *for all* $y \perp \ker T^*$; *moreover,* δ *can be chosen as* $\min\{\lambda \in \sigma([T^*T]^{1/2})\setminus\{0\}\}$;

(iii) There exists $\delta = \delta(T) > 0$ *such that if* $A \in L(H)$, $\|A\| \leq 1$ *and* $|\lambda| < \delta$, *then* $T+\lambda A$ *is semi-Fredholm and* ind$(T+\lambda A)$ = ind T, nul$(T+\lambda A) \leq$ nul T *and* nul $(T+\lambda A)^* \leq$ nul T^*; *moreover,* nul$(T+\lambda A)$ *and* nul$(T+\lambda A)^*$ *are constant for* $0 < |\lambda| < \delta$.

(iv) In particular, if $\{A_n\}_{n=1}^{\infty}$ *is a sequence of operators such that* $\|T - A_n\| \to 0$ $(n \to \infty)$, *then* A_n *is semi-Fredholm for all* $n \geq n_0$ *and* nul T $\geq$ lim sup$(n \to \infty)$ nul A_n, nul $T^* \geq$ lim sup$(n \to \infty)$ nul A_n^*;

(v) If $K \in K(H)$, *then* T+K *is semi-Fredholm and* ind(T+K) = ind T.

(vi) If B *is another semi-Fredholm operator and* ind T+ind B *is well-defined (i.e.,* {ind T, ind B} $\neq \{\infty,-\infty\}$ *or* $\{-\infty,\infty\}$), *then* TB *is a semi-Fredholm operator and* ind TB = ind T+ind B. *In particular,* T^k *is semi-Fredholm and* ind T^k = k(ind T), *for all* $k \geq 1$.

A semi-Fredholm operator T is a *Fredholm operator* if $-\infty <$ ind T $< \infty$. The well-known Atkinson's theorem asserts that T is Fredholm if and only if $\tilde{T}$ is invertible in $A(H)$ [119], [153]. Hence,

$$\rho_F(T) = \mathbb{C}\setminus\sigma_e(T) = \{\lambda \in \mathbb{C}: \ \lambda - T \text{ is Fredholm}\}$$

(the *Fredholm domain* of T) is an open subset of $\mathbb{C}$.

The *left (right) spectrum* of an element a of a Banach algebra A will be denoted by $\sigma_\ell(a)$ ($\sigma_r(a)$, resp) and its complement $\rho_\ell(a) = \mathbb{C}\setminus\sigma_\ell(a)$ ($\rho_r(a) = \mathbb{C}\setminus\sigma_r(a)$) is the *left (right,* resp) *resolvent set* of a. Thus, $\sigma_e(T) = \sigma_{\ell e}(T) \cup \sigma_{re}(T)$, where $\sigma_{\ell e}(T) = \sigma_\ell(\tilde{T})$ (*left essential spectrum*) and $\sigma_{re}(T) = \sigma_r(\tilde{T})$ (*right essential spectrum*). It is well-known that the intersection $\sigma_{\ell re}(T) = \sigma_{\ell e}(T) \cap \sigma_{re}(T)$ (some authors call $\sigma_{\ell re}(T)$ the *Wolf spectrum* of T) contains the boundary $\partial\sigma_e(T)$ of $\sigma_e(T)$

and therefore, it is a nonempty (compact) subset of $\mathbb{C}$. Its complement $\mathbb{C}\setminus\sigma_{\ell re}(T) = \rho_{\ell e}(T)\cup\rho_{re}(T)$, where $\rho_{\ell e}(T) = \mathbb{C}\setminus\sigma_{\ell e}(T)$ and $\rho_{re}(T) = \mathbb{C}\setminus\sigma_{re}(T)$, coincides with $\rho_{s-F}(T) = \{\lambda \in \mathbb{C}: \lambda-T$ is semi-Fredholm$\}$, the *semi-Fredholm domain* of T.

The following results are an immediate consequence of Theorem 1.13 and its proof (see [153,Chapter IV]).

COROLLARY 1.14. *Let* $T \in L(H)$; *then*

(i) $\rho_{s-F}(T)$ *is the disjoint union of the (possibly empty) open sets* $\{\rho^n_{s-F}(T)\}_{-\infty<n<\infty}$, $\rho^{\infty}_{s-F}(T)$ *and* $\rho^{-\infty}_{s-F}(T)$, *where*

$\rho^h_{s-F}(T) = \{\lambda \in \mathbb{C}: \lambda-T$ *is semi-Fredholm with* $\operatorname{ind}(\lambda-T) = h\}$, $-\infty\leq h\leq\infty$.

(ii) For each h, $-\infty \leq h \leq \infty$, $\rho^h_{s-F}(T)$ *is a lower semicontinuous function of* T *in the sense that, if* $\|T - T_n\| \to 0$ $(n \to \infty)$ *and* $\varepsilon > 0$, *then* $\rho^h_{s-F}(T) \subset [\rho^h_{s-F}(T_n)]_\varepsilon$ *for all* $n \geq n_0(\varepsilon)$.

(iii) $\rho^0_{s-F}(T)$ *contains the resolvent set* $\rho(T) = \mathbb{C}\setminus\sigma(T)$ *of* T, *and* $\sigma_0(T)$.

(iv) If $h \neq 0$, *then* $\rho^h_{s-F}(T)$ *is a bounded set.*

(v) If the minimal index of $\lambda-T$, $\lambda \in \rho_{s-F}(T)$, *is defined by*

$$\operatorname{min.ind}(\lambda-T) = \min\{\operatorname{nul}(\lambda-T),\operatorname{nul}(\lambda-T)^*\},$$

then the function $\lambda \to \operatorname{min.ind}(\lambda-T)$ *is constant on every component of* $\rho_{s-F}(T)$, *except for an at most denumerable subset* $\rho^s_{s-F}(T)$ *without limit points in* $\rho_{s-F}(T)$. *Furthermore, if* $\mu \in \rho^s_{s-F}(T)$ *and* λ *is a point of* $\rho_{s-F}(T)$ *in the same component as* μ *but* $\lambda \notin \rho^s_{s-F}(T)$, *then*

$$\operatorname{min.ind}(\mu-T) > \operatorname{min.ind}(\lambda-T).$$

(vi) If $\lambda \in \rho_{s-F}(T)$ *and* $\operatorname{nul}(\lambda-T) < \infty$ ($\operatorname{nul}(\lambda-T)^* < \infty$, *resp), then*

$$\lambda \notin \rho^s_{s-F}(T) \iff \ker(\lambda-T) \subset \cap_{n=1}^{\infty} \operatorname{ran}(\lambda-T)^n$$

$$(\ker(\lambda-T)^* \subset \cap_{n=1}^{\infty} \operatorname{ran}(\lambda-T)^{*n}, \textit{ resp.}).$$

$\rho^s_{s-F}(T)$ is the set of *singular points* of the semi-Fredholm domain $\rho_{s-F}(T)$ of T; $\rho^r_{s-F}(T) = \rho_{s-F}(T)\setminus\rho^s_{s-F}(T)$ is the set of *regular points*. It is completely apparent that $\rho^r_{s-F}(T)$ is open and contains $\rho(T)$. On the other hand, it is easily seen that $\rho^s_{s-F}(T)$ contains $\sigma_0(T)$. In Chapter III we shall return to the analysis of these sets.

1.5 On invariance and closures of subsets of $L(H)$

A subset R of $L(H)$ is called *invariant under unitary equivalence* if

$$T \in \mathcal{R} \Rightarrow \mathcal{U}(T) \subset \mathcal{R},$$

where $\mathcal{U}(\mathcal{H}) = \{U \in \mathcal{L}(\mathcal{H})$: U is unitary$\}$ and $\mathcal{U}(T) = \{UTU^*: U \in \mathcal{U}(\mathcal{H})\}$ is the *unitary orbit* of T.

It is completely apparent that $\mathcal{U}(T) \subset \mathcal{S}(T)$. We shall establish without proofs some very elementary facts that will be frequently used in the future.

PROPOSITION 1.15. *If $\mathcal{R}$ is a subset of $\mathcal{L}(\mathcal{H})$ invariant either under similarities, or under unitary equivalence, or under compact perturbations (i.e., $\mathcal{R}+\mathcal{K}(\mathcal{H}) = \mathcal{R}$), then $\mathcal{R}^-$ has the same property.*

1.6 Notes and remarks

Theorem 1.1 is just the quantitative version of [153,Theorem 3.16, p.212] (see also [177], or [132,Theorem 1]). Corollaries 1.3(ii) and 1.9(ii) are two elementary observations due to S. Grabiner [108] and D. A. Herrero [132], respectively. Proposition 1.10 is a mild improvement of a result due to D. A. Herrero [150,Lemma 4.3] (see also [44, Lemma 4.1]). The notion of "rank T = ∞-" for a compact operator T, not of finite rank, was introduced by J. Barría and D. A. Herrero in [44] in connection with the analysis of the similarity orbit of a nilpotent operator (See also Chapter VIII). The notion of "minimal index" is due to C. Apostol [10]. The fact that the singular points of $\rho_{s-F}(T)$ are isolated points of this set was discovered by I. C. Gohberg and M. G. Krein [107]. In the above mentioned article, C. Apostol proved that $\rho^s_{s-F}(T)$ is, precisely, the set of points of discontinuity of the function that maps $\lambda \in \rho_{s-F}(T)$ into the orthogonal projection of $\mathcal{H}$ onto $\ker(\lambda-T)$ (see also [25,Lemma 1.6 and Corollary 1.7]). This result will be analyzed in Section 3.3.

2 An apéritif: approximation problems in finite dimensional spaces

In this chapter we shall analyze several intrinsically finite dimensional problems, as well as infinite dimensional ones which can be solved through an essentially finite dimensional approach or by an argument in which the (finite or infinite) dimension of the underlying Hilbert space plays absolutely no role.

It will be convenient to introduce some notation: H will always denote a complex separable Hilbert space of dimension d, $0 \leq d \leq \infty$. If $0 \leq d < \infty$, then we shall also write $\mathbb{C}^d$ (with its canonical inner product) instead of H.

If A, B $\in L(H)$, $A \sim B$ ($A \simeq B$) will mean that A and B are *similar* (*unitarily equivalent*, resp.). $A \underset{\text{sim}}{\to} B$ will be used as an alternative way to indicate that $B \in S(A)^-$, i.e., that $\|B - W_n A W_n^{-1}\| \to 0$ $(n \to \infty)$ for a suitable sequence $\{W_n\}_{n=1}^{\infty}$ in $G(H)$. If $A \underset{\text{sim}}{\to} B$ and $B \underset{\text{sim}}{\to} A$ (equivalently, $S(A)^- = S(B)^-$), then we shall say that A and B are *asymptotically similar*. (In symbols: $A \# B$.) It is completely apparent that $\underset{\text{sim}}{\to}$ is a reflexive and transitive relation and that # is, indeed, an equivalence relation in $L(H)$.(Use Proposition 1.15. It is well-known and trivial that ~ and ≃ are also equivalence relations.)

If $A \in L(H_1)$ and $B \in L(H_2)$, where H_1 and H_2 are isomorphic Hilbert spaces (in symbols: $H_1 \simeq H_2$), then $A \# B$ will be understood as "up to a unitary mapping U from H_2 onto H_1", i.e., $S(A)^- = S(UBU^*)^-$. The same observation applies to the other relations.

The relation $\underset{\text{sim}}{\to}$ induces a partial order < in the quotient set $L(H)/\#$, defined by:

Let $[A] = \{T \in L(H)\colon\ T \# A\}$; $[B] < [A]$ if $A \underset{\text{sim}}{\to} B$.(Equivalently, $S(B)^- \subset S(A)^-$.)

Given a (finite or denumerable) uniformly bounded family $\{A_\nu\}_{\nu\in\Gamma}$ of operators such that $A_\nu \in L(H_\nu)$ for all ν in Γ, we shall denote by $\oplus_{\nu\in\Gamma} A_\nu$ the direct sum of the operators A_ν acting in the usual fashion on the orthogonal direct sum $H = \oplus_{\nu\in\Gamma} H_\nu$ of the spaces H_ν, i.e., if $x = \oplus_{\nu\in\Gamma} x_\nu$ ($\|x\| = [\sum_{\nu\in\Gamma}\|x\|^2]^{\frac{1}{2}} < \infty$) is a vector of H, then $[\oplus_{\nu\in\Gamma} A_\nu]x = \oplus_{\nu\in\Gamma} A_\nu x_\nu$. Clearly, $\|\oplus_{\nu\in\Gamma} A_\nu\| = \sup(\nu \in \Gamma)\ \|A_\nu\| < \infty$.

If $\Gamma = \{1,2,\ldots,n\}$, we shall also write $A_1 \oplus A_2 \oplus \ldots \oplus A_n$. If $A_\nu \simeq A \in L(H)$ for all ν in Γ and card $\Gamma = \alpha$ $(0 \leq \alpha \leq \infty)$, then $A^{(\alpha)}$ will denote the operator $\oplus_{\nu\in\Gamma} A_\nu$ acting on $\oplus_{\nu\in\Gamma} H_\nu = H^{(\alpha)}$ (orthogonal direct sum of α copies of H).

If M is a subspace of H, then $M^\perp = H \ominus M$ is the orthogonal complement of M in H.

Given f, g $\in H$, $f \otimes g \in L(H)$ is the rank one operator defined by $(f\otimes g)x = \langle x,g\rangle f$, where $\langle .,. \rangle$ denotes the inner product of H.

$F(H) = \{\sum_{j=1}^{n} f_j \otimes g_j : f_j, g_j \in H, j = 1,2,\ldots,n; n = 1,2,\ldots\}$

is the ideal of all finite rank operators acting on H.

Let $\{e_1, e_2, \ldots, e_k\}$ be the canonical orthonormal basis (ONB) of $\mathbb{C}^k$ and let $q_k \in L(\mathbb{C}^k)$ be the operator defined by

$$q_k = \sum_{j=1}^{k-1} e_j \otimes e_{j+1} \tag{2.1}$$

($k = 0,1,2,\ldots$; q_0 is the 0 operator acting on the trivial space $\{0\}$, q_1 is the 0 operator acting on the one-dimensional Hilbert space $\mathbb{C}^1 \simeq \mathbb{C}$, and q_k admits the matrix representation

$$q_k = \begin{pmatrix} 0 & 1 & 0 & \cdot & \cdot & \cdot & 0 & 0 \\ 0 & 0 & 1 & \cdot & \cdot & \cdot & 0 & 0 \\ 0 & 0 & 0 & \cdot & \cdot & \cdot & 0 & 0 \\ \cdot & \cdot & \cdot & \cdot & \cdot & & \cdot & \cdot \\ \cdot & \cdot & \cdot & & \cdot & \cdot & \cdot & \cdot \\ \cdot & \cdot & \cdot & & & \cdot & \cdot & \cdot \\ 0 & 0 & 0 & \cdot & \cdot & \cdot & 0 & 1 \\ 0 & 0 & 0 & \cdot & \cdot & \cdot & 0 & 0 \end{pmatrix} \quad (k \times k)$$

with respect to the canonical ONB, for $k = 2,3,\ldots$).

These operators will play a very important role throughout this monograph.

Finally, $sp(a) = \max\{|\lambda| : \lambda \in \sigma(a)\}$ will denote the *spectral radius* of $a \in A$ (a Banach algebra).

2.1 Closures of similarity orbits in finite dimensional spaces

As remarked in the introduction, for many approximation problems the "obvious" necessary conditions derived from the results of Chapter I turn out to be sufficient too. Here is a concrete example of this situation:

THEOREM 2.1. *Let* $T \in L(\mathbb{C}^d)$ *and let* $p(\lambda) = \Pi_{j=1}^{m} (\lambda-\lambda_j)^{k_j}$ $(\lambda_i \neq \lambda_j,$

for $i \neq j$*) be its minimal (monic) polynomial; then the closure of the similarity orbit of* T *is equal to*

$$S(T)^- = \{A \in L(\mathbb{C}^d): \text{ rank } q(A) \leq \text{rank } q(T) \textit{ for all } q|p\},$$

where $q|p$ *denotes a monic polynomial* q *dividing* p.

Furthermore, if $L \in L(\mathbb{C}^d)$, *then* $L \# T$ *if and only if* rank $q(L)$ = rank $q(T)$ *for all* $q|p$ *if and only if* $L \sim T$.

COROLLARY 2.2. *Let* $T \in L(\mathbb{C}^d)$*; then the following are equivalent:*

(i) $S(T)^-$ *is maximal with respect to inclusion (equivalently,* $[T]$ *is a maximal element of* $(L(\mathbb{C}^d)/\#,<)$*);*

(ii) $T \sim \oplus_{j=1}^m (\lambda_j + q_{k_j})$ *(*λ_j *and* k_j *as in Theorem 2.1,* $j = 1,2,\ldots,m$*);*

(iii) $\sum_{j=1}^m k_j = d$ *(where the* k_j*'s have the same meaning as in Theorem 2.1);*

(iv) T *is a cyclic operator;*

(v) $S(T)^- = \{A \in L(\mathbb{C}^d): \sigma(A) = \sigma(T)$ *and* $\dim H(\lambda;A) = \dim H(\lambda;T)$ *for all* λ *in* $\sigma(T)\}$.

COROLLARY 2.3. *Let* $T \in L(\mathbb{C}^d)$*; then the following are equivalent:*

(i) $S(T)^-$ *is minimal (equivalently,* $[T]$ *is a minimal element);*

(ii) k_j *(defined as in Theorem 2.1) is equal to* 1 *for all* $j = 1, 2,\ldots,m$*;*

(iii) T *is similar to a normal operator;*

(iv) $S(T)$ *is closed in* $L(\mathbb{C}^d)$*;*

(v) $T \in S(L)^-$ *for all* L *in* $L(\mathbb{C}^d)$ *such that* $\sigma(L) = \sigma(T)$ *and* $\dim H(\lambda;L) = \dim H(\lambda;T)$ *(*$\lambda \in \sigma(T)$*).*

Let $T, A \in L(\mathbb{C}^d)$ and assume that $\|A - W_n T W_n^{-1}\| \to 0$ $(n \to \infty)$ for a suitable sequence $\{W_n\}_{n=1}^{\infty}$ of operators in $G(\mathbb{C}^d)$. Clearly, $q(A)$ and $q(W_n T W_n^{-1}) = W_n q(T) W_n^{-1}$ are well-defined (for all $n = 1,2,\ldots$) for all $q|p$ and (by Proposition 1.7)

$$\|q(A) - W_n q(T) W_n^{-1}\| \to 0 \quad (n \to \infty).$$

Since rank $W_n q(T) W_n^{-1}$ = rank $q(T)$ (for all $n = 1,2,\ldots$), it follows from Proposition 1.12(i) that rank $q(A) \leq$ rank $q(T)$.

Hence the conditions of Theorem 2.1 are *necessary*. The *sufficiency* of these conditions will be proved in several steps.

The second statement of the theorem is a trivial consequence of the first one: It is obvious that $S(L)^- = S(T)^-$ if and only if $\sigma(L) = \sigma(T)$ and rank $q(L)$ = rank $q(T)$ for all $q|p$ (Use Proposition 1.15). On the other hand, a simple analysis of the Jordan forms of L and T shows that rank $q(L)$ = rank $q(T)$ for all $q|p$ if and only if L and T are similar. (In particular, this implies that they have the same spectrum.)

2.1.1 The nilpotent case

LEMMA 2.4. *If* $1 \le m \le k-1$, *then* $q_m \oplus q_k \underset{\text{sim}}{\to} q_{m+1} \oplus q_{k-1}$.

PROOF. Let $\{e_1, e_2, \ldots, e_m\}$ and $\{f_1, f_2, \ldots, f_k\}$ be the canonical ONB of $\mathbb{C}^m$ and $\mathbb{C}^k$, respectively, so that $q_m = \sum_{i=1}^{m-1} e_i \otimes e_{i+1}$ and $q_k = \sum_{j=1}^{k-1} f_j \otimes f_{j+1}$.

Given ε, $0 < \varepsilon < 1$, define $W_\varepsilon \in L(\mathbb{C}^m \oplus \mathbb{C}^k)$ by $W_\varepsilon e_i = e_i - (1/\varepsilon) f_{i+1}$ ($i = 1,2,\ldots,m$), $W_\varepsilon f_1 = \varepsilon f_1$ and $W_\varepsilon f_j = f_j$ ($j = 2,3,\ldots,k$). Straightforward computations show that W_ε is invertible, $W_\varepsilon^{-1} e_i = e_i + (1/\varepsilon) f_{i+1}$ ($i = 1,2,\ldots,m$), $W_\varepsilon^{-1} f_1 = (1/\varepsilon) f_1$ and $W_\varepsilon^{-1} f_j = f_j$ ($j = 2,3,\ldots,k$) and

$$Q_\varepsilon = W_\varepsilon (q_m \oplus q_k) W_\varepsilon^{-1} = Q + \varepsilon f_1 \otimes f_2,$$

where $Qf_1 = 0$, $Qe_1 = f_1$, $Qe_i = e_{i-1}$ ($i = 2,3,\ldots,m$), $Qf_2 = 0$ and $Qf_j = f_{j-1}$ ($j = 3,4,\ldots,k$).

It is immediate that $Q \simeq q_{m+1} \oplus q_{k-1}$ and that $Q = \lim(\varepsilon \to 0)\, Q_\varepsilon$ (unless otherwise stated, lim must always be understood as a limit in the *norm-topology*). □

LEMMA 2.5. *Let* $T \in L(\mathbb{C}^d)$ *be a nilpotent of order* m; *then*

$S(T)^- = \{A \in L(\mathbb{C}^d):\ \text{rank } A^j \le \text{rank } T^j$ *for* $j = 1,2,\ldots,m\}$.

PROOF. We have already observed that the condition "rank $A^j \le$ rank T^j for $j = 1,2,\ldots,m$" is necessary. (Observe that the minimal polynomial of T is $p(\lambda) = \lambda^m$.)

Assume that rank $A^j \le$ rank T^j for $j = 1,2,\ldots,m$. Clearly, we can directly assume that T and A are Jordan forms, i.e.,

$$T = q_{n_1}^{(\tau_1)} \oplus q_{n_2}^{(\tau_2)} \oplus \ldots \oplus q_{n_k}^{(\tau_k)} \quad \text{and } A = q_{m_1}^{(\alpha_1)} \oplus q_{m_2}^{(\alpha_2)} \oplus \ldots \oplus q_{m_h}^{(\alpha_h)}$$

($\tau_j, \alpha_i > 0$). We shall proceed by induction on

$$m(T,A) = d + \sum_{j=1}^{\infty} (\text{rank } T^j - \text{rank } A^j).$$

The case $m(T,A) = 1$ is trivial. Assume that $m(T_1, A_1) \le n$ implies that $T_1 \underset{\text{sim}}{\to} A_1$, whenever A_1 and T_1 are nilpotent operators acting on a finite dimensional space and satisfy rank $A_1^j \le$ rank T_1^j for all $j = 1,2,\ldots$, and let $A, T \in L(\mathbb{C}^d)$ be nilpotent operators such that rank $A^j \le$ rank T^j for all $j = 1,2,\ldots$, and $m(T,A) = n+1$.

If T and A have a common Jordan block q_r, then $T = q_r \oplus T_1$, $A = q_r \oplus A_1$, rank $A_1^j -$ rank $T_1^j =$ rank $A^j -$ rank $T^j \le 0$ for all $j = 1,2,\ldots,$

$m(T_1,A_1) \le (n+1)-r \le n$ and, by induction, $T_1 \underset{sim}{\to} A_1$. A fortiori, $T = q_r \oplus T_1 \underset{sim}{\to} q_r \oplus A_1 = A$.

If T and A have no common Jordan blocks and if $r \ge 1$ is the minimum index such that $m_h < n_r$, then T has the form

$$T = q_{n_{r-1}} \oplus q_{n_r} \oplus T'$$

(where $n_0 = 0$ and q_0 acts on a $\{0\}$-space, if $r = 1$) and we have

$$\text{rank } A^s < \text{rank } T^s \text{ for } n_{r-1} + 1 \le s \le m_h - 1.$$

Indeed, if rank A^s = rank T^s and if a_s (t_s, resp.) denotes the number of Jordan blocks of A (T, resp.) with order of nilpotency greater than or equal to s, then it is obvious that $a_s > t_s$ and this yields the contradiction

$$\text{rank } A^{s-1} = a_s + \text{rank } A^s > t_s + \text{rank } T^s = \text{rank } T^{s-1}.$$

Setting $T_1 = q_{n_{r-1}+1} \oplus q_{n_r - 1} \oplus T'$, we can check that rank $T_1{}^j$ = rank T^j for $j = 1,2,\ldots,n_{r-1}$ and rank $T_1{}^j$ = rank $T^j - 1$ for $n_{r-1}+1 \le j \le m_h - 1$, so that

$$\text{rank } A^j \le \text{rank } T_1{}^j \text{ for } j = 1,2,\ldots .$$

On the other hand, $m(T_1,A) \le m(T,A) - (\text{rank } T^{m_h-1} - \text{rank } T_1{}^{m_h-1}) = m(T,A) - 1 = n$, and consequently, $T_1 \underset{sim}{\to} A$. If $r > 1$, then $T \underset{sim}{\to} T_1$ by Lemma 2.4. If $r = 1$, then $q_{n_1} \simeq \sum_{j=1}^{n_1-1} e_j \otimes e_{j+1} \in L(\mathbb{C}^{n_1})$ is similar to $\varepsilon e_1 \otimes e_2 + \sum_{j=2}^{n_1-1} e_j \otimes e_{j+1}$ ($\varepsilon > 0$) and, letting $\varepsilon \to 0$, we conclude that $T \underset{sim}{\to} T_1$. In either case, $T \underset{sim}{\to} T_1 \underset{sim}{\to} A$, and therefore $A \in S(T)^-$. □

2.1.2 Proof of Theorem 2.1

Assume that $T, A \in L(\mathbb{C}^d)$ $(0 \le d < \infty)$, the minimal polynomial of T is p, $p(\lambda) = \Pi_{j=1}^m (\lambda - \lambda_j)^{k_j}$ ($\lambda_i \ne \lambda_j$, if $i \ne j$) and rank $q(A) \le$ rank $q(T)$ for all $q|p$.

We want to show that $T \underset{sim}{\to} A$. Clearly, we can directly assume (without loss of generality) that T and A are unitarily equivalent to their Jordan forms; let $T = \oplus_{j=1}^m (\lambda_j + Q_j)$, where $\mathbb{C}^d = \oplus_{j=1}^m H_j$ and Q_j is a Jordan nilpotent acting on the subspace H_j, $0 < \dim H_j \le d$, $j = 1,2,\ldots,m$. It is easily seen that $H(\lambda_r;T) = H_r = \ker(\lambda_r - T)^{k_r} = \text{ran } p_r(T)$ and $(H_r)^\perp = \text{ran}(\lambda_r - T)^{k_r} = \ker p_r(T)$, where $p_r(T) = \Pi_{j=1, j\ne r}^m (\lambda - \lambda_j)^{k_j}$ $(r = 1,2,\ldots,m)$, whence we readily conclude that $\dim H(\lambda_r;A) = \dim H_r$ for all $r = 1,2,\ldots,m$ and (since $d = \sum_{j=1}^m \dim H_j$) that $\sigma(A) = \sigma(T) = \{\lambda_1,\lambda_2,\ldots,\lambda_m\}$. Furthermore, since $\text{rank}(\lambda_r - A)^s p_r(A) \le \text{rank}(\lambda_r - T)^s p_r(T)$ for all $s = 1,2,\ldots,k_r$, $r = 1,2,\ldots,m$, it is not difficult to see that

$A \simeq \oplus_{r=1}^{m} (\lambda_r+R_r)$, where R_r is a Jordan nilpotent acting on H_r and satisfying the conditions

$$\text{rank } (R_r)^s \leq \text{rank } (Q_r)^s,\ s = 1,2,\ldots,k_r, r=1,2,\ldots,m.$$

By Lemma 2.5, $Q_r \underset{\text{sim}}{\to} R_r$ and, a fortiori, $\lambda_r+Q_r \underset{\text{sim}}{\to} \lambda_r+R_r$ for all $r = 1,2,\ldots,m$, whence it readily follows that

$$T = \oplus_{j=1}^{m} (\lambda_j+Q_j) \underset{\text{sim}}{\to} \oplus_{j=1}^{m} (\lambda_j+R_j) \simeq A.$$

The proof of Theorem 2.1 is complete now. □

REMARK 2.6. Since nul $q(T) = d - \text{rank } q(T)$, the conditions "rank $q(A) \leq \text{rank } q(T)$ for all $q|p$" can be replaced by the conditions "nul $q(A) \geq \text{nul } q(T)$ for all $q|p$".

PROOF OF COROLLARIES 2.2 AND 2.3. It is easily seen that T has the form of Corollary 2.2(ii) if and only if $S(T)^-$ is maximal; (v) => (i) is a trivial implication and (ii) => (v) follows from Theorem 2.1. Now Corollary 2.2 follows from the well-known algebraic fact that (ii), (iii) and (iv) are equivalent.

Similarly, $S(T)^-$ is minimal if and only if $k_j = 1$ for all $j = 1,2,\ldots,m$, if and only if T is similar to a normal operator acting on $\mathbb{C}^d$; (v) => (i) is a trivial implication and (ii) => (v) follows from Theorem 2.1. On the other hand, it easily follows from Theorem 2.1 and its proof that if $k_j = 1$ for all $j = 1,2,\ldots,m$, then the minimal polynomial of $A \in S(T)^-$ is necessarily equal to p and that $A \sim T$, i.e., $S(T)^- = S(T)$ is a closed subset of $L(\mathbb{C}^d)$.

Conversely, if $S(T)$ is closed and $T \underset{\text{sim}}{\to} A$, then $A \in S(T)^- = S(T)$ and therefore $A \sim T$. A fortiori, $A \underset{\text{sim}}{\to} T$, whence it readily follows that $[A] = [T] = S(T)$ is a minimal element of $(L(\mathbb{C}^d)/\#,<)$. The proof of Corollary 2.3 is complete now. □

2.1.3 The lattice $(N(\mathbb{C}^d)/\#,<)$

The partially ordered set (poset) $(L(H)/\#,<)$ is never a lattice (unless dim $H = 0$). However, it contains many interesting lattices. Consider the case when $H = \mathbb{C}^d$ is a finite dimensional space and let

$$T = \oplus_{j=1}^{n} q_j^{(\tau_j)} \in L(\mathbb{C}^d);$$

then rank $T^j = \tau_{j+1}+2\tau_{j+2}+\ldots+(n-j)\tau_n$ and this implies that

$$\tau_j = \text{rank } T^{j-1} - 2\text{rank } T^j + \text{rank } T^{j+1} \geq 0, \text{ for } j = 1,2,\ldots,n. \quad (2.2)$$

Let Σ_d be the set of all (d+1)-tuples $(m_0,m_1,m_2,\ldots,m_d)$ of non-negative integers such that

$$\begin{cases} d = m_0 > m_1 \geq m_2 \geq \ldots \geq m_d = 0, \text{ and} \\ m_{j-1} - 2m_j + m_{j+1} \geq 0 \text{ for all } j = 1,2,\ldots,d-1. \end{cases} \tag{2.3}$$

Observe that these two conditions imply that $m_j > m_{j+1}$, unless $m_j = 0$, and $m_{j-1} - m_j \geq m_j - m_{j+1}$ for $j = 1,2,\ldots,d-1$. It follows that $0 \leq m_j \leq d-j$ for all $j = 0,1,2,\ldots,d$.

Let $\mu = (m_0,m_1,\ldots,m_d)$ and $\mu' = (m_0',m_1',\ldots,m_d')$ be two elements of Σ_d; then we shall write

$$\mu \leq \mu' \text{ if } m_j \leq m_j' \text{ for all } j = 0,1,2,\ldots,d,$$

and

$$\mu \vee \mu' = (\max\{m_0,m_0'\},\max\{m_1,m_1'\},\ldots,\max\{m_d,m_d'\}).$$

It is easily seen that $(\Sigma_d,\leq)$ is a poset,

$$d = \max\{m_0,m_0'\} > \max\{m_1,m_1'\} \geq \ldots \geq \max\{m_d,m_d'\} = 0$$

and

$$\begin{aligned} 2\max\{m_j,m_j'\} &\leq \max\{m_{j-1}+m_{j+1},m_{j-1}'+m_{j+1}'\} \\ &\leq \max\{m_{j-1},m_{j-1}'\}+\max\{m_{j+1},m_{j+1}'\}, \end{aligned}$$

so that $\mu \vee \mu' \in \Sigma_d$. It is completely apparent that $\mu \vee \mu'$ is the least upper bound (l.u.b.) of μ and μ' with respect to the partial order $\leq$. Since Σ_d is finite, every subset of Σ_d has a l.u.b.. In particular,

$$\mu \wedge \mu' = \text{l.u.b.}\{\nu \in \Sigma_d: \ \nu \leq \mu \text{ and } \nu \leq \mu'\}$$

is the (unique) greatest lower bound of μ and μ'.

It readily follows that $(\Sigma_d,\leq)$ is a finite *lattice* with *supremum* $(d,d-1,d-2,\ldots,2,1,0)$ and *infimum* $(d,0,0,\ldots,0,0,0)$.

Given $\mu = (m_0,m_1,\ldots,m_d) \in \Sigma_d$, define $T_\mu \in N(\mathbb{C}^d)$ by

$$T_\mu = \oplus_{j=1}^{d} q_j^{(\tau_j)},$$

where $\tau_j = m_{j-1}-2m_j+m_{j+1}$ for $j = 1,2,\ldots,d-1$ and $\tau_d = m_{d-1}$; (2.2) and (2.3) guarantee that the mapping

$$\mu \to [T_\mu] \tag{2.4}$$

is a bijection from Σ_d onto $N(\mathbb{C}^d)/\#$ and, moreover, that rank $T_\mu^{\,j} = m_j$ for $j = 0,1,2,\ldots,d$. Combining these observations with Theorem 2.1, we obtain

THEOREM 2.7. *The mapping* (2.4) *defines an order-preserving bijection from* $(\Sigma_d,\leq)$ *onto* $(N(\mathbb{C}^d)/\#,\leq)$. *In particular,* $(N(\mathbb{C}^d)/\#,\leq)$ *is a finite lattice with supremum* $[q_d]$ *and infimum* $[0]$ $(0 = q_1^{(d)})$.

2.1.4 Closures of similarity orbits of finite rank operators

Let H be an infinite dimensional Hilbert space and let T, A $\in$ $F(H)$; then T and A are algebraic operators with nul T = nul A = ∞ and there exists a finite dimensional subspace $H(T,A)$ reducing both, A and T, such that $T|H(T,A)^{\perp} = A|H(T,A)^{\perp} = 0$ ($H(T,A)$ can always be defined so that dim $H(T,A) \leq 2$ rank T + 2 rank A).

Assume that A $\in S(T)^{-}$; then we can prove exactly as in the finite dimensional case that rank q(A) $\leq$ rank q(T) and, by using Proposition 1.12(ii), that nul q(A) $\geq$ nul q(T) for all q|p, where p is the minimal polynomial of T. Conversely, if A satisfies those conditions, then it is not difficult to check that $A|H(T,A)$ satisfies the same conditions with respect to $T|H(T,A)$ and therefore, by Theorem 2.1,

$$T|H(T,A) \underset{\text{sim}}{\to} A|H(T,A).$$

A fortiori, $T = T|H(T,A)\oplus 0 \underset{\text{sim}}{\to} A = A|H(T,A)\oplus 0$; hence, we have

COROLLARY 2.8. *Let* T $\in L(H)$ *be a (necessarily algebraic) finite rank operator with minimal polynomial* p, *then*

$$S(T)^{-} = \{A \in L(H): \text{ rank } q(A) \leq \text{rank } q(T) \textit{ and } \text{nul } q(A) \geq \text{nul } q(T) \textit{ for all } q|p\}.$$

Let L $\in L(H)$; *then* L # T *if and only if* L ~ T.

REMARK 2.9. Since H is infinite dimensional, the conditions "rank q(A) $\leq$ rank q(T) and nul q(A) $\geq$ nul q(T) for all q|p" of Corollary 2.8 still imply that $\sigma(A) = \sigma(T)$. However, the following example shows that those two conditions cannot be replaced by "$\sigma(A) = \sigma(T)$ and rank q(A) $\leq$ rank q(T) for all q|p": Let T be an orthogonal projection of rank 2 and let A be an orthogonal projection of rank 1; then $\sigma(A) = \sigma(T) = \{0,1\}$, and rank q(A) $\leq$ rank q(T) for all q|p ($p(\lambda) = \lambda(\lambda-1)$),but A cannot belong to $S(T)^{-}$, because $1 = \dim H(1;A) \neq \dim H(1;T) = 2$. Reversing the roles of A and T, we see that conditions of Corollary 2.8 cannot be replaced by "$\sigma(A) = \sigma(T)$ and nul q(A) $\geq$ nul q(T) for all q|p".

Let $NF(H) = N(H)\cap F(H)$ and $NF_k(H) = \{T \in NF(H): \text{ rank } T \leq k-1\}$. Since H is infinite dimensional,

$$NF_1(H) \subset NF_2(H) \subset \ldots \subset NF_k(H) \subset NF_{k+1}(H) \subset \ldots$$

is an infinite chain (all the inclusions are proper: no two sets in this chain coincide), and this chain naturally induces a chain of lattices

$$(NF_1(H)/\#,<) \subset (NF_2(H)/\#,<) \subset \ldots \subset (NF_k(H)/\#,<) \subset (NF_{k+1}(H)/\#,<) \subset \ldots,$$

where (as can be easily checked by using Theorem 2.7) $(NF_k(H)/\#,<)$ is order-isomorphic with $(\Sigma_k,\leq)$; moreover, the above chain is in correspondence with a chain of natural order-preserving inclusions between the lattices $\{(\Sigma_k,\leq)\}_{k=1}^{\infty}$ ($(\Sigma_1,\leq) \subset (\Sigma_2,\leq) \subset \ldots \subset (\Sigma_k,\leq) \subset (\Sigma_{k+1},\leq) \subset \ldots$). Let $(\Sigma,\leq)$ be the set union of the Σ_k's with the induced partial order, i.e., if μ, $\mu' \in \Sigma$, then μ, $\mu' \in \Sigma_k$ for all $k \geq k_o(\mu,\mu')$ and we define $\mu \leq \mu'$ in $(\Sigma,\leq)$ if $\mu \leq \mu'$ holds in $(\Sigma_{k_o},\leq)$.

COROLLARY 2.10. $(NF(H)/\#,<)$ *is a lattice with infimum* $[0]$ *and no maximal element. This lattice is order-isomorphic with* $(\Sigma,\leq)$.

REMARK 2.11. However (as we shall see in Chapter VIII), $(NF(H)/\#,<)$ does have a supremum in $(L(H)/\#,<)$.

2.2 The distance from the set of all non-zero orthogonal projections to $N(H)$

2.2.1 The limit case

Let $P(H)$ be the set of all non-zero orthogonal projections in $L(H)$. What is the exact value of the distance

$$\eta_\infty = \inf\{\|P-Q\|: \quad P \in P(H),\ Q \in N(H)\} \tag{2.5}$$

from $P(H)$ to $N(H)$ (H an infinite dimensional space)?

According to Example 1.5, $\|P-Q\| > \frac{1}{2}$ for all Q in $L(H)$ such that $\sigma(Q)\cap\{\lambda: \ |\lambda-1| \leq \frac{1}{2}\} = \emptyset$.

Since $\sigma(Q) = \{0\}$ for every nilpotent Q, the above inequality holds, in particular, for all Q in $N(H)$. Hence, $\eta_\infty \geq \frac{1}{2}$. It will be shown that $\eta_\infty = \frac{1}{2}$; more precisely, if

$$\eta_k = \inf\{\|P-Q\|: \quad P \in P(H),\ Q \in N_k(H)\}$$

and

$$\delta_k = \inf\{\|P-Q\|: \quad P \in P(\mathbb{C}^k),\ Q \in N(\mathbb{C}^k)\},$$

then it is completely apparent that $\frac{1}{2} \leq \eta_\infty \leq \eta_k \leq \delta_k \leq 1$ and η_k (δ_k) is non-increasing with k. We have the following

THEOREM 2.12. $\frac{1}{2} < \delta_k \leq \frac{1}{2} + \sin\frac{\pi}{(m+1)}$, *where* $m = [(k-1)/2]$ (= *the integral part of* $(k-1)/2$), *for all* $k \geq 3$.

In particular, $\lim(k\to\infty)\ \delta_k = \lim(k\to\infty)\ \eta_k = \eta_\infty = \frac{1}{2}$.

We shall need some auxiliary results.

LEMMA 2.13. *Let* $1 \le m < n < \infty$ *and let* $m < r < n$, s *be such that* $r+s = m+n$. *If* $\{e_i\}_{i=1}^{r}$ $(\{f_j\}_{j=1}^{s})$ *is an ONB of* $\mathbb{C}^r$ $(\mathbb{C}^s$, *resp.),*

$$T = (\textstyle\sum_{i=1}^{r-1} e_i \otimes e_{i+1}) + (\sum_{j=1}^{s-1} f_{j+1} \otimes f_j)$$

and

$$L = (\textstyle\sum_{i=m+1}^{r-1} e_i \otimes e_{i+1}) + f_1 \otimes e_{m+1} + (\sum_{j=1}^{s-1} f_{j+1} \otimes f_j) + (\sum_{i=1}^{m-1} e_i \otimes e_{i+1})$$

$(T, L \in L(\mathbb{C}^r \oplus \mathbb{C}^s))$, *then* $T \simeq q_r \oplus q_s$, $L \simeq q_n \oplus q_m$ *and there exists a unitary operator* $V \in L(\mathbb{C}^r \oplus \mathbb{C}^s)$ *such that*

$$\|VTV^* - L\| = \|T - V^*LV\| = s(m), \text{ where } s(m) = 2\sin\frac{\pi}{(m+1)},$$

$Ve_i = e_i$, *for* $m+1 \le i \le r$, *and* $Vf_j = f_j$, *for* $m+1 \le j \le s$.

PROOF. Define

$$g_t = \begin{cases} e_{m+1-t}, & m+1-r \le t \le 0, \\ (\cos\frac{t\pi}{2(m+1)})e_{m+1-t} + (\sin\frac{t\pi}{2(m+1)})f_t, & 1 \le t \le m, \\ f_t, & m+1 \le t \le s, \end{cases}$$

$$k_t = (-\sin\frac{t\pi}{2(m+1)})e_{m+1-t} + (\cos\frac{t\pi}{2(m+1)})f_t, \quad 1 \le t \le m,$$

and

$$R = (\textstyle\sum_{t=m+1-r}^{s-1} g_{t+1} \otimes g_t) + (\sum_{t=1}^{m-1} k_{t+1} \otimes k_t).$$

It is completely apparent that $(\{g_t\}_{t=m+1-r}^{s} \cup \{k_t\}_{t=1}^{m})$ is also an ONB of $\mathbb{C}^r \oplus \mathbb{C}^s$, the pair $\{g_t, k_t\}$ generates the same two-dimensional subspace as the pair $\{e_{m+1-t}, f_t\}$ (for each t such that $1 \le t \le m$), $T \simeq q_r \oplus q_s$ and $L \simeq q_n \oplus q_m \simeq R$. Now a straightforward computation shows that $L = VRV^*$, where V is the unitary operator defined by

$$Vg_t = \begin{cases} e_{m+1-t}, & m+1-r \le t \le 0 \\ f_t, & 1 \le t \le s \end{cases}$$

and

$$Vk_t = e_{m+1-t}, \qquad 1 \le t \le m.$$

On the other hand,

$$(T-R)g_t = 0, \qquad m+1-r \le t \le -1,$$

$$(T-R)g_0 = (1-\cos\frac{\pi}{2(m+1)})e_m - (\sin\frac{\pi}{2(m+1)})f_1,$$

$$(T-R)g_t = (\cos\frac{\pi}{2(m+1)} - 1)g_{t+1} - (\sin\frac{\pi}{2(m+1)})k_{t+1}, \quad 1 \le t \le m-1,$$

$$(T-R)g_m = (\sin\frac{m\pi}{2(m+1)} - 1)f_{m+1}$$

$$(T-R)g_t = 0, \qquad m+1 \le t \le s,$$

$$(T-R)k_t = (\sin\frac{\pi}{2(m+1)})g_{t+1} + (\cos\frac{\pi}{2(m+1)} - 1)k_{t+1}, \quad 1 \le t \le m-1,$$

and

$$(T-R)k_m = (\cos\frac{m\pi}{2(m+1)})f_{m+1}.$$

Since $\bigvee\{g_t,k_t\} = \bigvee\{e_{m+1-t},f_t\}$ $(1 \le t \le m$; $\bigvee\{\ldots\}$ denotes the *closed linear span of* $\{\ldots\})$, these two-dimensional subspaces are pairwise orthogonal and $(T-R)g_t \perp (T-R)k_t$ for $1 \le t \le m-1$, we conclude that

$$||T-R|| = \max\{[\sin^2\frac{\pi}{2(m+1)} + (\cos\frac{\pi}{2(m+1)} - 1)^2]^{\frac{1}{2}},$$

$$\max_{|a|^2+|b|^2=1} |a(\sin\frac{m\pi}{2(m+1)} - 1) + b\cos\frac{m\pi}{2(m+1)}|\}$$

$$= [\sin^2\frac{\pi}{2(m+1)} + (\cos\frac{\pi}{2(m+1)} - 1)^2]^{\frac{1}{2}} = 2\sin\frac{\pi}{(m+1)} = s(m).$$

Since $||VTV^* - L|| = ||T - V^*LV|| = ||T - R||$, we are done. □

The above proof admits a very simple geometric description: The action of q_r $(e_r \to e_{r-1} \to e_{r-2} \to \ldots \to e_2 \to e_1 \to 0)$ can be described by an arrow of length r, and the action of $T = q_r \oplus q_s$ by two arrows of lengths r and s, as follows

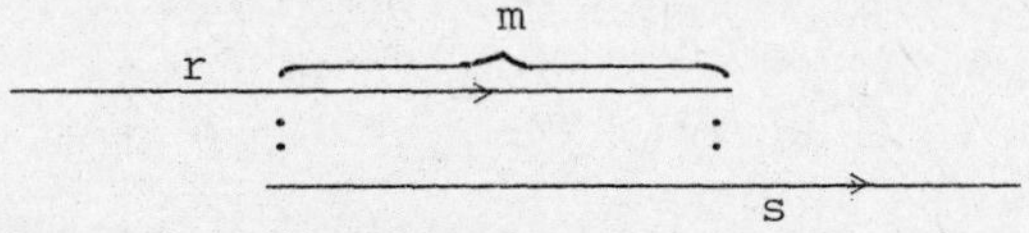

Similarly, R can be indicated as

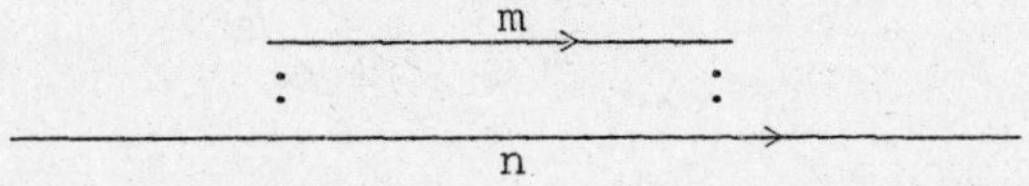

T maps e_{m+1} to e_m, the pair $\{e_{m+1-j},f_j\}$ onto the pair $\{e_{m-j},f_{j+1}\}$ of "parallel" vectors $(j = 1,2,\ldots,m-1)$ and the pair $\{e_1,f_m\}$ onto $\{0, f_{m+1}\}$. On the other hand, R maps e_{m+1} to $(\cos\pi/2(m+1))e_m+(\sin\pi/2(m+1))f_1$, which is equal to the result of a "slight twist" of the vector e_m in the two-dimensional space $\bigvee\{e_m,f_1\}$, the pair $\{e_{m+1-j},f_j\}$ onto a "slight twist" of the "parallel pair" $\{e_{m-j},f_{j+1}\}$ $(j = 1,2,\ldots,m-1)$ and a "slight twist" (in the reverse sense) of f_m to f_{m+1}.

The accumulation of all these "slight twists" after m+1 steps will map e_m to f_{m+1} and f_1 to 0. This fact will be described by the following scheme

(T → R)

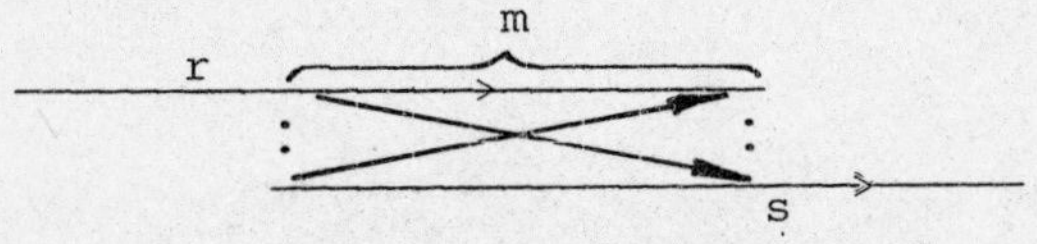

Conversely, if we take $(\{g_t\}_{t=m+1-r}^{s} \cup \{k_t\}_{t=1}^{m})$ as the original ONB of $\mathbb{C}^r\oplus\mathbb{C}^s \simeq \mathbb{C}^n\oplus\mathbb{C}^m$ and modify R to obtain T, then the corresponding scheme will be

(R → T)

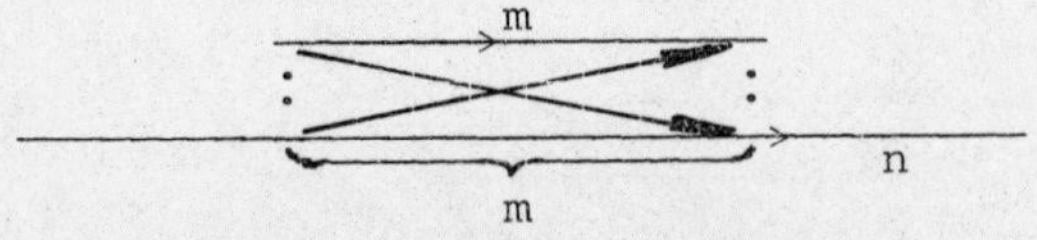

In this case, the twist will move g_0 to k_m to 0 and k_1 to $-g_{m+1}$.

In what follows, we shall freely use these schemes instead of analytic descriptions in the proofs.

COROLLARY 2.14. *If* $k \geq 3$, $1 \leq m \leq [(k-1)/2]$, $p = k-m$ *and* $U_p = \sum_{j=1}^{p-1} e_j \otimes e_{j+1} + e_p \otimes e_1 \in L(\mathbb{C}^p)$, *then* $U_p \simeq q_p + (q_p^*)^{p-1} \simeq V_p$, *where* V_p *is the unitary operator defined by* $V_p e_j = (\omega_p)^j e_j$ *(where* $-\omega_p$ *is a primitive p-th root of* 1*) and there exists a unitary operator* $V \in L(\mathbb{C}^k)$ *such that* $||Vq_kV^* - q_m \oplus U_p|| = s(m)$.

PROOF. It is straightforward to check that $U_p \simeq q_p + (q_p^*)^{p-1} \simeq V_p$ (Observe that the three operators are unitary and determinant$(\lambda - U_p)$ = determinant$(\lambda - V_p) = \lambda^p - (-1)^p$.)

Represent q_k by a "curled arrow" and modify the action of the operator in the subspace $\bigvee\{e_1, e_2, \ldots, e_m, e_{m+1}, e_{k-m}, e_{k-m+1}, \ldots, e_k\}$ as in Lemma 2.13

$(q_k \to q_m \oplus U_p)$

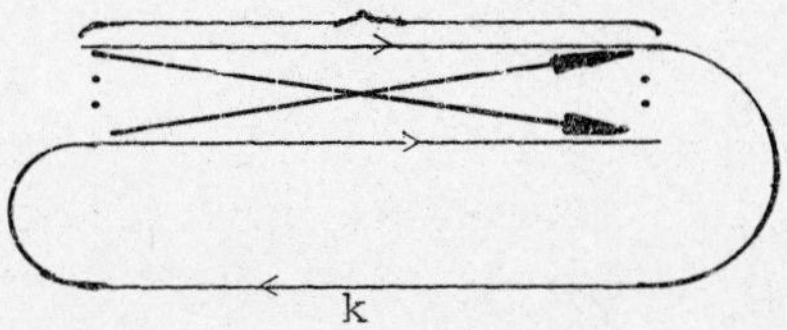

We conclude as in the above lemma that there exists an operator $L_{m,p} \simeq q_m \oplus U_p$ such that $||q_k - L_{m,p}|| = s(m)$. □

PROOF OF THEOREM 2.12. With the notation of Corollary 2.14: Let $P \in L(\mathbb{C}^k)$ be the orthogonal projection of $\mathbb{C}^k$ onto the one-dimensional kernel of $((-1) - q_m \oplus U_p)$, where $m = [(k-1)/2]$ and $p = k-m$; then

$$||P - (-\tfrac{1}{2})Vq_kV^*|| \leq ||P + (\tfrac{1}{2})q_m \oplus U_p|| + (\tfrac{1}{2})||Vq_kV^* - q_m \oplus U_p||$$

$$\leq \tfrac{1}{2} + \tfrac{1}{2}\, s(m) = \tfrac{1}{2} + \sin \frac{\pi}{(m+1)} .$$

Hence, $\delta_k \leq \frac{1}{2} + \sin \frac{\pi}{(m+1)}$. It readily follows that

$$\delta_k = \inf\{||P - Q||: \quad P \in P(\mathbb{C}^k),\ Q \in N(\mathbb{C}^k) \text{ and } ||Q|| \leq 3/2 + \sin \frac{\pi}{(m+1)}\}. \tag{2.6}$$

Since $P(\mathbb{C}^k) \times \{Q \in N(\mathbb{C}^k): \ ||Q|| \leq 3/2 + \sin \frac{\pi}{(m+1)}\}$ is a compact subset of $L(\mathbb{C}^k) \times L(\mathbb{C}^k)$, it is easily seen that the infimum in (2.6) is actually attained for some pair (P_{min}, Q_{min}) in this set. Thus, by our previous remarks (Example 1.5),

$$\delta_k = ||P_{min} - Q_{min}|| > \tfrac{1}{2}. \qquad \square$$

2.2.2 On the exact values of δ_k and η_k

PROPOSITION 2.15. *If* $2 \leq h = \dim H \leq \infty$, *then*

$$\inf\{\|P - Q\|: \quad P \in P(H),\ Q \in N_2(H)\} = \sqrt{2}/2.$$

Furthermore, the above infimum is actually attained by the operators

$$P_{min} = \begin{pmatrix} 1 & 0 \\ 0 & 0 \end{pmatrix} \oplus 0_{h-2} \in P(H) \text{ and } Q_{min} = \begin{pmatrix} \frac{1}{2} & -\frac{1}{2} \\ \frac{1}{2} & -\frac{1}{2} \end{pmatrix} \oplus 0_{h-2} \in N_2(H),$$

where the 2×2 *matrices act on a two-dimensional subspace* H_2 *of* H *and* 0_{h-2} *denotes the* 0 *operator acting on* $(H_2)^{\perp}$.

PROOF. It is straightforward to check that $P_{min} \in P(H)$, $Q_{min} \in N_2(H)$ and $\|P_{min} - Q_{min}\| = \sqrt{2}/2$.

On the other hand, if $Q \in N_2(H)$ and $P \in P(H)$, then either $Q = 0$ and $\|P - Q\| = \|P\| = 1$, or $Q \neq 0$.

Assume that $\|P - Q\| < \sqrt{2}/2$. If $x \in \ker Q$ and $y \perp \ker Q$, then $\|Px\| = \|(P-Q)x\| < (\sqrt{2}/2)\|x\|$ and $\|Py\| = \|P^*y\| = \|(P-Q)^*y\| < (\sqrt{2}/2)\|y\|$.

Let $z \in \operatorname{ran} P$ (i.e., $Pz = z$) be a unit vector and write $z = x+y$, $x \in \ker Q$, $y \perp \ker Q$. Since $\|z\|^2 = \|x\|^2 + \|y\|^2 = 1$, it follows that $\|x\|+\|y\| \leq \sqrt{2}$. We have

$$1 = \|Pz\| = \|Px + Py\| \leq \|Px\|+\|Py\| < (\sqrt{2}/2)(\|x\|+\|y\|) \leq 1,$$

a contradiction. Hence, $\|P - Q\| \geq \sqrt{2}/2$. □

It readily follows that $\delta_2 = \eta_2 = \sqrt{2}/2$. This is the only known value of δ_k or η_k $(k \geq 2)$. If $k = 3$,

$$P_3 = 1/3 \begin{pmatrix} 1 & 1 & 1 \\ 1 & 1 & 1 \\ 1 & 1 & 1 \end{pmatrix} \quad \text{and} \quad Q_3 = 2/3 \begin{pmatrix} 0 & 1 & 1 \\ 0 & 0 & 1 \\ 0 & 0 & 0 \end{pmatrix} = 2/3[q_3 + (q_3)^2],$$

then $P_3 \in P(\mathbb{C}^3)$ (P is a rank one projection), $Q_3 \in N(\mathbb{C}^3)$ and $\|P_3 - Q_3\|$ is equal to $2/3$. Thus, $\frac{1}{2} \leq \eta_3 \leq \delta_3 \leq 2/3$.

This result is certainly better than the poor estimate $\delta_3 \leq \frac{1}{2} + \sin \frac{1}{2}\pi = 3/2$ given by Theorem 2.12 and the same is true for the k-dimensional analogues $P_k = 1/k \sum_{i=1}^{k}\sum_{j=1}^{k} e_i \otimes e_j$ and $Q_k = (2/k)[q_k + (q_k)^2 + (q_k)^3 + \ldots + (q_k)^{k-1}]$ of P_3 and Q_3, respectively, *for small values of* k. However, there is some numerical evidence that the estimate $\frac{1}{2} + \sin\frac{\pi}{(m+1)}$ is much better than $\|P_k - Q_k\|$ for all $k \geq 20$.

CONJECTURE 2.16. $\delta_k = \eta_k$ for all $k \geq 2$, $\{k(\eta_k - \frac{1}{2})\}_{k=2}^{\infty}$ has a limit and

$$\tfrac{1}{2}\pi < \lim(k \to \infty)\ k(\eta_k - \tfrac{1}{2}) < 2\pi.$$

2.2.3 A companion problem: The distance from the set of all non-zero idempotents to $N(H)$

Clearly, $P(H)$ is not invariant under similarities. The smallest set invariant under similarities containing $P(H)$ is the set of all non zero idempotents acting on H, $E(H) = \{WPW^{-1}: P \in P(H), W \in G(H)\} = \{E \in L(H): E^2 = E \neq 0\}$. We shall see later that if R_1 and R_2 are two subsets of $L(H)$, where H is an infinite dimensional space, then $R_1^- \cap R_2^-$ is "very large", in general; namely, when R_j^- contains an operator R_j such that $\tilde{R}_j$ is not an algebraic element of $A(H)$, for j = 1,2. (See Chapter of the second monograph.) Very little is known about

$$\text{dist}[R_1, R_2] = \inf\{\|R_1 - R_2\|: R_j \in R_j, j = 1,2\}$$

for the case when one of the sets R_1^- or R_2^- is a subset of the algebraic operators (or, more generally, operators whose images are algebraic elements of the Calkin algebra).

In particular, what is the value of $\eta'_\infty = \text{dist}[E(H), N(H)]$ for the case when H is infinite dimensional? What is the answer if $H = \mathbb{C}^k$ or $N(H)$ is replaced by $N_k(H)$ (and dim $H = \infty$) for some $k \geq 2$?

Clearly, if

$$\delta'_k = \text{dist}[E(\mathbb{C}^k), N(\mathbb{C}^k)]$$

and

$$\eta'_k = \text{dist}[E(H), N_k(H)] \quad (H \text{ infinite dimensional}),$$

then $0 \leq \eta'_k \leq \delta'_k$, for all $k \geq 2$.

PROPOSITION 2.17. *(i)* $\delta'_k = 1/k$, *but* $\|E - Q\| > 1/k$ *for all* E *in* $E(\mathbb{C}^k)$ *and all* Q *in* $N(\mathbb{C}^k)$ $(k \geq 2)$.

(ii) $0 \leq \eta'_k \leq 1/k$ *and* $\eta'_\infty = 0$.

PROOF. (i) Let $\{e_j\}_{j=1}^k$ be the canonical ONB of $\mathbb{C}^k$ and let $P_k = (1/k)\sum_{i,j=1}^k e_i \otimes e_j$ and $W_{k,n} = \sum_{j=1}^k n^{k-j} e_j \otimes e_j$ $(n = 1,2,\ldots)$. Straightforward computations show that P_k is a rank one projection, $W_{k,n}$ is invertible,

$$E_{k,n} = W_{k,n} P_k W_{k,n}^{-1} = \sum_{i,j=1}^k (n^{j-i} k^{-1})\, e_i \otimes e_j \in E(\mathbb{C}^k)$$

and

$$Q_{k,n} = \sum_{1 \leq i < j \leq k} (n^{j-i} k^{-1})\, e_i \otimes e_j \in N(\mathbb{C}^k).$$

Since

$$\|E_{k,n} - Q_{k,n}\| = \|\sum_{1 \leq j \leq i \leq k} (n^{j-i} k^{-1})\, e_i \otimes e_j\|$$

$$= (1/k)\|\sum_{i=0}^{k-1} \sum_{r=1}^{k-i} n^{-r}\, e_{r+i} \otimes e_r\|$$

$$\leq (1/k)\{\|\sum_{r=1}^k e_r \otimes e_r\| + (1/n)\sum_{i=1}^{k-1}\|\sum_{r=1}^{k-i} e_{i+r} \otimes e_r\|\} < 1/k + 1/n,$$

we see that $\delta_k' \le 1/k$.

On the other hand, if $F \in E(\mathbb{C}^k)$ and $\{f_j\}_{j=1}^k$ is an ONB, then $\{1\} \subset \sigma(F) \subset \{0,1\}$ and

$$\text{trace } F = \sum_{j=1}^k \langle Ff_j, f_j\rangle = \text{rank } F \ge 1,$$

so that $\langle Ff_j, f_j\rangle \ge 1/k$ for (at least) one value of j, $1 \le j \le k$.

Given F as above and $Q \in N(\mathbb{C}^k)$, we can find an ONB $\{f_j\}_{j=1}^k$ so that Q is strictly upper triangular with respect to that basis, i.e., $\langle Qf_i, f_j\rangle = 0$ for $1 \le j \le i \le k$. In particular, we have

$$\|F - Q\| \ge \max_{1\le j\le k} |\langle (F-Q)f_j, f_j\rangle| = \max_{1\le j\le k} |\langle Ff_j, f_j\rangle| \ge 1/k,$$

so that $\|F - Q\| \ge 1/k$, i.e., $\delta_k' \ge 1/k$.

Furthermore, since $\langle Qf_i, f_j\rangle = 0$ for $1 \le j \le i \le k$, it easily follows that, either $\|F - Q\| > 1/k$ or $\langle Ff_i, f_j\rangle = 0$ for $1 \le j < i \le k$ and $\langle Ff_j, f_j\rangle = 1/k$ for all $j = 1,2,\ldots,k$; i.e., F admits an upper triangular matrix with $1/k$ in all the diagonal entries with respect to the ONB $\{f_j\}_{j=1}^k$. But the second possibility implies that $\sigma(F) = \{1/k\}$, a contradiction (recall that $k \ge 2$). Hence, $\|F - Q\| > \delta_k' = 1/k$ for all F in $E(\mathbb{C}^k)$ and all Q in $N(\mathbb{C}^k)$.

Since the second statement is a trivial consequence of the first one, we are done. □

Clearly, the trace argument cannot be applied in the case when H is infinite dimensional, but the above result suggests the following

CONJECTURE 2.18. $\eta_k' = \delta_k' = 1/k$, but $\|E - Q\| > 1/k$ for all E in $E(H)$ and all Q in $N_k(H)$ $(k \ge 2)$.

The last result of this section says that the above conjecture is true at least for $k = 2$.

PROPOSITION 2.19. *If* $2 \le h = \dim H \le \infty$, *then*

$$\eta_2' = \delta_2' = \inf\{\|E - Q\|: \ E \in E(H),\ Q \in N_2(H)\} = \tfrac{1}{2},$$

but this infimum cannot be attained for any pair (E,Q), $E \in E(H)$, $Q \in N_2(H)$.

PROOF. If $E = 1$, then it is clear that $\|E - Q\| = \|1 - Q\| \ge \text{sp}(1-Q) = 1$ for every quasinilpotent Q. Let

$$E = \begin{pmatrix} 1 & M \\ 0 & 0 \end{pmatrix} \begin{matrix} \text{ran } E \\ (\text{ran } E)^\perp \end{matrix}$$

be the matrix of $E \in E(H)\backslash\{1\}$ with respect to the decomposition $H = \text{ran } E \oplus (\text{ran } E)^\perp$ and let

$$Q = \begin{pmatrix} \cdot & \cdot \\ C & D \end{pmatrix}$$

be the matrix of $Q \in L(H)$ (with respect to the same decomposition).

It is immediate that $Q \in N_2(H)$ if and only if

$$A^2+BC = D^2+CB = AB+BD = CA+DC = 0.$$

Assume that $\|E-Q\| \le \frac{1}{2}$; then $\|1-A\| \le \frac{1}{2}$ and $\|D\| \le \frac{1}{2}$ and therefore $\sigma(A) \subset \{\lambda: |1-\lambda| \le \frac{1}{2}\}$ and $\sigma(D) \subset \{\lambda: |\lambda| \le \frac{1}{2}\}$. Thus, by the spectral mapping theorem,

$$\sigma(A^2) = \sigma(-BC) \subset \{\lambda: \operatorname{Re}\lambda \ge \tfrac{1}{4}\}$$

and

$$\sigma(D^2) = \sigma(-CB) \subset \{\lambda: |\lambda| \le \tfrac{1}{4}\}.$$

Since $\sigma(-BC)\backslash\{0\} = \sigma(-CB)\backslash\{0\}$ (see, e.g., [119], [153], [172])and A is invertible (recall that $\|1-A\| \le \frac{1}{2} < 1$), it readily follows that $\sigma(-BC) = \sigma(A^2) = \{\frac{1}{4}\} \subset \sigma(-CB) = \sigma(D^2) \subset \{0,\frac{1}{4}\}$ and $\sigma(A) = \{\frac{1}{2}\} \subset \sigma(D) \subset \{0,\frac{1}{2}\}$. Hence, $\|E-Q\| \ge \|1-A\| \ge sp(1-A) = \frac{1}{2}$.

Assume that $\|E-Q\| = \frac{1}{2}$. Since $\sigma(A) = \{\frac{1}{2}\} = \sigma_\ell(A)$, there exists a sequence $\{x_n\}_{n=1}^{\infty}$ of unit vectors in ran E such that $\|(A-\frac{1}{2})x_n\| \to 0$ $(n \to \infty)$. On the other hand, $A^2 = -BC$ and A invertible imply that $\|Cx\| \ge \varepsilon\|x\|$ for some $\varepsilon > 0$ and for all x in ran E, so that

$$\begin{aligned}\|E-Q\| &\ge \limsup(n\to\infty)\ \|(E-Q)x_n\| \\ &\ge \limsup(n\to\infty)\{\|(1-A)x_n\|^2+\|-Cx_n\|^2\}^{\frac{1}{2}} \\ &\ge (\tfrac{1}{4}+\varepsilon^2)^{\frac{1}{2}} > \tfrac{1}{2},\end{aligned}$$

a contradiction. Hence

$$\|E-Q\| > \tfrac{1}{2} \text{ for all } Q \text{ in } N_2(H).$$

On the other hand, it readily follows from Proposition 2.17 that

$$\inf\{\|E-Q\|:\ E \in E(H),\ Q \in N_2(H)\} \le \tfrac{1}{2}. \qquad \square$$

2.3 On the distance to $N_k(H)$

2.3.1 A general upper bound

LEMMA 2.20. *Suppose that* $T \in L(H)$, $\|T\| \le 1$ *and* $\|T^k\| \le \varepsilon$ *for some* $k \ge 2$ *and some* ε, $0 < \varepsilon \le 1$. *Let* $(T^*T)^{\frac{1}{2}} = \int_{[0,1]} \lambda\, dE$ *(spectral decomposition) and let* $P = E([0,\sqrt{\varepsilon}])$; *then*

$$\|[(1-P)T(1-P)]^{k-1}\| \le (k-1)\sqrt{\varepsilon}.$$

PROOF. Clearly, $\|TP\| \le \sqrt{\varepsilon}$, and

$$\varepsilon^2\|x\|^2 \geq \|T^k x\|^2 = \|(T^*T)^{\frac12}T^{k-1}x\|^2$$
$$= \|P(T^*T)^{\frac12}T^{k-1}x+(1-P)(T^*T)^{\frac12}T^{k-1}x\|^2$$
$$\geq \|(1-P)(T^*T)^{\frac12}T^{k-1}x\|^2 = \|(T^*T)^{\frac12}(1-P)T^{k-1}x\|^2$$
$$\geq \varepsilon\|(1-P)T^{k-1}x\|^2,$$

for all x in H. Hence, $\|(1-P)T^{k-1}\| \leq \sqrt{\varepsilon}$.

On the other hand,

$$(1-P)T^{k-1} = [(1-P)T]^{k-1}+(1-P)TPT[(1-P)T]^{k-3}+(1-P)T^2PT[(1-P)T]^{k-4}+ \dots +(1-P)T^{k-2}PT,$$

so that

$$\|[(1-P)T(1-P)]^{k-1}\| \leq \|[(1-P)T]^{k-1}\| \leq \|(1-P)T^{k-1}\|+ \{\|(1-P)TPT[(1-P)T]^{k-3}\|+\|(1-P)T^2PT[(1-P)T]^{k-4}\|+\dots \|(1-P)T^{k-2}PT\|\} \leq \|(1-P)T^{k-1}\|+(k-2)\|TP\| \leq (k-1)\sqrt{\varepsilon}. \square$$

THEOREM 2.21. *If* $T \in L(H)$, $\|T\| \leq 1$ *and* $\|T^k\| \leq \varepsilon$ *for some* $k \geq 2$ *and some* ε, $0 < \varepsilon \leq 1$, *then*

$$\operatorname{dist}[T, N_k(H)] \leq \phi_k(\varepsilon), \tag{2.7}$$

where $\phi_k(\varepsilon)$ *is a continuous, positive, non-decreasing function defined on* $(0,1]$ *such that* $\lim(\varepsilon \to 0)\ \phi_k(\varepsilon) = 0$.

Moreover, $\phi_k(\varepsilon)$ *can be inductively defined by* $\phi_2(\varepsilon) = (2\varepsilon)^{\frac12}$ *and* $\phi_k(\varepsilon) = \{\varepsilon+\phi_{k-1}((k-1)\sqrt{\varepsilon})^2\}^{\frac12}$, *for* $k = 3,4,\dots$.

PROOF. Let P be defined as in Lemma 2.20 and let

$$T = \begin{pmatrix} E_1 & T_1 \\ E_2 & T_2 \end{pmatrix} \begin{matrix} \operatorname{ran} P \\ \ker P \end{matrix}$$

be the matrix of T with respect to the decomposition $H = \operatorname{ran} P \oplus \ker P$.

If k = 2, define

$$T' = PT(1-P) = \begin{pmatrix} 0 & T_1 \\ 0 & 0 \end{pmatrix}.$$

Since

$$\begin{pmatrix} E_1 & 0 \\ E_2 & 0 \end{pmatrix}\Big|\ker P = 0, \quad \begin{pmatrix} 0 & 0 \\ 0 & T_2 \end{pmatrix}\Big|\operatorname{ran} P = 0$$

and $\ker P \perp \operatorname{ran} P$, it follows from Lemma 2.20 and its proof that

$$\|T - T'\| = \left\|\begin{pmatrix} E_1 & 0 \\ E_2 & T_2 \end{pmatrix}\right\| \leq \left\{\left\|\begin{pmatrix} E_1 & 0 \\ E_2 & 0 \end{pmatrix}\right\|^2 + \left\|\begin{pmatrix} 0 & 0 \\ 0 & T_2 \end{pmatrix}\right\|^2\right\}^{\frac12} \leq (\varepsilon+\varepsilon)^{\frac12} = (2\varepsilon)^{\frac12}.$$

Assume that $k \geq 3$ and (2.7) holds for $j \leq k-1$, with $\phi_2(\varepsilon) = (2\varepsilon)^{\frac12}$ and $\phi_j(\varepsilon) = \{\varepsilon+\phi_{j-1}((j-2)\sqrt{\varepsilon})^2\}^{\frac12}$, for $j = 3,4,..,k-1$.

Clearly, $\|T_2\| = \|(1-P)T(1-P)\| \leq 1$ and, by Lemma 2.20, $\|T_2{}^{k-1}\| \leq$

$(k-1)\sqrt{\varepsilon}$. Thus, by our inductive construction, we can find $T_2' \in N_{k-1}$ (ran P) such that $\|T_2-T_2'\| \le \phi_{k-1}((k-1)\sqrt{\varepsilon})$. Define

$$T' = \begin{pmatrix} 0 & T_1 \\ 0 & T_2' \end{pmatrix}.$$

It is easily seen that $T' \in N_k(H)$ and a formal repetition of our previous argument (for the case when $k = 2$) shows that

$$\|T - T'\| \le \phi_k(\varepsilon) = (\text{def}) \ \{\varepsilon+\phi_{k-1}((k-1)\sqrt{\varepsilon})^2\}^{\frac{1}{2}}. \qquad \square$$

COROLLARY 2.22. *Let* $p(\lambda) = \Pi_{j=1}^{m} (\lambda-\lambda_j)^{k_j}$ *($\lambda_i \ne \lambda_j$ if $i \ne j$) be a polynomial and let* $M_p(H) = \{A \in L(H): \ p(A) = 0\}$. *Given* $\varepsilon > 0$, *there exists* $\delta > 0$ *such that, if* $\|T\| \le 1$ *and* $\|p(T)\| \le \delta$, *then*

$$\text{dist}[T, M_p(H)] < \varepsilon.$$

PROOF. If $m = 1$, then the result follows from Theorem 2.21. Assume that $m \ge 2$. There exists $\varepsilon_1 > 0$ such that the m open disks $D_1, D_2, \ldots, D_m$ of radius ε_1 centered at $\lambda_1, \lambda_2, \ldots, \lambda_m$, respectively, are pairwise disjoint. Let $\varepsilon_o > 0$ be such that if $|p(\lambda)| < \varepsilon_o$, then λ is contained in one of these disks.

Let

$$P_j = \frac{1}{2\pi i} \int_{\partial D_j} (\lambda-T)^{-1} \, d\lambda.$$

We shall show that P_k is a bounded function of $\|p(T)\|$, $\|T\|$, ε_o, $p(\lambda)$ for small $\|p(T)\|$. Observe that $p(\lambda) - p(T) = (\lambda-T)q(\lambda,T)$ for a polynomial q in the variables λ, T. Thus

$$1 - p(T)p(\lambda)^{-1} = (\lambda-T)q(\lambda,T)p(\lambda)^{-1}.$$

Now, for λ such that $|\lambda-\lambda_j| = \varepsilon_1$ we have that $\|p(\lambda)^{-1}p(T)\|$ is small if $\|p(T)\|$ is. Let $E = p(\lambda)^{-1}p(T)$; then

$$(\lambda-T)^{-1} = p(\lambda)^{-1}q(\lambda,T)(1-E)^{-1} = p(\lambda)^{-1}q(\lambda,T)+J,$$

where $\|J\| \le \|p(\lambda)^{-1}q(\lambda,T)\| \cdot \|E\|/(1-\|E\|)$ is small since $\|p(\lambda)^{-1}q(\lambda,T)\|$ is bounded by a funtion of $\|T\|$ and $p(\lambda)$. Thus the P_j's are bounded ($j = 1, 2, \ldots, m$).

On the other hand,

$$\|(T-\lambda_j)^{k_j}P_j\| = \|\frac{1}{2\pi i} \int_{\partial D_j} p(\lambda)^{-1}(\lambda-\lambda_j)^{k_j}q(\lambda,T) + (\lambda-\lambda_j)^{k_j}J \, d\lambda\|$$

$$= \|\frac{1}{2\pi i} \int_{\partial D_j} (\lambda-\lambda_j)^{k_j} J d\lambda\| \le \|J\| \cdot \varepsilon_o^{k_j+1}.$$

Since each $(T-\lambda_j)^{k_j}P_j$ is small (provided $\|p(T)\|$ is small) we may use Theorem 2.21 to perturb TP_j to a T_j' in $L(\text{ran } P_j)$ such that $(T_j'-\lambda_j)$ is a nilpotent of order at most k_j.

If $H = \text{ran } P_j \dot{+} \ker P_j$, define $T_j = T_j' \dot{+} 0$ with respect to this decomposition ($j = 1,2,\ldots,m$) and $T' = \sum_{j=1}^{m} T_j$. Then $p(T') = 0$ and

$$\|T - T'\| \leq \max\{\|P_j\|\cdot\|T_j' - T|\text{ran } P_j\| : \ 1 \leq j \leq m\} < \varepsilon,$$

provided $\|p(T)\| < \delta$ for some $\delta > 0$ small enough. □

Let T, k and ε be as in Theorem 2.21 and let η, $0 < \eta \leq 1$, be the k-th root of ε (i.e., $\|T^k\| \leq \eta^k$) and define $\psi_k(\eta) = \phi_k(\varepsilon)$, $k = 2,3,\ldots$; then Theorem 2.21 implies that, if $k = 2$, then $\text{dist}[T,N_2(H)] \leq \sqrt{2\eta}$; if $k = 3$, then $\text{dist}[T,N_3(H)] \leq (\eta^3+4\eta^{3/2})^{\frac{1}{2}}, \ldots$, and $\text{dist}[T,N_k(H)] = O(\varepsilon^{2^{-k}}) = O(\eta^{k2^{-k}})$ for all $k \geq 2$.

These results can be slightly improved if, instead of $P = E([0,\sqrt{\varepsilon}])$, we choose $P = E([0,\{2\varepsilon^2/(k-2)\}^{1/3}])$ $(k \geq 3)$ in Lemma 2.20. With this choice of P, it is possible to show that $\psi_k(\eta)$ can actually be defined so that

$$\text{dist}[T,N_k(H)] \leq \psi_k(\eta) = O(\eta^{(9k/8)(2/3)^k})$$

($k = 3,4,\ldots$), but these results seem to be very far from the best possible ones.

CONJECTURE 2.23. There exists a continuous function $\psi(\eta)$, defined on $[0,1]$, such that $\psi(\eta) > 0$ on $(0,1]$, $\psi(0) = 0$ and $\text{dist}[T,N_k(H)] \leq \psi(\eta)$ for all T in $L(H)$ such that $\|T\| \leq 1$ and $\|T^k\| \leq \eta^k$ (i.e., the functions $\psi_k(\eta)$ can be replaced by a single one).

2.3.2 Two illustrative examples

The rough argument of the proof of Proposition 1.10 might suggest that those estimates are very poor. However, Lemma 2.13 shows that the "very poor" lower estimates given by Proposition 1.10 are actually the best possible except, perhaps, for a constant factor independent of k. Indeed, if T and L have the form of that lemma, then $\text{nul } T^{m+1} = 2(m+1) > \text{nul } L^{m+1} = 2m+1$. Thus, if W (U) is an invertible (unitary, resp.) operator, we can always find a unit vector $x = x(W)$ ($= x(U)$, resp.) such that $\|L^{m+1}x\| = 1$, but $(WTW^{-1})^{m+1}x = 0$ ($(UTU^*)^{m+1}x = 0$, resp.), whence it readily follows from Proposition 1.10(i) ((ii), resp.) that $\text{dist}[L, S(T)] \geq 2^{1/(m+1)} - 1$ ($\text{dist}[L,U(T)] \geq 1/(m+1)$, resp.).

An even more surprising example can be constructed on the same lines. We shall need the following auxiliary result (With the notation of Lemma 2.13):

COROLLARY 2.24. *(i) Let* $\{g_n\}_{-\infty<n<\infty}$ *be an ONB of* H *and let* U *be the bilateral shift defined by* $Ug_n = g_{n+1}$ *for all* $n \in \mathbb{Z}$.

Let $H_+ = \bigvee\{g_n\}_{n\geq 0}$ ($H_- = \bigvee\{g_n\}_{n<0}$) *and let* $A \in L(H_+)$ ($B \in L(H_-)$, *resp.*) *be the forward (backward, resp.) unilateral shift defined by* $Ag_n = g_{n+1}$ *for all* $n \geq 0$ ($Bg_{-1} = 0$ *and* $Bg_n = g_{n+1}$ *for all* $n < -1$).

Finally, let $q_m = \sum_{t=1}^{m-1} k_t \otimes k_{t+1} \in L(\mathbb{C}^m)$. *Then there exists a unitary operator* $V_m : H \to \mathbb{C}^m \oplus H$ *such that*

$$\|A\oplus B - V_m(q_m \oplus U)V_m^*\| = s(m), \quad V_m g_0 = k_1 \text{ and } V_m g_{-1} = -k_m.$$

(ii) If $k \geq 3$ *and* $1 \leq m \leq [(k-1)/2]$, *there exists a unitary operator* $W_{k,m} : (\mathbb{C}^m)^{(\infty)} \oplus H \to (\mathbb{C}^k)^{(\infty)}$ *such that*

$$\|q_k^{(\infty)} - W_{k,m}(q_m^{(\infty)} \oplus U)W_{k,m}^*\| = s(m).$$

PROOF. (i) For suitably defined $A' \simeq A$ and $B' \simeq B$, we have

$(q_m \oplus U \to A'\oplus B')$

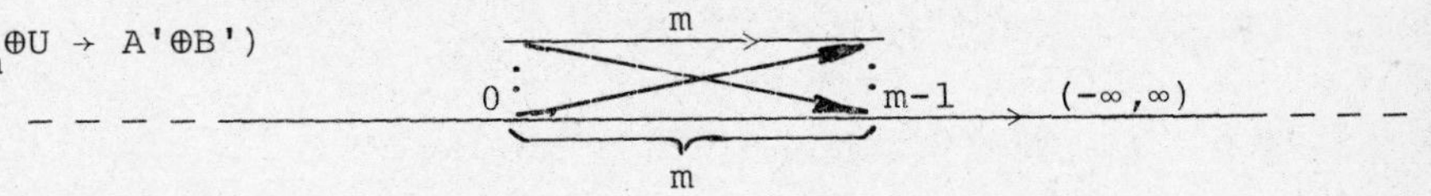

whence we obtain

$A'\oplus B'$:

$\|(q_m\oplus U) - (A'\oplus B')\| = s(m)$ and the unitary operator V_m can be chosen so that $V_m(A'\oplus B')V_m^* = A\oplus B$, $V_m g_0 = k_1$ and $V_m g_{-1} = -k_m$.

(ii) Since $r = k-2m \geq 1$, there exists $U' \simeq U$ such that

$(q_k^{(\infty)} \to q_m^{(\infty)} \oplus U')$

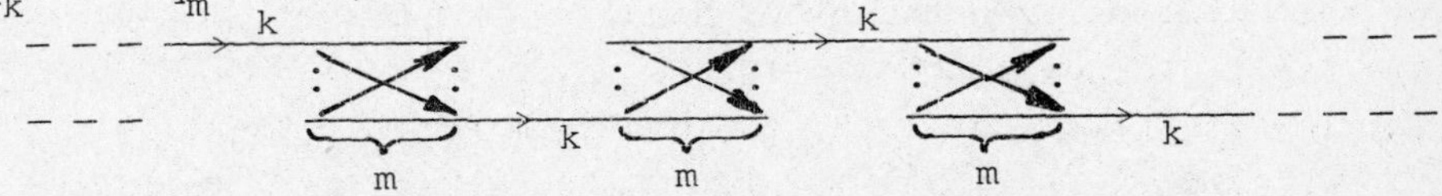

(the fact that $r \geq 1$ guarantees that we can consistently apply the argument of Lemma 2.13 to each step.), with $\|q_k^{(\infty)} - q_m^{(\infty)} \oplus U'\| = s(m)$, whence the result follows. □

EXAMPLE 2.25. Let S be a unilateral shift of multiplicity one and let $T \in L(H)$ be unitarily equivalent to $S^{(\infty)} \oplus S^{*(\infty)}$; then $\|T^k\| = 1$ for all $k \geq 1$, $\sigma(T) = \sigma_{\ell re}(T) = \{\lambda: \ |\lambda| \ \leq 1\}$ and

$$2^{1/k} - 1 \leq \text{dist}[T, N_k(H)] \leq 4 \sin \pi/([(k-1)/2]+1) < 8\pi/k,$$

for all $k \geq 3$.

PROOF. It is not difficult to see that $\|T^k\| = 1$ for all $k \geq 1$ and that $\text{nul}(\lambda-T) = \text{nul}(\lambda-T)^* = \infty$ for all λ in the open unit disk $D = \{\lambda: |\lambda| < 1\}$, whence it follows that $\sigma(T) = \sigma_{\ell re}(T) = D^-$.

Let $Q \in N_k(H)$. Since $Q^k = 0$, it follows from Proposition 1.10(iii) that $\|T - Q\| \geq 2^{1/k} - 1$ ($k = 1,2,3,\ldots$). On the other hand, if $k \geq 3$, $m =$

$[(k-1)/2]$ and the operators are defined with respect to a suitable ONB of H, it is not difficult to infer from Corollary 2.24(i) and (ii) that $((q_k{}^{(\infty)})^{(\infty)} \simeq q_k{}^{(\infty)})$:

$$\|(q_k{}^{(\infty)})^{(\infty)} - T\| \le \|(q_k{}^{(\infty)})^{(\infty)} - (q_m \oplus U')^{(\infty)}\| + \|(q_m \oplus U')^{(\infty)} - T\|$$
$$= \|q_k{}^{(\infty)} - q_m{}^{(\infty)} \oplus U'\| + \|q_m \oplus U' - S \oplus S^*\| = 2s(m)$$
$$= 4\{\sin \pi/([(k-1)/2]+1)\} < 8\pi/k.$$

Our next example shows that the infinite dimensional ampliations of finite dimensional examples can produce certain surprises. (An operator T in $L(H)$ is called an *ampliation* if $T \simeq A \otimes 1 \simeq A^{(\infty)}$ for some operator $A \in L(H_o)$, $1 \le \dim H_o \le \infty$.)

EXAMPLE 2.26. Let $k > h \ge 3$; then the operators $A = q_h{}^{(k)}$ and $B = q_k{}^{(h)}$ $(A, B \in L(\mathbb{C}^{kh}))$ satisfy

$$\text{dist}[B, S(A)] = 1; \tag{2.8}$$

however,

$$2^{1/k} - 1 \le \text{dist}[B^{(\infty)}, S(A^{(\infty)})] \le 4\{\sin \pi/([(k-1)/2]+1) + \sin \pi/([(h-1)/2]+1)\} < 8\pi(1/k+1/h). \tag{2.9}$$

PROOF. Observe that nul $A = k >$ nul $B = h$. Thus, $\text{dist}[B,S(A)] \ge 1$. (Use Proposition 1.10(i) as in our previous observations at the beginning of this section.) On the other hand, $A \sim \varepsilon A$ for all $\varepsilon > 0$, so that $0 \in S(A)^-$, whence we obtain (2.8).

The lower estimate of (2.9) follows from Proposition 1.10(iii) $([A^{(\infty)}]^h = 0$, $\|[B^{(\infty)}]^h\| = \|B^{(\infty)}\| = 1$.) The upper estimate follows from the proof of Example 2.25: For suitably chosen $T \simeq S^{(\infty)} \oplus S^{*(\infty)}$, $A' \simeq A$ and $B' \simeq B$, we have

$$\|B' - A'\| \le \|B' - T\| + \|T - A'\| \le 4\{\sin \pi/([(k-1)/2]+1) + \sin \pi/([(h-1)/2]+1)\} < 8\pi(1/k+1/h). \qquad \square$$

Example 2.26 suggests that, if H is infinite dimensional, then $N^1(H) = \{Q \in N(H): \|Q\| \le 1\}$ "looks like" $\mathbb{Q}+i\mathbb{Q}$ (where $\mathbb{Q}$ denotes the set of all rational numbers), in the following sense: Observe that $\mathbb{Q}+i\mathbb{Q} = \cup_{k=1}^{\infty} (\mathbb{Q}_k+i\mathbb{Q}_k)$, where $\mathbb{Q}_k = \{m/n: m \in \mathbb{Z}, 1 \le n \le k\}$, $\mathbb{Q}_k+i\mathbb{Q}_k$ is a "large" nowhere dense subset of $\mathbb{Q}+i\mathbb{Q}$,

$$\max\{\text{dist}[\lambda, \mathbb{Q}_k+i\mathbb{Q}_k]: \lambda \in \mathbb{Q}+i\mathbb{Q}\} < 1/(k\sqrt{2}) \to 0 \quad (k \to \infty)$$

and $(\mathbb{Q}+i\mathbb{Q})^-$ (in the complex plane) coincides with the much larger set $\mathbb{C}$, and $N^1(H) = \cup_{k=1}^{\infty} [N^1(H) \cap N_k(H)]$.

CONJECTURE 2.27. $\max\{\mathrm{dist}[Q, N^1(H)\cap N_k(H)]: Q \in N^1(H)\} \le C/k$ for some constant $C > 0$ independent of k.

An affirmative answer to this conjecture would provide some heuristic explanation to the wild structure of $N^1(H)^-$ (see Chapter V).

2.3.3 An example on approximation of normal operators by nilpotents

Let $k > 50$, $p = [\sqrt{k}/2\sqrt{\pi}]$, $\eta = p^{-1}$ and $r = [k\eta/2]$ and let $Q_k \in L(\mathbb{C}^k)$ be the operator defined by $Q_k = \sum_{j=1}^{k-1} \alpha_j \; e_{j+1}\otimes e_j$ with respect to the ONB $\{e_j\}_{j=1}^k$ of $\mathbb{C}^k$, where

$$\alpha_j = \begin{cases} n\eta, & \text{for } r(n-1) < j \le rn,\ n = 1,2,\dots,p, \\ n\eta, & \text{for } r(2p-n) < j \le r(2p-n+1),\ n = 1,2,\dots,p, \\ 0, & \text{for } 2rp < j \le k-1. \end{cases}$$

(Roughly speaking: Q_k is a truncated weighted shift; the weights α_j grow from η to 1 through p steps of length r and then go down from 1 to 0 through p steps of length r, so that the upper step has length 2r with weights equal to 1, i.e., $\alpha_j = 1$ for $r(p-1) < j \le r(p+1)$.)

In the first modification, we shall "ignore" the coördinates 1,2, ...,r(p-1) and r(p+1)+1,r(p+1)+2,...,k and apply Corollary 2.14 to the subspace $\vee\{e_{r(p-1)+1}, e_{r(p-1)+2}, \dots, e_{r(p+1)}\}$. It is easily seen that we can modify Q_k in order to obtain an operator $R_1' = T_1'\oplus U_{r+1}$, where $U_{r+1} \simeq e_1\otimes e_{r+1} + \sum_{j=1}^{r} e_{j+1}\otimes e_j$ is a unitary operator acting on a subspace of dimension r+1 and there exists an orthonormal system $\{f_1^1, f_2^1, \dots, f_{2r}^1\}$ such that $\{e_j\}_{j=1}^{r(p-1)} \cup \{f_j^1\}_{j=1}^{2r} \cup \{e_j\}_{j=r(p+1)+1}^{k}$ is an ONB of $\mathbb{C}^k$, $T_1'e_j = Q_k e_j$ for $j \notin (r(p-1), r(p+1)]$, $T_1'e_{r(p-1)} = f_1^1$, $T_1'f_h^1 = f_{h+1}^1$ for $h = 1,2,\dots, r-2$, $T_1'f_{r-1}^1 = (1-\eta)e_{r(p+1)+1}$ and $U_{r+1}f_h^1 = (\omega_{r+1})^h f_h^1$ for $h = r,r+1,\dots,2r$ ($-\omega_{r+1}$ is a primitive (r+1)-th root of 1); furthermore, $\|Q_k - R_1'\| = s(r-1)$.

Let T_1 be the operator obtained from T_1' by replacing each weight equal to 1 by $1-\eta$ and $R_1 = T_1\oplus U_{r+1}$; then $\|Q_k - R_1\| \le s(r-1)+\eta$.

Now we can apply the same argument to R_1 in order to obtain an operator $R_2 = T_2\oplus(1-\eta)U_{2r}\oplus U_{r+1}$, where U_{2r} is a unitary operator acting on a subspace of dimension 2r, whose eigenvalues are equal to minus the 2r 2r-th roots of 1, $T_2e_j = Q_k e_j$ for all $j \notin (r(p-2), r(p+2)]$, $T_2e_{r(p-2)} = (1-2\eta)f_1^2$, $T_2f_h^2 = (1-2\eta)f_{h+1}^2$ for $h = 1,2,\dots,r-2$, $T_2f_{r-1}^2 = (1-2\eta)e_{r(p+2)+1}$, $\{f_1^2, f_2^2, \dots, f_{r-1}^2\}$ is an orthonormal system that spans a subspace orthogonal to the span of the vector $(\{e_j\}_{j=1}^{r(p-2)} \cup \{e_j\}_{j=r(p+2)+1}^{k})$, $\|R_1 - R_2\| \le (1-\eta)[s(r-1)+\eta]$ and this second modification only affects the vectors in the subspace spanned by $(\{e_j\}_{j=r(p-2)+1}^{r(p-1)} \cup \{e_j\}_{j=r(p+1)+1}^{r(p+2)})$, so that $\|Q_k - R_2\| = \max\{\|Q_k - R_1\|, \|R_1 - R_2\|\} \le s(r-1)+\eta$, etc.

An easy inductive argument shows that after p-1 steps we finally obtain an operator

$$L_k = U_{r+1} \oplus [\oplus_{j=1}^{p-1} (1-j\eta)U_{2r}] \oplus q_1^{(k+1-(2p-1)r)}$$

such that $\|Q_k - L_k\| \le s(r-1)+\eta < 2\pi/r+[\sqrt{k}/2\sqrt{\pi}] < 5(\pi/k)^{1/2}$ for all $k > 50$. On the other hand, if $1 \le k \le 50$, then $5(\pi/k)^{1/2} > 1$. Thus, we have the following

PROPOSITION 2.28. *(i) For each* $k \ge 1$ *there exists a normal operator* $L_k \in L(\mathbb{C}^k)$ *such that* $\|L_k\| = 1$ *and* $\text{dist}[L_k, N(\mathbb{C}^k)] < 5(\pi/k)^{1/2}$.

(ii) If H is infinite dimensional, there exists a normal operator M *such that* $\sigma(M) = D^-$, *where* $D = \{\lambda: |\lambda| < 1\}$ *and* $\text{dist}[M, N_k(H)] < 5(\pi/k)^{1/2}$ *for all* $k = 1,2,\ldots$. *In particular,* $M \in N(H)^-$.

PROOF. (i) If $k > 50$, define L_k as above. If $1 \le k \le 50$, take $L_k = 1$, $Q_k = 0$.

(ii) Let $\{\lambda_m\}_{m=1}^{\infty}$ be an enumeration of all those points λ in D^- such that both $|\lambda_m|$ and $(\arg \lambda_m)/\pi$ are rational numbers (arg 0 is defined equal to 0) and let M be a diagonal normal operator with eigenvalues $\lambda_1, \lambda_2, \ldots, \lambda_m, \ldots$ such that $\text{nul}(\lambda_m - M) = \infty$ for all $m = 1,2,\ldots$, i.e., $M = (\text{diag}\{\lambda_1, \lambda_2, \ldots, \lambda_m, \ldots\})^{(\infty)}$.

Given k, it is easy to see that M can be written as $M \simeq (\oplus_{m=1}^{\infty} \lambda_m L_k)^{(\infty)}$, whence it readily follows that

$$\text{dist}[M, N_k(H)] \le (\sup_m |\lambda_m|)\|L_k - Q_k\| < 5(\pi/k)^{1/2}.$$

A fortiori, $M \in N(H)^-$. □

The result of Proposition 2.28(i) is, in a certain sense, the best possible. Observe that if $N_k \in L(\mathbb{C}^k)$ is normal and there exist $Q_k \in N(\mathbb{C}^k)$ and $\varepsilon_k > 0$ such that $\|N_k - Q_k\| < \varepsilon_k$, then (by Corollary 1.6 (i)) $\sigma(N_k)_{\varepsilon_k}$ is a connected set containing the origin. If the points of $\sigma(N_k)$ are more or less evenly distributed in a connected neighborhood Ω of the origin with smooth boundary (namely, $\Omega = D$), then $\sigma(N_k)_{\varepsilon_k}$ will include Ω and therefore $k\pi\varepsilon_k^2 \ge m_2(\Omega)$, where m_2 denotes the planar Lebesgue measure. Hence,

$$\varepsilon_k \ge [m_2(\Omega)/(\pi k)]^{1/2} = O(k^{-1/2}).$$

(On the other hand, if ε_k is too small, then $\sigma(N_k)_{\varepsilon_k}$ cannot be connected, a contradiction.)

CONJECTURE 2.29. There exists a constant $C > 0$ (independent of k) such that $\mathrm{dist}[N,N(\mathbb{C}^k)] \geq Ck^{-\frac{1}{2}}$ for every normal operator $N \in L(\mathbb{C}^k)$ such that $||N|| = 1$ $(k = 1,2,\ldots)$.

The following result provides some extra support to this conjecture. Observe that if $A \in L(\mathbb{C}^k)$ is hermitian and $0 \leq A \leq 1$, then the points of $\sigma(A)$ are not evenly distributed in any set of positive measure. (More precisely, $m_2(\sigma(A)_\varepsilon) \leq 2\varepsilon+\pi\varepsilon^2$ independently of k, and $2\varepsilon+\pi\varepsilon^2 \to 0$, as $\varepsilon \to 0$.)

PROPOSITION 2.30. *If* $A \in L(\mathbb{C}^k)$, $0 \leq A \leq 1$, *and* $1 \in \sigma(A)$, *then* $\mathrm{dist}[A,N(\mathbb{C}^k)] > (1/2\sqrt{k})$, $k = 1,2,\ldots$.

PROOF. Assume that $||A - Q|| \leq \varepsilon$ for some $Q \in N(\mathbb{C}^k)$, $Q = H+iJ$ (Cartesian decomposition); then $||A - H|| = ||\mathbb{R}\mathrm{e}(A-Q)|| \leq ||A - Q|| \leq \varepsilon$ and

$$\mathrm{trace}\ (H) = \mathrm{trace}\ (\mathbb{R}\mathrm{e}\ Q) = \mathbb{R}\mathrm{e}\ \mathrm{trace}\ (Q) = \mathbb{R}\mathrm{e}\ 0 = 0.$$

On the other hand, it is easily seen that $\sigma(A)_\varepsilon \supset [0,1]$ (Use Corollary 1.6(i)), so that

$$\mathrm{trace}\ (A) \geq 1+(1-2\varepsilon)+(1-4\varepsilon)+\ldots+(1-2n\varepsilon),$$

where $n = [1/2\varepsilon]$ (= integral part of $(1/2\varepsilon)$). It is clear that $1/2\varepsilon \geq n > 1/2\varepsilon-1$. Hence, $(n+1) > 1/2\varepsilon$ and

$$\mathrm{trace}\ (A) \geq (n+1) - 2\varepsilon\sum_{j=0}^{n-1} j = (n+1)-(n+1)n\varepsilon = (n+1)(1-n\varepsilon) > 1/4\varepsilon.$$

Let $A = \int \lambda\ dE$ and $H = \int \lambda\ dF$ (spectral decompositions). If $\alpha \in [0,1]$, $\varepsilon' > \varepsilon$ and rank $F((\alpha-\varepsilon',\infty))$ < rank $E([\alpha,\infty))$, then it is easily seen that there exists a unit vector $x \in \mathrm{ran}\ E([\alpha,\infty))\ominus\mathrm{ran}\ F((\alpha-\varepsilon',\infty))$ and therefore

$$||A - H|| \geq |\langle(A-H)x,x\rangle| > \alpha-(\alpha-\varepsilon) = \varepsilon, \qquad (2.10)$$

a contradiction.

Hence, rank $F([\alpha-\varepsilon,\infty)) \geq$ rank $E([\alpha,\infty))$ and, by symmetry, rank $F([\alpha,\infty)) \leq$ rank $E([\alpha-\varepsilon,\infty))$ for all $\alpha \in [0,1]$. It folows that

$$0 = \mathrm{trace}\ (H) > \mathrm{trace}\ (A) - k\varepsilon > 1/4\varepsilon - k\varepsilon.$$

Hence, $\varepsilon > 1/2\sqrt{k}$. By a compactness argument (exactly as in the proof of Theorem 2.12), we conclude that

$$\mathrm{dist}[A,N(\mathbb{C}^k)] = \min\{||A - Q||:\ Q^k = 0,\ ||Q|| \leq 2\} > 1/2\sqrt{k}.\square$$

2.3.4 On the distance to a similarity orbit

Let $T \in L(\mathbb{C}^d)$ be a *cyclic* operator with minimal polynomial p, $p(\lambda) = \Pi_{j=1}^m (\lambda-\lambda_j)^{k_j}$ ($\lambda_i \neq \lambda_j$, if $i \neq j$); then $\sum_{j=1}^m k_j = d$ and T is similar to the Jordan form $\oplus_{j=1}^m (\lambda_j+q_{k_j})$.

Let $A \in L(\mathbb{C}^d)$ be an operator with spectrum $\sigma(A) = \{\alpha_1,\alpha_2,\ldots,\alpha_n\}$

and dim $H(\alpha_j;A) = h_i$.(Clearly, $\sum_{j=1}^{n} h_i = d$.) Define $\mu_1 = \mu_2 = \ldots = \mu_{k_1} = \lambda_1$, $\mu_{k_1+1} = \mu_{k_1+2} = \ldots = \mu_{k_1+k_2} = \lambda_2, \ldots, \mu_{d-k_m+1} = \mu_{d-k_m+2} = \ldots = \mu_d = \lambda_m$ and $\beta_1 = \beta_2 = \ldots = \beta_{h_1} = \alpha_1, \beta_{h_1+1} = \beta_{h_1+2} = \ldots = \beta_{h_1+h_2} = \alpha_2, \ldots, \beta_{d-h_n+1} = \beta_{d-h_n+2} = \beta_{d-h_n+3} = \ldots = \beta_d = \alpha_n$; then A admits a representation as an upper triangular matrix of the form

$$A = \begin{pmatrix} \beta_1 & a_{12} & a_{13} & \cdot & \cdot & \cdot & a_{1d} \\ & \beta_2 & a_{23} & \cdot & \cdot & \cdot & a_{2d} \\ & & \beta_3 & \cdot & \cdot & \cdot & a_{3d} \\ & & & \cdot & & & \cdot \\ & & & & \cdot & & \cdot \\ & & 0 & & & \cdot & \cdot \\ & & & & & & \beta_d \end{pmatrix}$$

(with respect to a suitable ONB of $\mathbb{C}^d$). It is not difficult to conclude from Theorem 2.1 that

$$T \underset{sim}{\rightarrow} B = \begin{pmatrix} \mu_1 & a_{12} & a_{13} & \cdot & \cdot & \cdot & a_{1d} \\ & \mu_2 & a_{23} & \cdot & \cdot & \cdot & a_{2d} \\ & & \mu_3 & \cdot & \cdot & \cdot & a_{3d} \\ & & & \cdot & & & \cdot \\ & & & & \cdot & & \cdot \\ & & 0 & & & \cdot & \cdot \\ & & & & & & \mu_d \end{pmatrix}.$$

Moreover, the same result applies to any upper triangular representation of A. Hence, we have

COROLLARY 2.31. *Let* A *and* T *be as above; then*

$$\text{dist}[A, S(T)] \le \min_{\sigma \in \Sigma(k)} \max_{1 \le j \le k} |\mu_j - \beta_{\sigma(j)}|, \tag{2.11}$$

where $\Sigma(k)$ *denotes the set of all permutations of* k *elements.*

Unfortunately, the estimate (2.11) is very poor, in general. Namely, if L_k and Q_k have the form of Proposition 2.28(i), then q_k is cyclic, $q_k \underset{sim}{\rightarrow} Q_k$ (by Theorem 2.1) and

$$\text{dist}[L_k, S(q_k)] \le \|L_k - Q_k\| < 5(\pi/k)^{1/2} \to 0 \quad (k \to \infty).$$

However, $sp(L_k) = 1$ $(-1 \in \sigma(L_k))$ and $\sigma(q_k) = \{0\}$, so that the only information that we can obtain from Corollary 2.31 is that $\text{dist}[L_k, S(q_k)] \le 1$.

PROBLEM 2.32. Find a formula for $\text{dist}[A, S(T)]$ $(A, T \in L(\mathbb{C}^d))$.

We shall close this section with a partial answer to this problem.

COROLLARY 2.33. *If* $T = \oplus_{k=1}^{n} q_k^{(\tau_k)}$ *and* $A = \oplus_{k=1}^{n} q_k^{(\alpha_k)}$ *are finite rank operators*, rank T^j = rank A^j *for* $j = 1,2,\ldots,r$ *and* rank T^{r+1} < rank A^{r+1} *for some* $r \geq 2$, *then*

$$2^{1/(r+1)} - 1 \leq \text{dist}[A, S(T)] \leq 2\, s([(r-1)/2]).$$

PROOF. The lower estimate follows from Proposition 1.10(i).

In order to obtain the upper estimate, we can directly assume without loss of generality that $A, T \in L(\mathbb{C}^d)$ (for some d, $0 < d < \infty$). Then, our hypotheses and formula (2.2) imply that $\tau_j = \alpha_j$ for $j = 1,2,\ldots, r-1$, but $\tau_r < \alpha_r$. Since $r > 1$, this means, in particular, that T and A have exactly the same number of direct summands, which is equal to $\tau = \sum_{j=1}^{n} \tau_j$.

After eliminating all common direct summands, we can directly assume (without loss of generality) that $T = \oplus_{k=r+1}^{n} q_k^{(\tau_k)}$ and $A = q_r^{(\alpha_r)} \oplus \{\oplus_{k=r+1}^{n} q_k^{(\alpha_k)}\}$, and $\alpha_k \tau_k = 0$ for $k = r+1, r+2, \ldots, n$.

Let $m = [(r-1)/2]$ and let $n = d - \tau m$; then

$(A \to q_n \oplus_m^{(\tau)})$

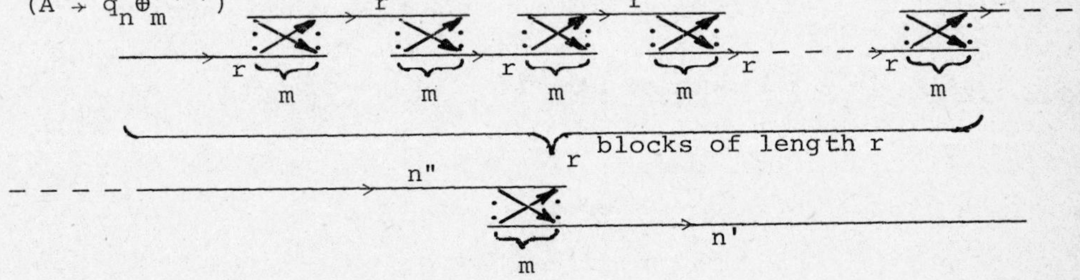

(where $r < n'' \leq n' \leq n$ and n' is the largest Jordan block of A), whence we obtain $\text{dist}[A, U(q_n \oplus q_m^{(\tau)})] \leq s(m)$.

Similarly, we have $\text{dist}[T, U(q_n \oplus q_m^{(\tau)})] \leq s(m)$. Combining these two inequalities, we obtain

$$\text{dist}[A, S(T)] \leq \text{dist}[A, U(T)] \leq \text{dist}[A, U(q_n \oplus q_m^{(\tau)})]$$
$$+ \text{dist}[T, U(q_n \oplus q_m^{(\tau)})] \leq 2\, s(m). \qquad \square$$

2.4 On the distance from a compact operator to $N(H)$

Let $\{f_j\}_{j=1}^{d}$ be an ONB of $\mathbb{C}^d$ and let $F = \sum_{j=1}^{d} \lambda_j f_j \otimes f_j + \sum_{1 \leq i < j \leq d} a_{ij} f_i \otimes f_j \in L(\mathbb{C}^d)$. Clearly, $\text{sp}(F) = \max\{|\lambda_j| : 1 \leq j \leq d\}$. Embed $\mathbb{C}^d$ in $\mathbb{C}^{kd}$ for some $k \geq 2$ and let $\{e_1^1, e_2^1, \ldots, e_k^1, e_1^2, e_2^2, \ldots, e_k^2, \ldots, e_1^d, \ldots, e_k^d\}$ be an ONB for $\mathbb{C}^{kd}$ such that f_j belongs to the subspace H_j spanned by the vectors $\{e_r^j\}_{r=1}^{k}$ ($j = 1,2,\ldots,d$). It follows from Theorem 2.12 that $e_1^j, e_2^j,$

$\ldots,e_k^j$ can be chosen in such a way that $\|f_j\otimes f_j - \frac{1}{2}\sum_{r=1}^{k-1} e_r^j\otimes e_{r+1}^j\| < \frac{1}{2} + 2\pi/k$. Define

$$G = \begin{pmatrix} G_1 & A_{12} & A_{13} & \cdot & \cdot & \cdot & A_{1d} \\ & G_2 & A_{23} & \cdot & \cdot & \cdot & A_{2d} \\ & & G_3 & \cdot & \cdot & \cdot & A_{3d} \\ & & & \cdot & & & \cdot \\ & & & & \cdot & & \cdot \\ & & 0 & & & \cdot & \cdot \\ & & & & & & G_d \end{pmatrix} \in L(\mathbb{C}^{kd}),$$

where $G_j = (\lambda_j/2)\sum_{r=1}^{k-1} e_r^j\otimes e_{r+1}^j$ and $A_{ij} = a_{ij}f_i\otimes f_j$; then

$$\|F\oplus 0_{((k-1)d)} - G\| = \max\{\|G_j - \lambda_j f_j\otimes f_j\|: \quad 1 \le j \le d\}$$

$$< \max\{|\lambda_j|/2: \quad 1 \le j \le d\}+2\pi/k = sp(F)/2 + 2\pi/k.$$

Let $K \in K(H)$ (H infinite dimensional) and let $\varepsilon > 0$ be given. Then there exists a finite rank operator F_ε such that $\|K - F_\varepsilon\| < \varepsilon/3$. Moreover, by the upper semicontinuity of the spectrum (Corollary 1.2), F_ε can be chosen so that $sp(F_\varepsilon) < sp(K)+\varepsilon/3$.

Since $F_\varepsilon \in F(H)$, there exists a finite dimensional subspace H_ε of H, $\dim H_\varepsilon = d \ge 1$, such that H_ε reduces F_ε and $F_\varepsilon|H_\varepsilon^\perp = 0$. Let $F = F_\varepsilon|H_\varepsilon$, let M_ε be a subspace of dimension kd containing H_ε for some k large enough to guarantee that $2\pi/k < \varepsilon/3$ and define $G \in L(M_\varepsilon)$ as above and $G_\varepsilon \in L(H)$ in such a way that $G_\varepsilon|M_\varepsilon = G$ and $G_\varepsilon|M_\varepsilon^\perp = 0$.

Then $G_\varepsilon \in F(H)$, $G_\varepsilon^{kd} = 0$ and

$$\|K - G\| \le \|K - F_\varepsilon\|+\|F_\varepsilon - G_\varepsilon\| < \varepsilon/3+\|F - G\| < \varepsilon/3+\tfrac{1}{2}\,sp(F)+\varepsilon/3$$
$$< \tfrac{1}{2}\,sp(K)+\varepsilon.$$

Since ε can be chosen arbitrarily small, we obtain the following upper bound:

PROPOSITION 2.34. *If* $K \in K(H)$ *(H an infinite dimensional space), then the distance from* K *to the set of all finite rank nilpotent operators cannot exceed* $\frac{1}{2}\,sp(K)$. *In particular, every compact quasinilpotent operator can be uniformly approximated by finite rank nilpotents.*

2.5 Notes and remarks

The problem of characterizing the closure of a similarity orbit in simple terms was raised by D. A. Herrero in [139]. This reference contains all the basic properties of the sets $S(a)^-$ (for a in a Banach algebra A), the notion of asymptotic similarity, several properties of the poset $(A/\#,<)$, the analysis of several relevant examples, etc. The results include the proof of the implications (ii) => (iv) => (i) of Corollary 2.3 (in the above mentioned more general setting [139,Proposition 1]).

Theorem 2.1 and Corollary 2.8 are due to J. Barría and D. A. Herrero [43,Theorem 1.1], who also proved that $(F(H)\cap N(H)/\#,<)$ is a lattice $(0 \leq \dim H \leq \infty)$. The concrete model $(\Sigma_d,\leq)$ for this lattice (Theorem 2.7) is an unpublished result of C. Apostol and D. Voiculescu [34].

In [129], J. H. Hedlund tried to determine the exact value of $\eta_\infty = \mathrm{dist}[P(H),N(H)]$ and showed that $1/4 \leq \eta_\infty \leq \eta_2 = \delta_2 = \sqrt{2}/2$ (Proposition 2.15). He also analyzed the norm of the difference $P_n - Q_n$ for several values of $n \leq 100$ and suggested that these operators could be used to prove that $\delta_k \to \frac{1}{2}$ $(k \to \infty)$.(Personal communication to the author.) The solution of the problem evolved as follows: 1) $\eta_\infty = \frac{1}{2}$ (D. A. Herrero [132,Corollary 9], by using infinite dimensional arguments; 2) δ_k does converge to $\frac{1}{2}$, as $k \to \infty$ (D. A. Herrero [133,Section 7(d)], modifying the previous argument); 3) $\delta_k \leq \frac{1}{2}+[1+(k-1)^{\frac{1}{2}}]/2k \simeq \frac{1}{2}+1/2\sqrt{k}$ (N. Salinas, [181,Lemma 3.3], by using an argument due to P. R. Halmos [125]); 4) $\delta_k > \frac{1}{2}$ and $\eta_k < \frac{1}{2}+(8 \log k)/k$ (D. A. Herrero, [149,Proposition 6.5]) and 5) $\delta_k < \frac{1}{2}+\sin \pi/([(k-1)/2]+1)$ (D. A. Herrero, [150,Corollary 5.2]).

P. R. Halmos and L. J. Wallen called an operator T in $L(H)$ a *power partial isometry* if T^k is a partial isometry for all $k \geq 1$ and proved that T has this property if and only if $T \simeq \{\oplus_{k=1}^{\infty} q_k^{(\tau_k)}\}\oplus S^{(\alpha)}\oplus S^{*(\beta)}\oplus V$, where S is a unilateral shift of multiplicity one and adjoint S* and V is a unitary operator [126]. Lemma 2.13 and its consequences (Corollaries 2.14, Sections 2.3.2 and 2.3.3, Corollary 2.33) are contained in the article [150], by D. A. Herrero, essentially devoted to obtain upper and lower estimates for the distance $\mathrm{dist}[A,U(T)]$ from a power partial isometry A to the unitary orbit of another power partial isometry T. However, as remarked in [150], the technique described by Lemma 2.13 is due to I. D. Berg [48] (see also [46]); this technique has also been exploited by C. Apostol and D. Voiculescu in a different context (unpublished results: Corollary 2.24(i) is due to these authors.)

Proposition 2.28(ii) is the key result of the article [132], where it has been obtained by a very different argument.

Proposition 2.30 is an unpublished observation of D. A. Herrero. According to Proposition 2.30 and [149,Corollary 6.1], if $A \in L(\mathbb{C}^k)$, $0 \leq A \leq 1$ and $1 \in \sigma(A)$, then

$$1/2\sqrt{k} < \mathrm{dist}[A,N(\mathbb{C}^k)] < 10/\log k,$$

for all k large enough.

The results of Section 2.2.3 are unpublished observations of D.

A. Herrero (However, the fact that dist[$E(\mathbb{C}^k)$, $N(\mathbb{C}^k)$] = 1/k was also independently proved by B. Aupetit and J. Zemánek in [200,Example 2.4]). It is worth to remark that the proof of Proposition 2.17(i) actually yields the following stronger result:

PROPOSITION 2.35. *For each* $k \geq 2$

$$\inf\{\|T - Q\|: \quad T \in L(\mathbb{C}^k),\ \sigma(T) = \{0,1\},\ Q \in N(\mathbb{C}^k)\} = 1/k,$$

but this infimum is never attained.

The results of Section 2.3.1 are due to S. L. Campbell and R. Gellar [64].

That every compact quasinilpotent operator is the norm limit of finite rank nilpotents is an observation of R. G. Douglas [123,Problem 7] and the improvement given here (Proposition 2.34) is due to D. A. Herrero [149,Proposition 6.6].

3 The main tools of approximation

In this chapter and the following one, we shall analyze the main "tools" of approximation. Some of these "tools" developed within the framework of Operator Theory: The Rosenblum equation $AX - XB = C$, Rota's universal model and its generalizations, the triangular representation of C. Apostol and the theorem of C. Apostol, C. Foiaş and D. Voiculescu on normal restrictions of small compact perturbations of operators.

These, as well as some of their consequences, will be developed here. Two other important results (the Brown-Douglas-Fillmore theorem and Voiculescu's "non-commutative Weyl-von Neumann theorem"), from the theory of C*-algebras, will be analyzed in Chapter IV.

3.1 The Rosenblum operator: $X \to AX - XB$

3.1.1 Linear operator equations

Let A be a Banach algebra with identity. The mapping $a \to L_a$, where $L_a \in L(A)$ is defined by $L_a(b) = ab$ $(b \in A)$ is called the *regular left representation* of A. It is well-known that this mapping is an isometric isomorphism from A into $L(A)$.

Similarly, the *regular right representation* (defined by $a \to R_a \in L(A)$, where $R_a(b) = ba$) is an isometric *anti-isomorphism* (i.e., $ab \to R_b R_a$). In particular, $\sigma(L_a) = \sigma(R_a) = \sigma(a)$ for all a in A.

Given $a, b \in A$, $\tau_{ab} \in L(A)$ is defined by $\tau_{ab}(c) = ac - cb$, i.e., $\tau_{ab} = L_a - R_b$. It is easily seen that L_a and R_b commute and that

$$\tau_{(a-\lambda)(b-\mu)} = \tau_{ab} - (\lambda-\mu),$$

$(\lambda, \mu \in \mathbb{C})$, so that

$$\|\tau_{ab}\| = \|\tau_{(a-\lambda)(b-\lambda)}\| \leq \inf_{\mu\in\mathbb{C}}\{\|L_{(a-\mu)}\|+\|R_{(b-\mu)}\|\}$$
$$= \inf_{\mu\in\mathbb{C}}\{\|a-\mu\|+\|b-\mu\|\}.$$

LEMMA 3.1. *Let X be a complex Banach space and let* $A, B \in L(X)$ *be two operators such that* AB = BA; *then*

$$\sigma(A+B) \subset \sigma(A) + \sigma(B) = \{\alpha+\beta: \quad \alpha \in \sigma(A),\ \beta \in \sigma(B)\}$$

and

$$\sigma(AB) \subset \sigma(A)\sigma(B) = \{\alpha\beta: \quad \alpha \in \sigma(A),\ \beta \in \sigma(B)\}.$$

PROOF. Let $\mathcal{B} \subset L(X)$ be a maximal abelian algebra containing A and B. Clearly, $1 \in \mathcal{B}$ and $(\lambda-A)^{-1}$ ($(\mu-B)^{-1}$) belongs to $\mathcal{B}$ for all $\lambda \notin \sigma(A)$ ($\mu \notin \sigma(B)$, resp.).

Let $M_{\mathcal{B}}$ be the set of all multiplicative linear functionals on $\mathcal{B}$; then the Gelfand theory implies that

$$\sigma(A+B) = \{\phi(A+B): \quad \phi \in M_{\mathcal{B}}\} = \{\phi(A)+\phi(B): \quad \phi \in M_{\mathcal{B}}\}$$
$$\subset \{\phi(A): \quad \phi \in M_{\mathcal{B}}\}+\{\phi(B): \quad \phi \in M_{\mathcal{B}}\} = \sigma(A) + \sigma(B)$$

and

$$\sigma(AB) = \{\phi(AB): \quad \phi \in M_{\mathcal{B}}\} = \{\phi(A)\phi(B): \quad \phi \in M_{\mathcal{B}}\}$$
$$\subset \{\phi(A): \quad \phi \in M_{\mathcal{B}}\}.\{\phi(B): \quad \phi \in M_{\mathcal{B}}\} = \sigma(A).\sigma(B).$$ □

COROLLARY 3.2. $\sigma(\tau_{ab}) \subset \sigma(a) - \sigma(b)$.

PROOF. Since $L_aR_b = R_bL_a$, it follows from Lemma 3.1 that $\sigma(\tau_{ab}) = \sigma(L_a - R_b) \subset \sigma(L_a) - \sigma(R_b)$. On the other hand, $\sigma(L_a) = \sigma(a)$ and $\sigma(R_b) = \sigma(b)$, whence the result follows. □

We shall see later (Corollary 3.20) that when $A = L(H)$ this inclusion is actually an equality.

3.1.2 Approximate point spectrum of a sum of commuting operators

The *approximate point spectrum* of $A \in L(X)$ is the set

$$\sigma_\pi(A) = \{\lambda \in \mathbb{C}: \quad \lambda\text{-}A \text{ is not bounded below}\}$$

and

$$\sigma_\delta(A) = \{\lambda \in \mathbb{C}: \quad \lambda\text{-}A \text{ is not onto}\}$$

is the *approximate defect spectrum* of A.

It is completely apparent that $\sigma_p(A) \subset \sigma_\pi(A) \subset \sigma_\ell(A)$ and $\sigma_\delta(A) \subset \sigma_r(A)$. Furthermore, if $A^\dagger$ denotes the Banach space adjoint of A, then it is not difficult to see that

$$\sigma_\pi(A) = \sigma_\delta(A^\dagger) \text{ and } \sigma_\delta(A) = \sigma_\pi(A^\dagger).$$

(If X is a Hilbert space, $\sigma_\pi(A) = \sigma_\ell(A) = \sigma_r(A^*)^*$ and $\sigma_\delta(A) = \sigma_r(A) = \sigma_\ell(A^*)^*$, where $\Omega^* = \{\bar{\lambda}: \quad \lambda \in \Omega\}$ for each $\Omega \subset \mathbb{C}$.)

LEMMA 3.3. *Given any Banach space X, there is an isometric imbedding of X into a larger Banach space X', and a mapping $A \to A'$ of $L(X)$ into $L(X')$ which is an isometric isomorphism such that every A' is an extension of A and $\sigma_p(A') = \sigma_\pi(A') = \sigma_\pi(A)$.*

PROOF. Let $\ell^\infty(X)$ be the Banach space of all bounded sequences of elements of X with the norm $\|\{x_n\}_{n=1}^\infty\| = \sup_n \|x_n\|_X$ and let $c_o(X) = \{\{x_n\} \in \ell^\infty(X): \|x_n\| \to 0 \ (n \to \infty)\}$. It is easily seen that $c_o(X)$ is a subspace of $\ell^\infty(X)$.

We define $X' = \ell^\infty(X)/c_o(X)$ and the imbedding of X into X' by $x \to [\{x_n\}]$ (= the coset of $\{x_n\}$), where $x_n = x$ for all $n = 1,2,\ldots$. Clearly, this mapping is an isometric isomorphism of X into X'.

Similarly, given A in $L(X)$, we define $A' \in L(X')$ by $A'[\{x_n\}] = [\{Ax_n\}]$; then $A \to A'$ defines an isometric isomorphism from $L(X)$ into $L(X')$.

If $\lambda \in \sigma_\pi(A)$, then there exists a sequence $\{x_n\}_{n=1}^\infty$ of unit vectors in X such that $\|(\lambda-A)x_n\| \to 0 \ (n \to \infty)$. It is not difficult to check that $A'[\{x_n\}] = \lambda[\{x_n\}]$, whence it follows that $\sigma_\pi(A) \subset \sigma_p(A') \subset \sigma_\pi(A')$.

On the other hand, if $\|(\lambda-A)x\| \geq c\|x\|$ for some $c > 0$ and for all x in X, then it can be easily seen that $\|(\lambda-A')[\{x_n\}]\| \geq c\|[\{x_n\}]\|$ for all $[\{x_n\}] \in X'$. Hence, $\sigma_\pi(A') \subset \sigma_\pi(A)$ and therefore

$$\sigma_p(A') = \sigma_\pi(A') = \sigma_\pi(A). \qquad \square$$

THEOREM 3.4. *If $A, B \in L(X)$ and $AB = BA$, then*

(i) $\sigma_\pi(A+B) \subset \sigma_\pi(A)+\sigma_\pi(B)$.

(ii) $\sigma_\pi(AB) \subset \sigma_\pi(A).\sigma_\pi(B)$.

PROOF. (i) Let $\lambda \in \sigma_\pi(A+B)$; then, by Lemma 3.3, $\lambda \in \sigma_p((A+B)') = \sigma_p(A'+B')$ and $A'B' = B'A'$. It readily follows that the subspace

$$M = \ker(A'+B'-\lambda)$$

is invariant under A' and B', and $B'|M = (\lambda-A')|M$.

Thus, if $\alpha \in \sigma_\pi(A'|M)$ then $\lambda-\alpha \in \sigma_\pi(B'|M)$ and therefore

$$\lambda = \alpha+(\lambda-\alpha) \in \sigma_\pi(A'|M)+\sigma_\pi(B'|M) \subset \sigma_\pi(A')+\sigma_\pi(B') = \sigma_\pi(A)+\sigma_\pi(B).$$

(ii) Take $\lambda \in \sigma_\pi(AB)$; then $N = \ker(A'B'-\lambda)$ is invariant under A' and B'. If $\lambda = 0$ then $\sigma_\pi(A')$ or $\sigma_\pi(B')$ contains 0, so $0 \in \sigma_\pi(A)\sigma_\pi(B)$. If $\lambda \neq 0$, then $A'|N$ has the two sided inverse $\lambda^{-1}B'|N$. Choose α in $\partial\sigma(A'|N)$; then $\lambda/\alpha \in \partial\sigma(B'|N)$. But this implies that $\alpha \in \sigma_\pi(A'|N) \subset \sigma_\pi(A')$, $\lambda/\alpha \in \sigma_\pi(B'|N) \subset \sigma_\pi(B')$, giving in turn

$$\lambda = \alpha(\lambda/\alpha) \in \sigma_\pi(A').\sigma_\pi(B') = \sigma_\pi(A).\sigma_\pi(B). \qquad \square$$

The duality between approximate point spectrum and approximate defect spectrum mentioned above, gives the following

COROLLARY 3.5. *If* A, B $\in L(X)$ *and* AB = BA, *then*

(i) $\sigma_\delta(A+B) \subset \sigma_\delta(A)+\sigma_\delta(B)$.

(ii) $\sigma_\delta(AB) \subset \sigma_\delta(A).\sigma_\delta(B)$.

LEMMA 3.6. *Let* H *be a Hilbert space and let* $A \in L(H)$; *then*

(i) $\sigma_\pi(L_A) = \sigma_\delta(R_A) = \sigma_\ell(A)$.

(ii) $\sigma_\delta(L_A) = \sigma_\pi(R_A) = \sigma_r(A)$.

PROOF. (i) Let $\lambda \in \sigma_\pi(A) = \sigma_\ell(A)$; then there exists a sequence $\{x_n\}_{n=1}^\infty$ of unit vectors such that $||(\lambda-A)x_n|| \to 0$ $(n \to \infty)$. It follows that $||(\lambda-L_A)(x_n \otimes x_n)|| = ||(\lambda-A)x_n \otimes x_n|| = ||(\lambda-A)x_n|| \to 0$, so that $\lambda-L_A$ cannot be bounded below. Hence, $\sigma_\ell(A) \subset \sigma_\pi(L_A)$.

Conversely, if $\lambda \in \sigma_\pi(L_A)$, then there exists a sequence $\{X_n\}_{n=1}^\infty$ in $L(H)$ such that $||X_n|| = 1$ for all $n = 1,2,...$, and $||(\lambda-A)X_n|| \to 0$ $(n \to \infty)$. For each n, choose $y_n \in H$ such that $1 \le ||y_n|| \le 2$ and $||X_n y_n|| = ||X_n|| = 1$; then $||(\lambda-A)(X_n y_n)|| \to 0$ $(n \to \infty)$, whence we conclude that $\lambda \in \sigma_\pi(A)$. It follows that $\sigma_\ell(A) = \sigma_\pi(L_A)$.

If $\lambda-R_A$ is onto, then there exists B in $L(H)$ such that $(\lambda-R_A)B = B(\lambda-A) = 1$, so that $\lambda \notin \sigma_\ell(A)$. Hence, $\sigma_\delta(R_A) \supset \sigma_\ell(A)$.

Conversely, if $\lambda-A$ is left invertible and $C(\lambda-A) = 1$, then $(\lambda-R_A) R_C = R_{(\lambda-A)}R_C = R_{C(\lambda-A)} = R_1 = 1$, i.e., $\lambda-R_A$ is right invertible, whence we conclude that $\sigma_\delta(R_A) \subset \sigma_r(R_A) \subset \sigma_\ell(A)$. Hence, $\sigma_\delta(R_A) = \sigma_\ell(A)$.

The second statement follows by the same arguments. □

From Theorem 3.4, Corollary 3.5 and Lemma 3.6, we immediately obtain the following

COROLLARY 3.7. *Let* A, B $\in L(H)$; *then*

(i) $\sigma_\pi(\tau_{AB}) \subset \sigma_\ell(A) - \sigma_r(B)$.

(ii) $\sigma_\delta(\tau_{AB}) \subset \sigma_r(A) - \sigma_\ell(B)$.

We shall see later (Section 3.1.4) that these inclusions are actually equalities.

3.1.3 Local one-side resolvents in $L(H)$

Let $T \in L(H)$ and let Ω be an open subset of $\rho_r(T)$. A continuous function $F:\Omega \to L(H)$ satisfying the equations

$$(\lambda-T)F(\lambda) \equiv 1 \quad (\lambda \in \Omega) \quad \text{and} \quad F(\lambda) - F(\mu) = (\mu-\lambda)F(\lambda)F(\mu) \quad (\lambda,\mu \in \Omega)$$

will be called a *right resolvent* for T (defined on Ω). It is completely apparent that such a function F is an analytic function defined on

Ω with derivative $F'(\lambda) = -F^2(\lambda)$.

Similarly, if Φ is an open subset of $\rho_\ell(T)$ and $G:\Phi \to L(H)$ is a continuous function satisfying $G(\lambda)(\lambda-T) \equiv 1$ $(\lambda \in \Phi)$ and $G(\lambda)-G(\mu) = (\mu-\lambda)G(\lambda)G(\mu)$ $(\lambda, \mu \in \Phi)$, then G is called a *left resolvent* for T (defined on Φ; clearly, G is analytic on Φ). Moreover, if $F(\lambda)$ is a right resolvent for T^* (defined on $\Omega = \Phi^*$), then $G(\lambda) = F(\bar{\lambda})^*$ is a left resolvent for T (defined on Φ).

For any $\lambda \in \rho_r(T)$, $\mu \in \rho_\ell(T)$, we shall put

$$\begin{cases} R_r(\lambda,T) = (\lambda-T)^*[(\lambda-T)(\lambda-T)^*]^{-1}, \\ R_\ell(\mu,T) = [(\mu-T)^*(\mu-T)]^{-1}(\mu-T)^*. \end{cases} \tag{3.1}$$

It is easily seen that $R_r(.,T):\rho_r(T) \to L(H)$ $(R_\ell(.,T):\rho_\ell(T) \to L(H))$ is continuous on this domain and satisfies the equation $(\lambda-T)R_r(\lambda,T) \equiv 1$, $\lambda \in \rho_r(T)$ $(R_\ell(\mu,T)(\mu-T) \equiv 1$, $\mu \in \rho_\ell(T)$, resp.). However, $R_r(.,T)$ $(R\ (.,T))$ does not satisfy, in general, the resolvent equation, i.e., it is not a right (left, resp.) resolvent for T.

THEOREM 3.8. *Let* $T \in L(H)$. *Given* $\varepsilon > 0$, *there exists a right resolvent for* T *defined on* $\rho_F(T) \cap \rho_r(T)$ *except for an at most denumerable set* $S \subset \rho_r(T)$ *which does not accumulate in* $\rho_r(T)$, *such that*

$$S \subset [\partial\rho_r(T)]_\varepsilon = \{\lambda \in \mathbb{C}: \ \text{dist}[\lambda, \partial\rho_r(T)] \le \varepsilon\}.$$

Applying the above theorem to T^*, we obtain the following dual result.

COROLLARY 3.9. *Let* $T \in L(H)$. *Given* $\varepsilon > 0$, *there exists a left resolvent for* T *defined on* $\rho_F(T) \cap \rho_\ell(T)$ *except for an at most denumerable set* $S' \subset \rho_\ell(T)$ *which does not accumulate in* $\rho_\ell(T)$, *such that* S' *is included in* $[\partial\rho_\ell(T)]_\varepsilon$.

The proof of Theorem 3.8 will be given in a series of lemmas. If $T \in L(H)$ and N is a subspace of H, then T_N will denote the *compression* of T to N, i.e., the restriction of $P_N T$ to N, where P_N denotes the orthogonal projection of H onto N. If N is invariant under T, then $T_N = T|N$.

LEMMA 3.10. *Let* $T \in L(H)$ *and let* M *be an invariant subspace of* T *such that* $0 \in \rho_{re}(T|M)$. *Then* ran $T_{M^\perp}$ *is closed if and only if* ran T *is closed.*

PROOF. Since $0 \in \rho_{re}(T|M)$, $P_{\ker(T|M)^*}$ will be a finite rank projection. Thus, if we write

$$N = \{z \in M^{\perp}: \ P_{\ker(T|M)^*}P_M Tz = 0\}$$

we have $N \subset M^{\perp}$, $\dim (M^{\perp}\ominus N) < \infty$ and, consequently, ran T is closed if and only if $T(M+N)$ is closed in H. This reduces our problem to showing that

$$T(M+N) \text{ is closed } \iff T_{M^{\perp}}N \text{ is closed.}$$

Observe that $P_{\ker(T|M)^*}P_M TN = 0$ implies that $P_M TM \subset TM$, whence we obtain that

$$T(M+N) \subset TM+P_{M^{\perp}}TN \subset T_M M+T_{M^{\perp}}N = TM+T_{M^{\perp}}N.$$

On the other hand, given $y \in M$ and $z \in N$, we can choose $y_o \in M$ such that $T_M y_o = -P_M Tz$, whence we obtain $T_M y+T_M\ z = T(y+y_o+z) \in T(M+N)$. Hence, we have the equality $T(M+N) = TM+T_{M^{\perp}}N$, where the sum in the right side is orthogonal and TM is closed.

It readily follows that $T(M+N)$ is closed if and only if $T_{M^{\perp}}N$ is closed. □

COROLLARY 3.11. *Let* $T \in L(H)$ *be an operator with closed range and let* M *be an invariant subspace of* T *such that* $(TM)^{-} = M$ *and* 0 *belongs to* $\rho_{\ell e}(T_M)$. *Then* $0 \in \rho_r(T_M)$.

PROOF. Since T* has closed range and $0 \in \rho_{re}(T_{M^{\perp}}^*)$, it follows from Lemma 3.10 that T_M^* has closed range. Moreover, since T_M has dense range, T_M^* must be injective, so that $0 \in \rho_{\ell}(T_M^*)$; equivalently, $0 \in \rho_r(T_M)$. □

LEMMA 3.12. *Let* $T \in L(H)$ *be an operator with closed range and let* M *be an invariant subspace of* T *such that* $\ker T \subset M$ *and* $0 \in \rho_r(T_M)$. *Then* $0 \in \rho_{\ell}(T_{M^{\perp}})$ *and the set* $\{\lambda \in \mathbb{C}: \ \ker(\lambda-T) \subset M\}$ *is a neighborhood of the origin.*

PROOF. Since $\ker T \subset M$ and T_M is right invertible, 0 cannot be an eigenvalue of $T_{M^{\perp}}$. Using Lemma 3.10 we see that ran $T_{M^{\perp}}$ is closed and therefore $0 \in \rho_{\ell}(T_{M^{\perp}})$.

For each λ in the *open set* $\rho_{\ell}(T_{M^{\perp}})$ and $x_{\lambda} \in \ker(\lambda-T)$, we have

$$0 = P_{M^{\perp}}(\lambda-T)x_{\lambda} = (\lambda-T_{M^{\perp}})P_{M^{\perp}}x_{\lambda}.$$

Since $\lambda-T_{M^{\perp}}$ is injective, it follows that $P_{M^{\perp}}x_{\lambda} = 0$, i.e., $x_{\lambda} \in M$. We conclude that

$$\rho_{\ell}(T_M\ ) \subset \{\lambda \in \mathbb{C}: \ \ker(\lambda-T) \subset M\}. \qquad \square$$

LEMMA 3.13. *Let* $T \in L(H)$, $\{\lambda_k\}_{k=1}^{n} \subset \sigma_p(T)$ *(*$\lambda_i \neq \lambda_j$ *if* $i \neq j$*) and let* $\{m_k\}_{k=1}^{n}$ *be natural numbers; then*

$$\sum_{k=1}^{n} \ker(\lambda_k - T)^{m_k} = \ker \Pi_{k=1}^{n} (\lambda_k - T)^{m_k}.$$

PROOF. The inclusion '$\subset$' is obvious and the converse inclusion is trivial for n = 1. We shall proceed by induction over n. Assume that we have

$$\sum_{k=1}^{n} \ker(\lambda_k - T)^{m_k} \supset \ker \Pi_{k=1}^{n} (\lambda_k - T)^{m_k}$$

and let $\lambda_{n+1} \in \sigma_p(T) \setminus \{\lambda_1, \lambda_2, .., \lambda_n\}$ and m_{n+1} be a natural number.

Given $x \in \ker \Pi_{k=1}^{n+1} (\lambda_k - T)^{m_k}$, we have that $(\lambda_{n+1} - T)^{m_{n+1}} x \in \ker \Pi_{k=1}^{n} (\lambda_k - T)^{m_k}$ and, by our inductive assumption, we can find $y_k \in \ker(\lambda_k - T)^{m_k}$ $(k = 1,2,...,n)$ such that $(\lambda_{n+1} - T)^{m_{n+1}} x = \sum_{k=1}^{n} y_k$. Define

$$x_k = \left\{\sum_{i=1}^{m_k - 1} (\lambda_{n+1} - \lambda_k)^{-i-1} (\lambda_k - T)^i\right\}^{m_{n+1}} y_k,$$

$k = 1,2,...,n$, and

$$x_{n+1} = x - \sum_{k=1}^{n} x_k.$$

Using the relations

$$(\lambda_{n+1} - T)^{m_{n+1}} x_k = \left[\sum_{i=0}^{m_k - 1} (\lambda_{n+1} - \lambda_k)^{-i} (\lambda_k - T)^i - \sum_{i=0}^{m_k - 1} (\lambda_{n+1} - \lambda_k)^{-i-1} (\lambda_k - T)^{i+1}\right]^{m_{n+1}} y_k = y_k,$$

we see that $x_k \in \ker(\lambda_k - T)^{m_k}$, $k = 1,2,...,n,n+1$, whence the inclusion

$$\sum_{k=1}^{n+1} \ker(\lambda_k - T)^{m_k} \supset \ker \Pi_{k=1}^{n+1} (\lambda_k - T)^{m_k}$$

follows. □

LEMMA 3.14. *Let* $T \in L(H)$ *and let* $x \in H$ *be such that the set of zeroes of the function* $\lambda \to P_{\ker(\lambda - T)} x$, *has an accumulation point* $\zeta \in \rho_r(T)$; *then* $P_{\ker(\lambda - T)} x = 0$, *for all* λ *in the component* Ω_ζ *of* $\rho_r(T)$ *containing the point* ζ.

PROOF. Let $\Delta = \{\lambda \in \Omega_\zeta: P_{\ker(\lambda - T)} x = 0\}$ and assume that $\Delta \neq \Omega_\zeta$. Since Δ has an accumulation point in Ω_ζ, we can find $\mu \in \Omega_\zeta \cap \partial\Delta$ and a sequence $\{\mu_k\}_{k=1}^{\infty} \subset \Delta$ such that $\mu_k \neq \mu$, $\lim(k \to \infty)\ \mu_k = \mu$.

Observe that $\lambda \to P_{\ker(\lambda - T)} = 1 - R_r(\lambda;T)(\lambda - T)$ (where $R_r(\lambda;T)$ is defined by (3.1)) is continuous at μ. Thus, if $M = \bigvee\{\ker(\mu_k - T)\}_{k=1}^{\infty}$, then $\ker(\mu - T) \subset M$ and therefore $\operatorname{ran}(\mu - T_M)$ is closed. But $\operatorname{ran}(\mu - T_M)$ is obviously dense; hence $\mu \in \rho_r(T_M)$.

Let $\Omega_o = \{\lambda \in \Omega_\zeta: \ker(\lambda - T) \subset M\}$. By Lemma 3.12, Ω_o is a neighborhood of μ. Since $x \in M$, we see that $P_{\ker(\lambda - T)} x = 0$ for all $\lambda \in \Omega_o$, contradicting the assumption that $\mu \in \partial\Delta$. This implies that $\Delta = \Omega_\zeta$ and the proof is complete. □

LEMMA 3.15. *Let* $T \in L(H)$, $x \in H$ *and* $\zeta \in \rho_r(T)$ *be such that* $P_{\ker(\zeta-T)^k}x = 0$, *for all* $k = 1,2,\ldots$, *then* $P_{\ker(\lambda-T)}x = 0$ *for all* λ *in the component* Ω_ζ *of* $\rho_r(T)$ *containing* ζ.

PROOF. Let $M = \vee\{\ker(\zeta-T)^k\}_{k=1}^{\infty}$. Since $\zeta-T$ is surjective, we see that $(\zeta-T)\ker(\zeta-T)^{k+1} \subset \ker(\zeta-T)^k$ $(k = 0,1,2,\ldots)$, so that $\operatorname{ran}(\zeta-T_M)$ is dense.

On the other hand, $\ker(\zeta-T) \subset M$. Hence, $\operatorname{ran}(\zeta-T_M)$ is closed and it follows that $\zeta \in \rho_r(T_M)$.

Since $x \in M$, it follows from Lemma 3.12 that the set $\Omega_o = \{\lambda \in \Omega_\zeta : \ker(\lambda-T) \subset M\}$ is a neighborhood of ζ, and therefore $P_{\ker(\lambda-T)}x \equiv 0$ $(\lambda \in \Omega_o)$. By Lemma 3.14, $P_{\ker(\lambda-T)}x = 0$ for all λ in Ω_ζ. □

PROPOSITION 3.16. *Let* $T \in L(H)$ *and let* σ *be a compact subset of* $\rho_r(T)\cap\sigma_p(T)$; *then there exists* x *in* H *such that*

$$P_{\ker(\lambda-T)}x \neq 0 \textit{ for all } \lambda \in \sigma.$$

PROOF. Let $\{\Omega_i\}_{i=1}^{h}$ be the components of $\rho_r(T)$ having nonempty intersection with σ and set $\sigma_i = \sigma\cap\Omega_i$. For each i $(i = 1,2,\ldots,h)$, we choose $\mu_i \in \sigma_i$ and $y_i \in \ker(\mu_i-T)$, $y_i \neq 0$. Then it is possible to find non-zero complex numbers α_i $(i = 1,2,\ldots,h)$ such that the vector $y = \sum_{i=1}^{h} \alpha_i y_i$ satisfies $P_{\ker(\mu_i-T)}y \neq 0$, $i = 1,2,\ldots,h$.

Let $\Delta = \{\lambda \in \sigma:\ P_{\ker(\lambda-T)}y = 0\}$. If $\Delta = \emptyset$, we take $x = y$. If $\Delta \neq \emptyset$, we can apply Lemmas 3.14 and 3.15 in order to show that Δ is a finite set, $\Delta = \{\lambda_k\}_{k=1}^{n}$ and that there exists natural numbers $\{m_k\}_{k=1}^{n}$ such that

$$y \in \{\ker(\lambda_k-T)^{m_k}\}^{\perp},\ y \notin \{\ker(\lambda_k-T)^{m_k+1}\}^{\perp},\ k = 1,2,..,n.$$

By Lemma 3.13, $y \in \{\ker\Pi_{k=1}^{n}(\lambda_k-T)^{m_k}\}^{\perp}$. On the other hand, since $\operatorname{ran}\{\Pi_{k=1}^{n}(\lambda_k-T)^{m_k}\}^*$ is closed, we can find $x \in H$ such that $y = \{\Pi_{k=1}^{n}(\lambda_k-T)^{m_k}\}^* x$.

Assume that $P_{\ker(\mu-T)}x = 0$ for some $\mu \in \sigma$; then $x = (\mu-T)^*z$ for some $z \in H$ and therefore $y = (\mu-T)^*\{\Pi_{k=1}^{n}(\lambda_k-T)^{m_k}\}^* x$. But this is impossible: if $\mu \neq \lambda_k$ (for all $k = 1,2,\ldots,n$), we have $P_{\ker(\mu-T)}y = 0$; if $\mu = \lambda_j$ (for some j, $1 \le j \le n$), then $y \in \{\ker(\lambda_j-T)^{m_j+1}\}^{\perp}$. In either case, this contradicts our assumptions on y. □

LEMMA 3.17. *Let* $T \in L(H)$, *and let* Ω_1 *be a component of* $\rho_r(T)$ *such that* $\operatorname{nul}(\lambda-T) \equiv 1$, $\lambda \in \Omega_1$. *For any* $\varepsilon > 0$ *there exists a right resolvent* R *of* T *on* Ω_1 *except for an at most denumerable set* S_1, *which does not accumulate in* Ω_1, *and satisfies* $S_1 \subset [\partial\Omega_1]_\varepsilon$.

PROOF. By Lemmas 3.14 and 3.15 and Proposition 3.16, there exists a vector $y \in H$ such that $P_{\ker(\lambda-T)}y \neq 0$ for all $\lambda \in \Omega_1 \setminus S_1$, where S_1 is an at most denumerable subset which does not accumulate in Ω_1 and such that $S_1 \subset [\partial\Omega_1]_\varepsilon$. (Take $\sigma = \{\lambda \in \Omega_1: \text{dist}[\lambda, \partial\Omega_1] \geq \varepsilon\}$, in Proposition 3.16.) Fix $\lambda_o \in \Omega_1 \setminus S_1$ and let z be a non-zero vector in $\ker(\lambda_o - T)$. If R_o is a fixed right inverse of $\lambda_o - T$, then any right inverse of $\lambda_o - T$ is of the form $R_t = R_o + z \otimes t$, where $t \in H$.

Choose $t = -\langle y,z\rangle^{-1} R_o^* y$ (Since $P_{\ker(\lambda_o - T)}y \neq 0$, it readily follows that $\langle y,z\rangle \neq 0$, so that t is well-defined.); then

$$\langle y, R_t x\rangle = \langle R_t^* y, x\rangle = \langle R_o^* y - (z\otimes t)^* y, x\rangle = \langle R_o^* y - \langle y,z\rangle\langle y,z\rangle^{-1} R_o^* y, x\rangle = 0$$

for all x in H, so that $y \perp \text{ran}\, R_t$.

For $\lambda \in \Omega_1$ the following identity holds:

$$(\lambda - T)R_t = ((\lambda - \lambda_o)R_t + 1).$$

This shows that $\lambda \to -(\lambda-\lambda_o)^{-1}$ is a mapping from Ω_1 into the component of $\rho_F(R_t)$ which contains the point at infinity. Also, for $\lambda \in \Omega_1$, the Fredholm index of $(\lambda-\lambda_o)R_t + 1$ is zero. Suppose that $\ker[(\lambda_1 - \lambda_o)R_t + 1] \neq \{0\}$ for some $\lambda_1 \in \Omega_1 \setminus S_1$. Then it follows from the above identity that, for some $x \neq 0$, $(\lambda_1 - T)R_t x = 0$. In this case, $R_t x \in \ker(\lambda_1 - T)$ and since the last space is one-dimensional, $\ker(\lambda_1 - T) \subset \text{ran}\, R_t$. This contradicts $y \in \text{ran}\, R_t$ and $P_{\ker(\lambda_1 - T)}y \neq 0$.

It readily follows that the operator $(\lambda-\lambda_o)R_t + 1$ is invertible for all λ in $\Omega_1 \setminus S_1$. The operator valued function

$$R(\lambda) = R_t[(\lambda-\lambda_o)R_t + 1]^{-1}$$

is a right resolvent for T in $\Omega_1 \setminus S_1$. This completes the proof. □

LEMMA 3.18. *Let* $T \in L(H)$ *and let* Ω_n *be a component of* $\rho_r(T)$ *such that* $\text{nul}(\lambda - T) = n$, $\lambda \in \Omega_n$. *For any* $\varepsilon > 0$, *there exists a right resolvent* F *of* T *on* Ω_n *except for an at most denumerable set* S_n, *which does not accumulate on* Ω_n, *and satisfies* $S_n \subset [\partial\Omega_n]_\varepsilon$.

PROOF. We proceed by induction on n. The result is clear if n = 0, for then Ω_n is a component of the resolvent set $\rho(T)$ of T (and $F(\lambda) = (\lambda - T)^{-1}$, $\lambda \in \Omega_n$) and the case n = 1 is contained in the preceding lemma. Suppose the result has been obtained in the case n = k-1. Let Ω_k be a component of $\rho_r(T)$ such that $\text{nul}(\lambda - T) = k$, $\lambda \in \Omega_k$. It follows from Proposition 3.16 that for any $\varepsilon > 0$ there exists a vector $y \in H$ for which $P_{\ker(\lambda-T)}y \neq 0$, for all $\lambda \in \Omega_k \setminus S'$, where S' is an at most denumerable set which does not accumulate in Ω_k and satisfies $S' \subset [\partial\Omega_k]_\varepsilon$.

Let $M_\lambda = \ker(\lambda - T) \cap \{y\}^\perp$, for $\lambda \in \Omega_k$, and let $M = \bigvee\{M_\lambda\}_{\lambda\in\Omega_k}$. Obviously, M is invariant under T and relative to the decomposition $H = M \oplus M^\perp$,

$$T = \begin{pmatrix} T_M & A \\ 0 & T_{M^\perp} \end{pmatrix}.$$

It is easy to establish that $\lambda - T_M$ is onto for $\lambda \in \Omega_k$ and clearly $(\lambda - T_{M^\perp})$ is onto for $\lambda \in \Omega_k$. It follows that for $\lambda \in \Omega_k \backslash S'$, $\text{nul}(\lambda - T_M) = k-1$ and $\text{nul}(\lambda - T_{M^\perp}) = 1$. By Lemma 3.17 and our inductive hypothesis T_M has a right resolvent $R(\lambda)$ on $\Omega_k \backslash S'$ and $T_{M^\perp}$ has a right resolvent $G(\lambda)$ on $\Omega_k \backslash S''$, where S'' is a (possibly empty) finite or denumerable subset of Ω_k which does not accumulate on Ω_k and satisfies $S'' \subset [\partial\Omega_k]_\varepsilon$.

Define

$$F(\lambda) = \begin{pmatrix} R(\lambda) & -R(\lambda)AG(\lambda) \\ 0 & G(\lambda) \end{pmatrix}, \ \lambda \in \Omega_k \backslash S_k$$

(with respect to the above decomposition), where $S_k = S' \cup S''$. It is easily seen that $F(\lambda)$ is a right resolvent for T on $\Omega_k \backslash S_k$, S_k is at most denumerable, S_k does not have any accumulation point in Ω_k and $S_k \subset [\partial\Omega_k]_\varepsilon$. □

PROOF OF THEOREM 3.8. Clearly, it suffices to define a right resolvent F on each component of $\rho_r(T)$, except for an at most denumerable subset S with the desired properties. Let Ω be a component of $\rho_r(T)$ such that $\text{nul}(\lambda - T) = n(\Omega) \geq 0$, $\lambda \in \Omega$. If $n(\Omega) = 0$, then $\Omega \subset \rho(T)$ and the only possible definition for F is $F(\lambda) = (\lambda - T)^{-1}$ = the resolvent of T restricted to Ω.(This is true, in particular, for the unbounded component of $\rho_r(T)$.) If $1 \leq n(\Omega) < \infty$, then Ω is a bounded component of $\rho_r(T)$. If Ω intersects the compact set $\Delta = \{\lambda \in \rho_r(T) \backslash \rho(T)\colon \ \text{dist}[\lambda, \partial\rho_r(T)] \geq \varepsilon\}$, then we define F on $\Omega \backslash S(\Omega)$, where $S(\Omega)$ is an at most denumerable subset of Ω which does not accumulate in Ω and satisfies $S(\Omega) \subset (\partial\Omega)_\varepsilon$ by using Lemma 3.18. If $\Omega \cap \Delta = \emptyset$, then we can use the same arguments as in that lemma in order to construct a right resolvent F on $\Omega \backslash S(\Omega)$, where $S(\Omega)$ is an at most denumerable subset which does not accumulate in Ω (the condition $S(\Omega) \subset \Omega \subset (\partial\Omega)_\varepsilon$ is trivially satisfied in this case).

It is completely apparent that this defines a right resolvent for T on $\rho_r(T) \backslash S$, where $S = \cup\{S(\Omega)\colon \ \Omega$ is a component of $\rho_r(T) \backslash \rho(T)\}$ is an at most denumerable subset of $\rho_r(T)$ with the desired properties. □

3.1.4 The left and the right spectra of τ_{AB}

THEOREM 3.19. *Let* $A, B \in L(H)$; *then*

(i) $\sigma_\delta(\tau_{AB}) = \sigma_r(\tau_{AB}) = \sigma_r(A) - \sigma_\ell(B)$.

(ii) $\sigma_\pi(\tau_{AB}) = \sigma_\ell(\tau_{AB}) = \sigma_\ell(A) - \sigma_r(B)$.

PROOF. (i) By Corollary 3.7 and our observations at the beginning of Section 3.1.2, $\sigma_\delta(\tau_{AB}) \subset \sigma_r(A) - \sigma_\ell(B)$ and $\sigma_\delta(\tau_{AB}) \subset \sigma_r(\tau_{AB})$.

Assume that $\mu \in \sigma_r(A) - \sigma_\ell(B)$ (i.e., μ can be written as $\mu = \alpha-\beta$, where $\alpha \in \sigma_r(A)$ and $\beta \in \sigma_\ell(B)$) and that $\tau_{AB}-\mu$ is onto. Then, given C $\in L(H)$, there exists $X \in L(H)$ such that $(\tau_{AB}-\mu)(X) = \tau_{A-\alpha,B-\beta}(X) = (A-\alpha)X -X(B-\beta) = C$. Since $\mathrm{ran}(\tau_{AB}-\mu)$ is closed, there is a constant $m > 0$ such that $\|(\tau_{AB}-\mu)(T)\| \geq m\ \mathrm{dist}[T,\ker(\tau_{AB}-\mu)]$, for all T in $L(H)$. In particular, $X = X(C)$ can be chosen so that $(m/2)\|X\| \leq \|C\|$.

Since $\alpha \in \sigma_r(A)$ and $\beta \in \sigma_\ell(B)$, we can find unit vectors x, y in H such that $\|(A-\alpha)^*x\| < m/4$ and $\|(B-\beta)y\| < m/4$. Clearly, we can find $C \neq 0$ such that $\langle Cy,x\rangle = \|C\|$; for this C and $X = X(C)$ chosen as above, we have

$$0 < \|C\| = \langle Cy,x\rangle = \langle(\tau_{AB}-\mu)(X)y,x\rangle = \langle[(A-\alpha)X-X(B-\beta)]y,x\rangle$$
$$\leq |\langle Xy,(A-\alpha)^*x\rangle|+|\langle(B-\beta)y,X^*x\rangle|$$
$$\leq (2/m)\|C\|\{\|(A-\alpha)^*x\|+\|(B-\beta)y\|\} < \|C\|,$$

a contradiction.

We conclude that $\sigma_\delta(\tau_{AB}) = \sigma_r(A) - \sigma_\ell(B) \subset \sigma_r(\tau_{AB})$.

Assume that $\mu \notin \sigma_r(A) - \sigma_\ell(B)$; equivalently, $\sigma_r(A-\mu)\cap\sigma_\ell(B) = \emptyset$. Since $\sigma_r(A-\mu)$ and $\sigma_\ell(B)$ are nonempty, compact, and disjoint, there exists Cauchy domains Ω_A, Ω_B and Ω such that $\sigma_r(A-\mu) \subset \Omega_A \subset (\Omega_A)^- \subset \Omega$, $\sigma_\ell(B) \subset \Omega_B$ and $(\Omega_B)^-\cap\Omega^- = \emptyset$. It follows from [1], [2] (see also [185]) that there exists an analytic function $R_{A-\mu}(\lambda)$ defined on $\rho_r(A-\mu)$ such that $(A-\mu-\lambda)R_{A-\mu}(\lambda) \equiv 1$ on this domain. (If $\mathbb{C}\setminus\Omega_A \subset \rho_F(A-\mu)$, then we can choose $R_{A-\mu}(\lambda)$ as the right resolvent constructed in Theorem 3.8.); similarly, there exists an analytic function $L_B(\lambda)$ defined on $\rho_\ell(B)$ (If $\mathbb{C}\setminus\Omega_B \subset \rho_F(B)$, then we can choose $L_B(\lambda)$ as the left resolvent given by Corollary 3.9.) such that $L_B(\lambda)(B-\lambda) \equiv 1$ on this domain.

Clearly, $R_{A-\mu}$ and L_B are analytic in a neighborhood of $\partial\Omega$. Let Ω_0 be a component of Ω. Since $\sigma_\ell(B)\cap\Omega^- = \emptyset$, it follows from Cauchy's theorem that $\int_{\partial\Omega_0} L_B(\lambda)\, d\lambda = 0$.

A fortiori,

$$\frac{1}{2\pi i}\int_{\partial\Omega} L_B(\lambda)\, d\lambda = 0.$$

Let $\partial'\Omega = \partial\Omega\cap\rho(A)$ and $\partial''\Omega = \partial\Omega\setminus\rho(A)$. It is not difficult to check that $\partial'\Omega$ is the (positively oriented) boundary of a Cauchy domain containing $\sigma(A)$, while $\partial''\Omega$ is the (*negatively oriented*) boundary of a Cauchy domain $\Omega'' \subset \rho_r(A)\cap\sigma(A)$ such that $(\Omega'')^-\cap\Omega^- = \partial''\Omega$. Hence, by Cauchy's theorem and the properties of the analytic right inverse (Observe that $R_{A-\mu}(\lambda) = [\lambda-(A-\mu)]^{-1}$ for all $\lambda \notin \rho_r(A)\cap\sigma(A)$!), we have

$$-\int_{\partial''\Omega} R_{A-\mu}(\lambda)\, d\lambda = 0,$$

and

$$\frac{1}{2\pi i}\int_{\partial\Omega} R_{A-\mu}(\lambda)d\lambda = \frac{1}{2\pi i}\int_{\partial'\Omega} R_{A-\mu}(\lambda)d\lambda = \frac{1}{2\pi i}\int_{\partial'\Omega}[\lambda-(A-\mu)]^{-1}d\lambda = 1.$$

Given Y in $L(H)$, let

$$\phi(Y) = X = \frac{-1}{2\pi i}\int_{\partial\Omega} R_{A-\mu}(\lambda)YL_B(\lambda)d\lambda.$$

It is completely apparent that ϕ defines a bounded linear mapping from $L(H)$ into itself. We shall verify that ϕ is a right inverse for $\tau_{AB}-\mu$. Indeed,

$$\begin{aligned}(\tau_{AB}-\mu)(X) &= (A-\mu)X-XB = \\ &= \frac{-1}{2\pi i}\int_{\partial\Omega}(A-\mu)R_{A-\mu}(\lambda)YL_B(\lambda)d\lambda+\frac{1}{2\pi i}\int_{\partial\Omega}R_{A-\mu}(\lambda)YL_B(\lambda)Bd\lambda \\ &= \frac{-1}{2\pi i}\int_{\partial\Omega}(A-\mu-\lambda)R_{A-\mu}(\lambda)YL_B(\lambda)d\lambda+\frac{1}{2\pi i}\int_{\partial\Omega}R_{A-\mu}(\lambda)YL_B(B-\lambda)d\lambda \\ &= \frac{-1}{2\pi i}Y\int_{\partial\Omega}L_B(\lambda)d\lambda+\frac{1}{2\pi i}\left(\int_{\partial\Omega}R_{A-\mu}(\lambda)d\lambda\right)Y = Y,\end{aligned}$$

i.e., $(\tau_{AB}-\mu)\circ\phi = 1$ on $L(H)$.

Hence, $\sigma_r(\tau_{AB}) \subset \sigma_r(A) - \sigma_\ell(B)$ and the proof of (i) is complete.

(ii) By Corollary 3.7 and our observations at the beginning of Section 3.1.2, $\sigma_\pi(\tau_{AB}) \subset \sigma_\ell(A) - \sigma_r(B)$ and $\sigma_\pi(\tau_{AB}) \subset \sigma_\ell(\tau_{AB})$.

Assume that $\mu \in \sigma_\ell(A) - \sigma_r(B)$, i.e., $\mu = \alpha-\beta$ for some $\alpha \in \sigma_\ell(A)$ and some $\beta \in \sigma_r(B)$. This means that, given $\varepsilon > 0$, we can find unit vectors x, y in H such that $\|(A-\alpha)x\| < \varepsilon$ and $\|(B-\beta)^*y\| < \varepsilon$; then $x\otimes y \in L(H)$, $\|x\otimes y\| = 1$ and

$$\begin{aligned}\|(\tau_{AB}-\mu)(x\otimes y)\| &= \|\tau_{A-\alpha,B-\beta}(x\otimes y)\| = \|(A-\alpha)(x\otimes y)-(x\otimes y)(B-\beta)\| \\ &= \|(A-\alpha)x\otimes y - x\otimes(B-\beta)^*y\| \le \|(A-\alpha)x\|.\|y\|+\|x\|.\|(B-\beta)^*y\| \\ &< 2\varepsilon.\end{aligned}$$

Since ε can be chosen arbitrarily small, we conclude that $\tau_{AB}-\mu$ is not bounded below, i.e., $\mu \in \sigma_\pi(\tau_{AB})$. It readily follows that $\sigma_\pi(\tau_{AB}) = \sigma_\ell(A) - \sigma_r(B) \subset \sigma_\ell(\tau_{AB})$.

Now assume that $\sigma_\ell(A-\mu)\cap\sigma_r(B) = \emptyset$ and let Φ_A, Φ_B and Φ be three Cauchy domains such that $\sigma_\ell(A-\mu) \subset \Phi_A \subset (\Phi_A)^- \subset \Phi$, $\sigma_r(B) \subset \Phi_B$ and $(\Phi_B)^- \cap\Phi^- = \emptyset$. By [1], [2] (see also [185], Theorem 3.8 and Corollary 3.9), there exist an analytic function $R_B(\lambda)$ defined on $\rho_r(B)$ and an analytic function $L_{A-\mu}(\lambda)$ defined on $\rho_\ell(A-\mu)$, such that $(B-\lambda)R_B(\lambda) \equiv 1$ for all λ in $\rho_r(B)$ and $L_{A-\mu}(\lambda)(A-\lambda) \equiv 1$ for all λ in $\rho_\ell(A-\mu)$. Proceeding as in the last part of the proof of (i), we can easily check that

$$\psi(Y) = \frac{1}{2\pi i}\int_{\partial\Phi} L_{A-\mu}(\lambda)YR_B(\lambda)d\lambda$$

defines a left inverse of $\tau_{AB}-\mu$. It readily follows that $\sigma_\ell(T_{AB}) \subset \sigma_\ell(A) - \sigma_r(B)$. □

From Theorem 3.19 and its proof, we obtain

COROLLARY 3.20 (Rosenblum's theorem). *If* A, B $\in L(H)$, *then*

(i) $\sigma(\tau_{AB}) = \sigma(A) - \sigma(B)$;

(ii) If $\mu \notin \sigma(A) - \sigma(B)$, *then there exists a Cauchy domain* Ω *such that* $\sigma(A-\mu) \subset \Omega$, $\sigma(B) \cap \Omega^{-} = \emptyset$ *and*

$$(\tau_{AB}-\mu)^{-1}(X) = \frac{-1}{2\pi i} \int_{\partial\Omega} (\lambda - A + \mu)^{-1} X (\lambda - B)^{-1} \, d\lambda$$

By taking A = 0 or B = 0 in Theorem 3.19 and using Lemma 3.6, we obtain

COROLLARY 3.21. *If* A $\in L(H)$, *then*

(i) $\sigma_r(L_A) = \sigma_\delta(L_A) = \sigma_\ell(R_A) = \sigma_\pi(R_A) = \sigma_r(A)$.

(ii) $\sigma_\ell(L_A) = \sigma_\pi(L_A) = \sigma_r(R_A) = \sigma_\delta(R_A) = \sigma_\ell(A)$.

3.1.5 Rosenblum-Davis-Rosenthal corollary

The following is the most important consequence of Theorem 3.19 for the purposes of approximation.

COROLLARY 3.22. *Let* H_1 *and* H_2 *be two Hilbert spaces, let* $A \in L(H_1)$, $B \in L(H_2)$ *and* $C \in L(H_2, H_1)$ *and assume that* $\sigma_r(A) \cap \sigma_\ell(B) = \emptyset$. *Then the operators*

$$\begin{pmatrix} A & C \\ 0 & B \end{pmatrix} \quad \text{and} \quad A \oplus B$$

(acting on $H = H_1 \oplus H_2$.*) are similar.*

PROOF. Assume that H_1 and H_2 are infinite dimensional spaces; then we can identify them via a unitary mapping of H_1 onto H_2, i.e., we can directly assume that $H_1 = H_2 = H_o$ and $H = H_o^{(2)}$. By Theorem 3.19 (i), τ_{AB} is onto and therefore there exists $X \in L(H_o)$ such that $AX - XB = -C$. Then

$$\begin{pmatrix} A & C \\ 0 & B \end{pmatrix}\begin{pmatrix} 1 & X \\ 0 & 1 \end{pmatrix} = \begin{pmatrix} 1 & X \\ 0 & 1 \end{pmatrix}\begin{pmatrix} A & 0 \\ 0 & B \end{pmatrix}$$

and $\begin{pmatrix} 1 & X \\ 0 & 1 \end{pmatrix}$ is invertible $\left(\begin{pmatrix} 1 & X \\ 0 & 1 \end{pmatrix}^{-1} = \begin{pmatrix} 1 & -X \\ 0 & 1 \end{pmatrix}\right)$.

Hence,

$$A \oplus B = \begin{pmatrix} 1 & X \\ 0 & 1 \end{pmatrix}^{-1}\begin{pmatrix} A & C \\ 0 & B \end{pmatrix}\begin{pmatrix} 1 & X \\ 0 & 1 \end{pmatrix}.$$

If A acts on a finite dimensional space and B acts on an infinite dimensional space, define $\alpha = \|A\| + \|B\| + 1$ and consider the operators $\alpha \oplus A \in L(H_2 \oplus H_1)$, B and $C' = \begin{pmatrix} 0 \\ C \end{pmatrix} \in L(H_2, H_2 \oplus H_1)$. It follows from the first part of the proof that

$$\alpha\oplus\begin{pmatrix} A & C \\ 0 & B\end{pmatrix} = \begin{pmatrix} \alpha\oplus A & C' \\ 0 & B\end{pmatrix} = \begin{pmatrix} \alpha & 0 & 0 \\ 0 & A & C \\ 0 & 0 & B\end{pmatrix} \sim (\alpha\oplus A)\oplus B = \alpha\oplus(A\oplus B).$$

Assume that $W = \begin{pmatrix} W_{11} & W_{12} \\ W_{21} & W_{22}\end{pmatrix}$ is invertible, where $W_{11} \in L(H_2)$, $W_{22} \in L(H_1\oplus H_2)$, $W_{12} \in L(H_1\oplus H_2, H_2)$ and $W_{21} \in L(H_2, H_1\oplus H_2)$, and

$$W^{-1}\left\{\alpha\oplus\begin{pmatrix} A & C \\ 0 & B\end{pmatrix}\right\}W = \alpha\oplus(A\oplus B);$$

then

$$\left\{\alpha\oplus \begin{pmatrix} A & C \\ 0 & B\end{pmatrix}\right\}W-W[\alpha\oplus(A\oplus B)] = \left(\begin{array}{c:c} 0 & \alpha W_{12}-W_{12}(A\oplus B) \\ \hdashline \begin{pmatrix} A & C \\ 0 & B\end{pmatrix}W_{21}-W_{21}\alpha & \begin{pmatrix} A & C \\ 0 & B\end{pmatrix}W_{22}-W_{22}(A\oplus B)\end{array}\right) = 0,$$

so that $\alpha W_{12}-W_{12}(A\oplus B) = 0$ and $\begin{pmatrix} A & C \\ 0 & B\end{pmatrix}W_{21}-W_{21}\alpha = 0$.

Since

$$\sigma(\alpha)\cap\sigma\left(\begin{pmatrix} A & C \\ 0 & B\end{pmatrix}\right) \subset \sigma(\alpha)\cap[\sigma(A)\cup\sigma(B)] = \sigma(\alpha)\cap\sigma(A\oplus B) = \emptyset,$$

it readily follows from Corollary 3.20 that $W_{12} = 0$ and $W_{21} = 0$.

Hence $W = W_{11}\oplus W_{22}$, W_{11} and W_{22} are invertible operators, and (consider the (2,2)-entry of the above 2 × 2 matrix!)

$$W_{22}^{-1}\begin{pmatrix} A & C \\ 0 & B\end{pmatrix} W_{22} = A\oplus B,$$

i.e., $\begin{pmatrix} A & C \\ 0 & B\end{pmatrix}$ is similar to $A\oplus B$.

The cases when H_2 is finite dimensional and H_1 is infinite dimensional or both, H_1 and H_2 are finite dimensional spaces can be similarly analyzed to reach the same conclusion:

$$\begin{pmatrix} A & C \\ 0 & B\end{pmatrix} \sim A\oplus B. \qquad \square$$

3.1.6 The maximal numerical range of an operator

It was observed in Section 3.1.1 that $\|\tau_{AB}\| \le \min\{\|A-\lambda\|+\|B-\lambda\|: \lambda \in \mathbb{C}\}$. In order to complement our previous results about τ_{AB}, it will be shown that this inequality is actually an equality. The concept of *maximal numerical range* of an operator plays a central role here. Recall that the *numerical range* of $T \in L(H)$ is the set defined by

$$W(T) = \{\langle Tx,x\rangle: \ x \in H, \|x\| = 1\}.$$

The classical Toeplitz-Hausdorff theorem asserts that $W(T)$ is a convex set [119,Problem 166]. The maximal numerical range of T is the set

$$W_0(T) = \{\lambda \in \mathbb{C}: \ \langle Tx_n,x_n\rangle \to \lambda, \text{ where } \|x_n\| = 1 \text{ and } \|Tx_n\| \to \|T\|\}.$$

LEMMA 3.23. $W_0(T)$ *is a nonempty, closed, convex subset of* $W(T)^-$.

PROOF. Everything but convexity is obvious. Let $\lambda, \mu \in W_0(T)$. Without loss of generality, we can assume that $\|T\|=1$. Let $\{x_n\}_{n=1}^{\infty}$, $\{y_n\}_{n=1}^{\infty}$ be two sequences in H such that $\|x_n\|=\|y_n\|=1$ for all $n=1,2,\ldots$, $\|Tx_n\| \to 1$, $\|Ty_n\| \to 1$, $\langle Tx_n,x_n\rangle \to \lambda$ and $\langle Ty_n,y_n\rangle \to \mu$ $(n\to\infty)$. Consider $T_n = P_n TP_n$, where P_n is the projection of H onto $\vee\{x_n,y_n\}$ and let η be a point of the segment $[\lambda,\mu]$ joining λ and μ. Then for each n, it is possible, by the Toeplitz-Hausdorff theorem, to choose α_n, β_n such that $\langle Tu_n,u_n\rangle = \langle T_nu_n,u_n\rangle \to \eta$ and $\|u_n\|=1$, where $u_n=\alpha_nx_n+\beta_ny_n$. Observe that $|\langle x_n,y_n\rangle| \le \theta < 1$ and therefore $\max\{|\alpha_n|,|\beta_n|\} \le M = (1-\theta^2)^{-\frac{1}{2}}$ (otherwise $\|\alpha_nx_n+\beta_ny_n\|<1$) for all n sufficiently large. We have

$$1 \ge \|Tu_n\|^2 = \langle T^*Tu_n,u_n\rangle = \|u_n\|^2 - \langle(1-T^*T)u_n,u_n\rangle$$
$$\ge 1 - M\{\|(1-T^*T)x_n\| + \|(1-T^*T)y_n\|\}.$$

Since

$$0 \le \|(1-T^*T)x_n\| = \|x_n\|^2 - 2\|Tx_n\|^2 + \|T^*Tx_n\|^2$$
$$= (1-\|Tx_n\|^2) - (\|Tx_n\|^2 - \|T^*(Tx_n)\|^2) \le 1 - \|Tx_n\|^2 \to 0 \quad (n\to\infty)$$

and, similarly, $\|(1-T^*T)y_n\| \to 0$ $(n\to\infty)$, we conclude that $\|Tu_n\| \to 1$ and $\langle Tu_n,u_n\rangle \to \eta$ $(n\to\infty)$. Therefore, $W_0(T)$ is a convex set. □

If $A \in L(H)$, the operator $\delta_A = \tau_{AA}$ is the *inner derivation* of $L(H)$ induced by A.

LEMMA 3.24. *Let* $\mu \in W_0(T)$, *then* $\|\delta_T\| \ge 2(\|T\|^2 - |\mu|^2)^{\frac{1}{2}}$.

PROOF. Note that $\|\delta_T\| = \sup\{\|TA-AT\|: A \in L(H), \|A\|=1\}$. Since $\mu \in W_0(T)$, there exist $x_n \in H$ such that $\|x_n\|=1$, for all $n=1,2,\ldots$, $\|Tx_n\| \to \|T\|$ and $\langle Tx_n,x_n\rangle \to \mu$. Set $Tx_n = \alpha_nx_n+\alpha_ny_n$, where y_n is a unit vector orthogonal to x_n. Set $V_n = x_n\otimes x_n - y_n\otimes y_n$; then

$$\|(TV_n-V_nT)x_n\| = 2|\beta_n| \ge 2(\|T\|^2 - |\alpha_n|^2)^{\frac{1}{2}} - \varepsilon_n, \text{ where } \varepsilon_n \to 0 \ (n\to\infty).$$

Since $\alpha_n \to \mu$ $(n\to\infty)$, the proof is complete. □

PROPOSITION 3.25. *Let* $T \in L(H)$; *then the following are equivalent*

(i) $0 \in W_0(T)$;

(ii) $\|\delta_T\| = 2\|T\|$;

(iii) $\|T\|^2 + |\lambda|^2 \le \|T+\lambda\|^2$ *for all* $\lambda \in \mathbb{C}$;

(iv) $\|T\| \le \|T+\lambda\|$ *for all* $\lambda \in \mathbb{C}$.

PROOF. (i) => (ii) It follows from Lemma 3.24 that, if $0 \in W_0(T)$, then $\|\delta_T\| \ge 2\|T\|$. Therefore, $\|\delta_T\| = 2\|T\|$.

(ii) => (i) If $\|\delta_T\| = 2\|T\|$, then there exist $x_n \in H$ and $A_n \in L(H)$

such that $\|x_n\|=\|A_n\|=1$ and $\|(TA_n-A_nT)x_n\| \to 2\|T\|$ $(n \to \infty)$.

It readily follows that $\|A_nx_n\| \to 1=\|A_n\|$, $\|Tx_n\| \to \|T\|$ and $\|TA_nx_n\| \to \|T\|$. Passing, if necessary, to a subsequence, we can directly assume that $\langle Tx_n,x_n\rangle \to \mu$ and $\langle TA_nx_n,A_nx_n\rangle \to \gamma$ $(n \to \infty)$. Since $\|A_nx_n\| \to 1=\|x_n\|$, it is completely apparent that $\mu, \gamma \in W_0(T)$. Moreover, since $\|(TA_n-A_nT)x_n\| \to 2\|T\|$ and H is uniformly convex, the norm of $z_n = TA_nx_n+A_nTx_n$ tends to 0, as $n \to \infty$. Thus, we have

$$\langle TA_nx_n,A_nx_n\rangle = -\langle A_nTx_n,A_nx_n\rangle + \langle z_n,A_nx_n\rangle$$
$$= -\langle Tx_n,A_n{*}A_nx_n\rangle + \langle z_n,A_nx_n\rangle = -\langle Tx_n,x_n\rangle + \varepsilon_n,$$

where $\varepsilon_n \to 0$, as $n \to \infty$. (Indeed, it follows as in the last step of the proof of Lemma 3.24 that $\|A_n{*}A_nx_n - x_n\| \to 0$, as $n \to \infty$.) Thus,

$$\gamma = \lim(n \to \infty)\langle TA_nx_n,A_nx_n\rangle = -\lim(n \to \infty)\langle Tx_n,x_n\rangle = -\mu.$$

Since both μ and $-\mu$ belong to $W_0(T)$, it follows from Lemma 3.23 that $0 \in W_0(T)$.

(i) => (iii) If $0 \in W_0(T)$, then there exists a sequence $\{x_n\}_{n=1}^{\infty}$ of unit vectors such that

$$\|T+\lambda\|^2 \geq \|(T+\lambda)x_n\|^2 = \|Tx_n\|^2 + 2\mathrm{Re}\ \bar{\lambda}\langle Tx_n,x_n\rangle + |\lambda|^2 \to \|T\|^2 + |\lambda|^2$$

for all $\lambda \in \mathbb{C}$.

(iv) => (i) Assume that $\|T\| \leq \|T+\lambda\|$ for all $\lambda \in \mathbb{C}$, but $0 \notin W_0(T)$. By rotating T, we may assume that $\mathrm{Re}\ W_0(T) \geq c > 0$. Let $M = \{x \in H\colon \|x\| = 1,\ \mathrm{Re}\ \langle Tx,x\rangle \leq c/2\}$ and let $M = \sup\{\|Tx\|\colon x \in M\}$; then $M < \|T\|$.

Let $\mu = \min\{c/2, (\|T\|-M)/2\}$. If $x \in M$, then $\|(T-\mu)x\| \leq \|Tx\|+\mu \leq M+\mu < \|T\|$. If $x \notin M$, $\|x\|=1$, let $Tx = (a+ib)x+y$, where $y \perp x$; then

$$\|(T-\mu)x\|^2 = (a-\mu)^2+b^2+\|y\|^2 = \|Tx\|^2+(\mu^2 - 2a\mu) < \|T\|^2$$

because $a > \mu > 0$. Thus, $\|T-\mu\| < \|T\|$, contrary to the hypothesis.

Since (iii) => (iv) is a trivial implication, we are done.□

THEOREM 3.26. *Let* $T \in L(H)$; *then* $\|\delta_T\| = 2\min\{\|T-\lambda\|\colon \lambda \in \mathbb{C}\}$.

PROOF. We have already observed that $\|\delta_T\| \leq 2\inf\{\|T-\lambda\|\colon \lambda \in \mathbb{C}\}$. By an elementary argument of compactness (observe that $\|T-\lambda\| \to \infty$, as $|\lambda| \to \infty$), it is easily seen that the above infimum is actually attained at some point $\mu \in \mathbb{C}$, i.e., this infimum is a minimum. But $\|T-\mu\| \leq \|(T-\mu)+\lambda\|$ for all $\lambda \in \mathbb{C}$ implies that $\|\delta_T\| = \|\delta_{(T-\mu)}\| = 2\|T-\mu\|$. (Use Proposition 3.25: (iv) => (ii).) □

3.1.7 The norm of the operator τ_{AB}

LEMMA 3.27. *Let* $T \in L(H)$. *The mapping* $\lambda \to W_0(T+\lambda)$ *is upper semi-continuous.*

PROOF. We can assume, without loss of generality, that $\|T\| = 1$. Suppose that $\mathrm{Re}\, W_0(T) \le a$ and let $\varepsilon > 0$. Let $M = \sup\{\|Ax\|: \|x\| = 1, \mathrm{Re}\langle Tx,x\rangle \ge a+\varepsilon\}$; then $M < 1$. It is clear that $\|T+\lambda\| \ge 1 - |\lambda|$. However, for $y \in H$, $\|y\| = 1$ and $\mathrm{Re}\langle Ty,y\rangle \ge a+\varepsilon$, we see that

$$\|(T+\lambda)y\|^2 \le M^2+2|\lambda|+|\lambda|^2.$$

Thus, for $|\lambda| < (1-M^2)/4$, it follows that $\mathrm{Re}\, W_0(T+\lambda) < a+\varepsilon$.

By Lemma 3.23, $W_0(T+\lambda_o)$ is closed and convex (for fixed $\lambda_o \in \mathbb{C}$). Hence, $W_0(T+\lambda_o)$ coincides with the intersection of all the open half-planes containing it. Thus, we can find finitely many open halfplanes $S_1, S_2, \ldots, S_m$ such that

$$W_0(T+\lambda_o)_\varepsilon \subset \cap_{j=1}^m S_j \subset W_0(T+\lambda_o)_{2\varepsilon}.$$

By the first part of the proof, we can find $\delta = \delta(\varepsilon,\lambda_o) > 0$ such that

$$W_0(T+\lambda)_\varepsilon \subset \cap_{j=1}^m S_j \subset W_0(T+\lambda_o)_{2\varepsilon}$$

provided $|\lambda - \lambda_o| < \delta$.

Hence, $\lambda \to W_0(T+\lambda)$ is an upper semicontinuous mapping. □

We define the *normalized maximal numerical range* $W_N(T)$ of an operator $T \in L(H)$, $T \neq 0$, to be the set $W_0(T/\|T\|)$. From Lemma 3.27 we obtain the following

COROLLARY 3.28. *If* $\|T+\lambda\| \neq 0$ *for all* λ, *then the map* $\lambda \to W_N(T+\lambda)$ *is upper semicontinuous.*

LEMMA 3.29. *Let* $A, B \in L(H)$ *be two non-zero operators; then the following are equivalent*

(i) $\|\tau_{AB}\| = \|A\|+\|B\|$;

(ii) $W_N(A) \cap W_N(-B) \neq \emptyset$.

PROOF. The proof is very similar to that of Proposition 3.25 (i) => (ii), and so we shall only sketch a portion. Let $\lambda \in W_N(A) \cap W_N(-B)$ and $\varepsilon > 0$; then there exist $x, y \in H$ such that $\|x\| = \|y\| = 1$ and $\langle Ax,x\rangle = \lambda\|A\|+\varepsilon_A$ and $\langle By,y\rangle = -\lambda\|B\|+\varepsilon_B$, $0 \le \varepsilon_A, \varepsilon_B < \varepsilon$. Since

$$\langle Ax,x\rangle/\|A\| = -\langle By,y\rangle/\|B\|+\varepsilon',$$

it is possible to define an operator U of norm $1+\varepsilon''$ which sends x to y and $-By/\|B\|$ to $Ax/\|A\|$ (where ε', ε'' are small if ε is small). The rest of the proof is virtually unchanged. □

LEMMA 3.30. *Assume that* A, B $\in L(H)$ *and neither* A *nor* B *is a scalar multiple of the identity. Then*

$$\min\{\|A-\lambda\|+\|B-\lambda\|:\ \lambda\in\mathbb{C}\} = \|A-\mu\|+\|B-\mu\|$$

if and only if $W_N(A-\mu)\cap W_N(-(B-\mu)) \neq \emptyset$.

PROOF. Assume that $W_N(A-\mu)\cap W_N(-(B-\mu)) \neq \emptyset$. Then $\|\tau_{AB}\| = \|\tau_{A-\mu,B-\mu}\| = \|A-\mu\|+\|B-\mu\|$. Since it is obvious that $\|\tau_{AB}\|$ cannot be larger than $\min\{\|A-\lambda\|+\|B-\lambda\|:\ \lambda\in\mathbb{C}\}$, we see that the condition is necessary.

In order to prove the sufficiency, we can directly assume that $\mu = 0$. Thus, given $\lambda\in\mathbb{C}$ and $\varepsilon > 0$, there exist unit vectors x, y in H such that $\|(A+\lambda)x\|+\|(B+\lambda)y\| \geq \|A\|+\|B\|-\varepsilon$. After some algebraic transformations, we find that $\mathrm{Re}\,\bar{\lambda}(\langle Ax,x\rangle/\|A\|+\langle By,y\rangle/\|B\|) \leq K(|\lambda|^2+\varepsilon)$, where K is a constant independent of λ and ε.

Assume that $W_N(A)\cap W_N(-B) = \emptyset$. Then, $d_H[W_N(A),W_N(-B)] = \delta > 0$, and (by upper semicontinuity; Corollary 3.28) $d_H[W_N(A+\lambda),W_N(-B-\lambda)] > \delta/2$, for λ small. Thus, by convexity and upper semicontinuity, any choice of x, y which satisfies the above conditions, must satisfy the inequality $|\langle(A+\lambda)x,x\rangle/\|A+\lambda\| + \langle(B+\lambda)y,y\rangle/\|B+\lambda\|\,| \geq \delta/4$ for λ small. But then we are led to the inequality $|\lambda|\delta/8 \leq K|\lambda|^2$ for a suitable choice of arg λ and $|\lambda|$ small, which is impossible. Thus, $\mu = 0$ was not minimal, which completes the proof. □

THEOREM 3.31. *Let* A, B $\in L(H)$; *then*

$$\|\tau_{AB}\| = \min\{\|A-\lambda\|+\|B-\lambda\|:\ \lambda\in\mathbb{C}\}.$$

PROOF. Clearly, $\|\tau_{AB}\| \leq \min\{\|A-\lambda\|+\|B-\lambda\|:\ \lambda\in\mathbb{C}\}$. If A or B is a multiple of the identity, the rest of the proof is trivial. Let $\mu\in\mathbb{C}$ be any point such that the above minimum is attained at μ. By Lemmas 3.29 and 3.30, $\|\tau_{AB}\| = \|\tau_{A-\mu,B-\mu}\| = \|A-\mu\|+\|B-\mu\|$.

The proof is complete now. □

REMARK 3.32. It is completely apparent from Proposition 3.25(iii) that there exists exactly one $\mu\in\mathbb{C}$ such that $\|T-\mu\| = \min\{\|T-\lambda\|:\ \lambda\in\mathbb{C}\}$. However, simple examples show that, in general, the μ of Lemma 3.30 is not unique.

3.2 Generalized Rota's universal model

Let $T\in L(H)$. Since $\sigma(T)$ is a compact set, it has a fundamental system of open neighborhoods which are analytic Cauchy domains (i.e., Cauchy domains whose boundaries consists of pairwise disjoint regular analytic Jordan curves; see definition in Section 1.1).

Let $\Gamma = \partial\Omega$, where Ω is an analytic Cauchy domain containing $\sigma(T)$, and let $L^2(\Gamma)$ be the Hilbert space of (equivalent classes of) complex functions on Γ which are square integrable with respect to $(1/2\pi)$-times the arc-length measure on Γ; $M(\Gamma)$ will stand for the "multiplication by λ" operator acting on $L^2(\Gamma)$. The subspace $H^2(\Gamma)$ spanned by the rational functions with poles outside Ω^- is invariant under $M(\Gamma)$. By $M_+(\Gamma)$ and $M_-(\Gamma)$ we shall denote the restriction of $M(\Gamma)$ to $H^2(\Gamma)$ and its compression to $L^2(\Gamma)\ominus H^2(\Gamma)$, respectively, i.e.

$$M(\Gamma) = \begin{pmatrix} M_+(\Gamma) & Z \\ 0 & M_-(\Gamma) \end{pmatrix} \begin{matrix} H^2(\Gamma) \\ L^2(\Gamma)\ominus H^2(\Gamma) \end{matrix} \tag{3.2}$$

with respect to the above decomposition. (Here and in what follows, we write

$$A = \begin{pmatrix} A_{11} & A_{12} & . & . & . & A_{1k} \\ A_{21} & A_{22} & . & . & . & A_{2k} \\ . & . & & & & . \\ . & . & & & & . \\ . & . & & & & . \\ A_{k1} & A_{k2} & . & . & . & A_{kk} \end{pmatrix} \begin{matrix} H_1 \\ H_2 \\ . \\ . \\ . \\ H_k \end{matrix} \tag{3.3}$$

as an alternative way to indicate that A admits such a $k \times k$ operator matrix decomposition with respect to the orthogonal direct sum $H = \oplus_{j=1}^{k} H_j$, where $A_{ij}: H_j \to H_i$, $1 \leq i,j \leq k$. Such a decomposition is clearly unique.)

It is well-known [36], [73], [74, Sections 9] that $H^2(\Gamma)$ is a reproducing kernel space and that it can be realized as a space of analytic functions defined on Ω; furthermore, it can be easily checked that

$$\begin{cases} \sigma(M(\Gamma)) = \sigma_e(M(\Gamma)) = \sigma_e(M_+(\Gamma)) = \sigma_e(M_-(\Gamma)) = \Gamma, \\ \sigma(M_+(\Gamma)) = \sigma(M_-(\Gamma)) = \Omega^-, \\ \text{nul}(\lambda - M_+(\Gamma)) = \text{nul}(\lambda - M_-(\Gamma))^* = 0 \text{ and } \text{ind}(\lambda - M_+(\Gamma)) = \\ \text{ind}(\lambda - M_-(\Gamma))^* = -1 \text{ for all } \lambda \text{ in } \Omega, \\ M(\Gamma) \text{ is normal and} \\ \|r(M(\Gamma))\| = \|r(M_+(\Gamma))\| = \|r(M_-(\Gamma))\| = \max\{|r(\lambda)| : \lambda \in \Omega\} \\ \text{for each rational function } r \text{ with poles outside } \Omega^-. \end{cases} \tag{3.4}$$

The Hilbert space completion $H^2(\Gamma)\otimes H$ of the algebraic tensor product of $H^2(\Gamma)$ and H can be regarded as a space of analytic H-valued functions defined on Ω, or as a space of (equivalent classes of) weakly measurable square integrable functions on Γ, in which case it will also be denoted by $H^2(\Gamma;H)$.

THEOREM 3.33. *Let* $T \in L(H)$ *and* Ω *be as above, and let* $R = \text{ran}(M_+(\Gamma)\otimes 1 - 1\otimes T)$; *then* R *is a subspace of* $H^2(\Gamma;H)$ *invariant under* $M_+(\Gamma)\otimes 1$, *the restriction of* $M_+(\Gamma)\otimes 1$ *to* R *is similar to* $M_+(\Gamma)\otimes 1$ *and the compression of* $M_+(\Gamma)\otimes 1$ *to* $R^\perp$ *is similar to* T.

PROOF. Since $M_+(\Gamma)\otimes 1 - 1\otimes T$ is equal to multiplication by $\lambda - T$ in $H^2(\Gamma;H)$ and $\lambda - T$ is invertible on Γ, it is clear that R is closed. Moreover, the similarity of $M_+(\Gamma)\otimes 1$ and $M_+(\Gamma)\otimes 1|R$ is implemented by $M_+(\Gamma)\otimes 1 - 1\otimes T$. All we have to do is find an operator $L: H^2(\Gamma;H) \to H$ with range H and kernel R such that $L\circ(M_+(\Gamma)\otimes 1) = T\circ L$. We shall define for $f \in H^2(\Gamma;H)$ the element $Lf \in H$ by

$$Lf = \frac{1}{2\pi i}\int_{\partial\Omega}(\lambda - T)^{-1} f(\lambda)\,d\lambda.$$

It is easily seen that $L\circ(M_+(\Gamma)\otimes 1) = L\circ(1\otimes T) = T\circ L$, so that $R \subset \ker L$. On the other hand, if $f_x(\lambda) \equiv x$ $(x \in H)$, then $Lf_x = x$, so that ran L is equal to H.

Let Ω_1 be an analytic Cauchy domain containing Ω such that the restrictions of functions analytic in a neighborhood of Ω_1^- to Γ are dense in $H^2(\Gamma)$ (By Runge's approximation theorem, it suffices to choose Ω_1 so that every bounded component of $\mathbb{C}\backslash\Omega$ contains a component of $\mathbb{C}\backslash\Omega_1$) and let $e_o(\lambda) \equiv 1$, $e_o \in H^2(\Gamma)$.

Let $S \in L(H^2(\Gamma;H))$ be the operator defined by $Sf = f - e\otimes Lf$. Since $Lf = L(e\otimes Lf)$, it follows that ran $S = \ker L$. Thus, in order to complete the proof, it will be sufficient to show that $S(f|\Gamma) \in R$ for every H-valued function f analytic in a neighborhood of Ω_1.

Because of the analyticity of $(\lambda - T)^{-1}f(\lambda)$ in a neighborhood of $\Omega_1^-\backslash\Omega$, we have

$$Lf = \frac{1}{2\pi i}\int_{\partial\Omega_1}(\lambda - T)^{-1} f(\lambda)\,d\lambda.$$

It follows that for $\zeta \in \Omega$

$$(S(f|\Gamma))(\zeta) = \frac{1}{2\pi i}\int_{\partial\Omega_1}[(\lambda-\zeta)^{-1} - (\lambda - T)^{-1}]f(\lambda)\,d\lambda$$
$$= (\zeta - T)\,\frac{1}{2\pi i}\int_{\partial\Omega_1}(\lambda-\zeta)^{-1}(\lambda - T)^{-1} f(\lambda)\,d\lambda.$$

Since

$$\int_{\partial\Omega_1}(\lambda-\zeta)^{-1}(\lambda - T)^{-1} f(\lambda)\,d\lambda$$

is an analytic function of ζ $(\zeta \in \Omega_1)$, it readily follows that

$$S(f|\Gamma) = (M_+(\Gamma)\otimes 1 - 1\otimes T)\left(\frac{1}{2\pi i}\int_{\partial\Omega_1}(\lambda-\zeta)^{-1}(\lambda - T)^{-1} f(\lambda)\,d\lambda\right) \in R.$$

The proof is complete now. □

REMARK 3.34. According to the proof of Theorem 3.33, the diagram

$$\begin{array}{ccc} H^2(\Gamma;H) \supset R^{\perp} & \xrightarrow{\;T' = (M_+(\Gamma)\otimes 1)\;} & R^{\perp} \\ \Big\downarrow L|R^{\perp} & & \Big\downarrow L|R^{\perp} \\ H & \xrightarrow{\quad T \quad} & H \end{array}$$

is commutative, i.e., $T = (L|R^{\perp})^{-1} \circ T' \circ (L|R^{\perp})$.

Since $Lf_x = x$ (where $f_x(\lambda) \equiv x$, $x \in H$), it follows that $L|R^{\perp}$ is bounded below by

$$\inf\{\|f_x\|: \|x\| = 1\} = ([\text{length}(\Gamma)]/2\pi)^{\frac{1}{2}}.$$

On the other hand, the Cauchy-Schwartz inequality implies that

$$\|L\| \leq ([\text{length}(\Gamma)]/2\pi)^{\frac{1}{2}} \max\{\|(\lambda-T)^{-1}\|: \lambda \in \Gamma\},$$

so that

$$\|L|R^{\perp}\| . \|(L|R^{\perp})^{-1}\| \leq \max\{\|(\lambda-T)^{-1}\|: \lambda \in \Gamma\}. \qquad (3.5)$$

Let r be a rational function with poles outside Ω^- and let

$$M_+(\Gamma)\otimes 1 = \begin{pmatrix} M_+(\Gamma)\otimes 1|R & B \\ 0 & T' \end{pmatrix} \begin{matrix} R \\ R^{\perp}, \end{matrix}$$

where T' is similar to T; then

$$\|r(T')\| \leq \|r(M_+(\Gamma)\otimes 1)\| = \|r(M_+(\Gamma))\| = \max\{|r(\lambda)|: \lambda \in \Gamma\},$$

whence we obtain the following

COROLLARY 3.35. *(i) Given* $T \in L(H)$ *and a bounded open neighborhood* Φ *of* $\sigma(T)$, Φ *is a spectral set (in the sense of von Neumann) for some operator* $T' \sim T$, *i.e.,*

$$\|r(T')\| \leq \max\{|r(\lambda)|: \lambda \in \Phi^-\}$$

for all rational functions r *with poles outside* Φ^-.

(ii) In particular, if $\sigma(T) = \{0\}$ *and* $\varepsilon > 0$, *then there exists* $T' \sim T$ *such that* $\|T'\| < \varepsilon$ *and* $T \underset{\text{sim}}{\to} 0$.

LEMMA 3.36. *Let* $H = H_1 \oplus H_2 \oplus \ldots \oplus H_k$ *and assume that* $A \in L(H)$ *admits an upper triangular operator matrix with respect to this decomposition, i.e.,*

$$A = \begin{pmatrix} A_{11} & A_{12} & A_{13} & \cdot & \cdot & \cdot A_{1,k-1} & A_{1k} \\ & A_{22} & A_{23} & \cdot & \cdot & \cdot A_{2,k-1} & A_{2k} \\ & & A_{33} & \cdot & \cdot & \cdot A_{3,k-1} & A_{3k} \\ & & & \cdot & & \cdot & \cdot \\ & & & & \cdot & \cdot & \cdot \\ & & 0 & & & \cdot & \cdot \\ & & & & & A_{k-1,k-1} & A_{k-1,k} \\ & & & & & & A_{kk} \end{pmatrix} \begin{matrix} H_1 \\ H_2 \\ H_3 \\ \cdot \\ \cdot \\ \cdot \\ H_{k-1} \\ H_k \end{matrix};$$

then $A \underset{\text{sim}}{\to} A_{11} \oplus A_{22} \oplus A_{33} \oplus \ldots \oplus A_{k-1,k-1} \oplus A_{kk}$.

PROOF. Let $W_n = n \oplus n^2 \oplus n^3 \oplus \ldots \oplus n^{k-1} \oplus n^k$ (with respect to the same decomposition). Clearly, W_n is invertible with inverse $W_n^{-1} = n^{-1} \oplus n^{-2} \oplus n^{-3} \oplus \ldots \oplus n^{-k+1} \oplus n^{-k}$, and a straightforward computation shows that

$$\|W_n A W_n^{-1} - \oplus_{j=1}^{k} A_{jj}\| \le [n^{-1} + n^{-2} + \ldots + n^{-k+1}]\|A\| \to 0 \ (n \to \infty). \ \square$$

It follows from (3.2) and Theorem 3.33 that

$$M(\Gamma) \otimes 1 = \begin{pmatrix} M_+(\Gamma) \otimes 1 & Z \otimes 1 \\ 0 & M_-(\Gamma) \otimes 1 \end{pmatrix} \begin{matrix} H^2(\Gamma; H) \\ L^2(\Gamma) \otimes H \ominus H^2(\Gamma; H) \end{matrix}$$

$$= \begin{pmatrix} M_+(\Gamma) \otimes 1 | R & B & Z_1 \\ 0 & T' & Z_2 \\ 0 & 0 & M_-(\Gamma) \otimes 1 \end{pmatrix} \begin{matrix} R \\ H^2(\Gamma; H) \ominus R \\ L^2(\Gamma) \otimes H \ominus H^2(\Gamma; H) \end{matrix},$$

where $\begin{pmatrix} Z_1 \\ Z_2 \end{pmatrix} = Z \otimes 1$, $M_+(\Gamma) \otimes 1 | R \sim M_+(\Gamma) \otimes 1$ and $T' \sim T$. Combining this observation with Lemma 3.36, we obtain

COROLLARY 3.37. *Let* $T \in L(H)$ *and let* Ω *be an analytic Cauchy domain containing* $\sigma(T)$ *with* $\partial\Omega = \Gamma$; *then*

$$M(\Gamma) \underset{\text{sim}}{\to} M_+(\Gamma) \oplus M_-(\Gamma), \quad M(\Gamma) \otimes 1 \underset{\text{sim}}{\to} (M_+(\Gamma) \otimes 1) \oplus (M_-(\Gamma) \otimes 1),$$

$$M(\Gamma) \otimes 1 \underset{\text{sim}}{\to} (M_+(\Gamma) \otimes 1) \oplus T \oplus (M_-(\Gamma) \otimes 1) \text{ and } M_+(\Gamma) \otimes 1 \underset{\text{sim}}{\to} (M_+(\Gamma) \otimes 1) \oplus T.$$

3.3 Apostol's triangular representation

Given $T \in L(H)$, let $H_r(T) = \bigvee\{\ker(\lambda - T) : \lambda \in \rho_{s-F}^r(T)\}$, let $H_\ell(T) = \bigvee\{\ker(\lambda - T)^* : \lambda \in \rho_{s-F}^r(T)\}$ and let $H_o(T)$ be the orthogonal complement of $H_r(T) + H_\ell(T)$. Denote the compression of T to $H_r(T)$, $H_\ell(T)$ and $H_o(T)$ by T_r, T_ℓ and T_o, respectively. Recall that $A \in L(H)$ is a *triangular* operator if it admits an upper triangular matrix; i.e.,

$$A = \begin{pmatrix} a_{11} & a_{12} & a_{13} & \cdot & \cdot & \cdot \\ & a_{22} & a_{23} & \cdot & \cdot & \cdot \\ & & a_{33} & \cdot & \cdot & \cdot \\ & & & \cdot & \cdot & \cdot \\ & 0 & & & \cdot & \cdot \\ & & & & & \cdot \end{pmatrix}, \tag{3.6}$$

with respect to a suitable ONB.

Apostol's triangular representation and its basic properties are established in the following

THEOREM 3.38. *(i)* $H_r(T)$ *is orthogonal to* $H_\ell(T)$, *so that*

$$H = H_r(T) \oplus H_o(T) \oplus H_\ell(T). \tag{3.7}$$

(ii) $H_r(T)$ *and* $H_r(T)\oplus H_o(T)$ *are invariant under* T, *so that* T *admits a* 3 × 3 *upper triangular operator matrix representation*

$$T = \begin{pmatrix} T_r & * & * \\ 0 & T_o & * \\ 0 & 0 & T_\ell \end{pmatrix} \tag{3.8}$$

with respect to the above decompostion, where

(iii) $T_r = T|H_r(T)$ *is a triangular operator,* $\sigma(T_r) = \sigma_\ell(T_r) = \sigma_{\ell re}(T_r) \cup [\rho_r(T_r) \setminus \rho(T_r)]$, $\sigma(T_r)$ *is a perfect set, every component of* $\sigma_{\ell re}(T_r)$ *intersects the set* $\sigma_p(T_r)^-$ *and* $\sigma_p(T_r^*) = \emptyset$, *so that* $\rho^s_{s-F}(T_r) = \emptyset$ *and* $\min.\mathrm{ind}(\lambda - T_r) = 0$ *for all* $\lambda \in \rho_{s-F}(T_r)$, *and*

(iv) T_ℓ *is the adjoint in* $L(H_\ell(T))$ *of the triangular operator* $T^*|H_\ell(T)$, $\sigma(T_\ell) = \sigma_r(T_\ell) = \sigma_{\ell re}(T_\ell) \cup [\rho_\ell(T_\ell) \setminus \rho(T_\ell)]$, $\sigma(T_\ell)$ *is a perfect set, every component of* $\sigma_{\ell re}(T_\ell^*)$ *intersects the set* $\sigma_p(T_\ell^*)^-$ *and* $\sigma_p(T_\ell)$ $\emptyset$, *so that* $\rho^s_{s-F}(T_\ell) = \emptyset$ *and* $\min.\mathrm{ind}(\lambda - T_\ell) = 0$ *for all* $\lambda \in \rho_{s-F}(T_\ell)$.

Furthermore,

(v) $\lambda \to P_{\ker(\lambda - T)}$ $(\lambda \in \rho_{s-F}(T))$ *is a continuous function for* $\lambda \in \rho^r_{s-F}(T)$ *and discontinuous for* $\lambda \in \rho^s_{s-F}(T)$;

(vi) $\rho_{s-F}(T) \subset \rho_r(T_r) \cap \rho_\ell(T_\ell)$;

(vii) $\rho^r_{s-F}(T) \subset \rho(T_o)$;

(viii) $\rho^s_{s-F}(T) \subset \sigma_o(T_o)$;

(ix) $\sigma_o(T) \subset \rho(T_r) \cap \rho(T_\ell) \cap \sigma_o(T_o)$;

(x) *If* $\Lambda = \{\lambda_1, \lambda_2, \ldots, \lambda_m\}$ *is a finite subset of* $\rho^s_{s-F}(T)$, *then* $T \sim T_\Lambda \oplus T'_\Lambda$, *where* T_Λ *acts on a finite dimensional subspace,* $\sigma(T_\Lambda) = \Lambda$ *and* $\Lambda \subset \rho^r_{s-F}(T'_\Lambda)$.

We shall need several auxiliary results.

LEMMA 3.39. *If* $B \in L(H)$, $\lambda \in \sigma(B)$ *and* $\|(\lambda - B)x\| \geq M\|x\|$ *for all* $x \in$

H, *then* $||A - B|| \geq M$ *for every triangular operator* A.

PROOF. Assume that A has an upper triangular matrix of the form (3.6) with respect to the ONB $\{e_n\}_{n=1}^{\infty}$. Let P_n be the orthogonal projection of H onto $\bigvee\{e_1,e_2,\ldots,e_n\}$ and let y be a unit vector orthogonal to ran$(\lambda-B)$ (so that $(\lambda-B)^*y = 0$); clearly, $P_nAP_n = AP_n$.

Given ε, $0 < \varepsilon < 1$, there exists $n_o = n_o(\varepsilon)$ such that $||P_n(\lambda-B)^*y|| < \varepsilon$ for all $n \geq n_o$. $P_n(\lambda-B)^*P_n$ can be regarded as an operator acting on the finite dimensional space $M_n = \text{ran } P_n$. It follows from the finite dimensionality of M_n that this subspace contains a unit vector z_n such that $||[P_n(\lambda-B)^*P_n]^*z_n|| = ||P_n(\lambda-B)P_nz_n|| < \varepsilon$. We have $P_nz_n = z_n$, and

$$\begin{aligned}
||A - B|| &\geq \liminf(n \to \infty)\ ||(1-P_n)(A-B)P_n|| \\
&\geq \liminf(n \to \infty)\ ||(1-P_n)(\lambda-B)P_n|| \\
&\geq \liminf(n \to \infty)\ ||(1-P_n)(\lambda-B)P_nz_n|| \\
&\geq \liminf(n \to \infty)\ ||(\lambda-B)z_n|| - \limsup(n \to \infty)\ ||P_n(\lambda-B)P_nz_n|| \\
&\geq M - \varepsilon.
\end{aligned}$$

Since ε can be chosen arbitrarily small, we conclude that $||A - B|| \geq M$. □

COROLLARY 3.40. *If* A *is a triangular operator with matrix* (3.6) *with respect to the ONB* $\{e_n\}_{n=1}^{\infty}$, *then*

(i) $\sigma(A) = \sigma_\ell(A) = \sigma_{\ell re}(A) \cup \sigma_p(A)$;

(ii) Every clopen subset of $\sigma(A)$ *intersects* $d(A) = \{a_{nn}\}_{n=1}^{\infty}$ *and every component of* $\sigma(A)$ *intersects* $d(A)^-$;

(iii) Every isolated point of $\sigma(A)$ *belongs to* $d(A)$;

(iv) If $\ker(\lambda-A)^* \neq \{0\}$, *then* $\lambda \in d(A)$.

PROOF. (i) The equality $\sigma_\ell(A) = \sigma_{\ell re}(A) \cup \sigma_p(A)$ is trivial. On the other hand (since A is triangular) Lemma 3.39 implies that $\sigma(A) = \sigma_\ell(A)$.

(ii) Let σ_o be a clopen subset of $\sigma(A)$ such that $\sigma_o \cap d(A) = \emptyset$ and let Ω be a Cauchy domain containing $d(A)^-$ such that $\sigma_o \cap \Omega^- = \emptyset$. It is easily seen that, if

$$E = \frac{1}{2\pi i}\int_{\partial\Omega} (\lambda-A)^{-1}\, d\lambda,$$

then $Ee_n = e_n$ for all $n = 1,2,\ldots$.

Since E is an idempotent and ran $E \supset \bigvee\{e_n\}_{n=1}^{\infty} = H$, it readily follows that $\sigma_o = \emptyset$.

Since $\sigma(A)$ is a compact Hausdorff space, every component of $\sigma(A)$ is the intersection of all the clopen subsets containing it. Let σ be a component of $\sigma(A)$ and let $\{\sigma_\alpha : \alpha \in \Lambda\}$ be the family of all clopen subsets of $\sigma(A)$ containing σ. It follows from the first part of the proof that $\sigma_\alpha \cap d(A) \neq \emptyset$ for all α in Λ. A fortiori, $\sigma \cap d(A)^- = (\cap_{\alpha\in\Lambda}\ \sigma_\alpha) \cap d(A)^- =$

$\cap_{\alpha\in\Lambda}[\sigma_\alpha \cap d(A)^-] \neq \emptyset$ (by an obvious argument of compactness).

(iii) This is a trivial consequence of (ii).

(iv) Assume that $(\lambda-A)^*y = 0$ for some $\lambda \in \sigma(A)$ and some unit vector $y \in H$; then

$$0 = \langle(\lambda-A)^*y, e_n\rangle = \langle y, (\lambda-A)e_n\rangle, \text{ for all } n = 1,2,\ldots,$$

so that $y \perp (\lambda-A)M_n$, where $M_n = \vee\{e_1,e_2,\ldots,e_n\}$, for all $n = 1,2,\ldots$.

If $\lambda \notin d(A)$, then $(\lambda-A)M_n = M_n$ $(n = 1,2,\ldots)$ and therefore $y \perp \vee\{M_n\}_{n=1}^{\infty} = H$; so that, $y = 0$, a contradiction. Hence, $\lambda \in d(A)$. □

PROPOSITION 3.41. *Let* $T \in L(H)$ *and* $\mu \in \rho_{s-F}(T)$; *then*

$$Q = \lim(\lambda \to \mu)\, P_{\ker(\lambda-T)}$$

exists (in the norm topology). Q is the orthogonal projection onto $\ker(\mu-T) \cap \{\cap_{n=1}^{\infty} \operatorname{ran}(\mu-T)^n\}$.

PROOF. Assume first that $\mu-T$ is onto. By Corollary 1.14(v), $\lambda-T$ is onto for all λ in some neighborhood Ω of μ. It follows that $P_{\ker(\lambda-T)} = 1-R_r(\lambda,T)(\lambda-T)$ (where $R_r(\lambda,T)$ is defined by (3.1), $\lambda \in \Omega$) converges in the norm to $1 - R_r(\lambda,T)(\lambda-T) = P_{\ker(\lambda-T)}$ as $\lambda \to \mu$.

Put $M = \cap_{n=1}^{\infty} \operatorname{ran}(\mu-T)^n$. Because $(\mu-T)$ is semi-Fredholm, M is closed (i.e., is a subspace) and $(\mu-T)M = M$. Hence $(\mu-T_M)M = M$.

For $\lambda \neq \mu$ clearly $(\mu-T)\ker(\lambda-T) = \ker(\lambda-T)$ and so $\ker(\lambda-T) \subset M$. It follows that

$$P_{\ker(\lambda-T)} = P_{\ker(\lambda-T_M)} P_M$$

and the first paragraph shows that this has the desired limit. □

COROLLARY 3.42. *Let* $T \in L(H)$ *and* $\mu \in \rho_{s-F}(T)$. *The following conditions are equivalent*

(i) μ *is a regular point of* $\rho_{s-F}(T)$.

(ii) $P_{\ker(\lambda-T)}$ *is continuous at* $\lambda = \mu$.

PROOF. (i) => (ii) If $\operatorname{nul}(\lambda-T)$ is a finite constant in some neighborhood of μ then Corollary 1.14(vi) shows that $\ker(\lambda-T) \subset \cap_{n=1}^{\infty} \operatorname{ran}(\lambda-T)^n$.

By Proposition 3.41, $P_{\ker(\lambda-T)}$ is continuous at $\lambda = \mu$.

If $\operatorname{nul}(\lambda-T)^*$ is a finite constant in a neighborhood of μ then Corollary 1.14(vi) shows that

$$\ker(\mu-T)^* \subset \cap_{n=1}^{\infty} \operatorname{ran}(\mu-T)^{*n} \tag{3.9}$$

CLAIM: Since $\mu-T$ is semi-Fredholm, (3.9) is equivalent to

$$\ker(\mu-T) \subset \cap_{n=1}^{\infty} \operatorname{ran}(\mu-T)^n. \tag{3.10}$$

Indeed, if $\ker(\mu-T) \subset \cap_{n=1}^{\infty} \operatorname{ran}(\mu-T)^n$, then

$\operatorname{ran}(\mu-T)^* = [\ker(\mu-T)]^\perp \supset [\operatorname{ran}(\mu-T)^n]^\perp = \ker(\mu-T)^{*n}$, $n = 1,2,\ldots$.

Thus, if $y \in \ker(\mu-T)^*$, then $y = (\mu-T)^* y_1$ for some $y_1 \in H$. Therefore $\operatorname{ran}(\mu-T)^* \supset \ker(\mu-T)^{*2} \supset (\mu-T)^{*-1}[\{y\}] \neq \emptyset$, so that $y = (\mu-T)^{*2} y_2$ for some $y_2 \in H$ and $\operatorname{ran}(\mu-T)^* \supset \ker(\mu-T)^{*3} \supset (\mu-T)^{*-2}[\{y\}] \neq \emptyset$.

By induction, there exists y_n in H such that $y = (\mu - T)^{*n} y_n$, $n = 1,2,\ldots$. Hence, $y \in \cap_{n=1}^{\infty} \operatorname{ran}(\mu-T)^{*n}$.

This proves that (3.10) implies (3.9). The converse implication follows by taking adjoints.

Since (3.9) and (3.10) are equivalent, we conclude (as in the first part of the proof) that $P_{\ker(\lambda-T)}$ is continuous at $\lambda = \mu$.

(ii) => (i) If $P_{\ker(\lambda-T)}$ is continuous at $\lambda = \mu$, then (3.10) holds. Hence (3.9) also holds and (i) follows from Corollary 1.14(vi).

□

Let $H = \oplus_{j=1}^{k} H_j$, where $H_j \simeq H$ for all $j = 1,2,\ldots,k$, and let $A \in L(H)$ with matrix of the form (3.3) with respect to this decomposition. Observe that A is compact if and only if A_{ij} is compact for all i and j. On the other hand, since $H_j \simeq H$ (i.e., H_j is infinite dimensional) for all j, H_j can be identified with H via a fixed unitary map ($j = 1, 2,\ldots,k$).

Then $L(H)$ can be identified with the algebra of all $k \times k$ matrices with entries in $L(H)$, $K(H)$ can be identified with the ideal of all $k \times k$ matrices with entries in $K(H)$ and (by taking the corresponding quotients) $A(H)$ can also be identified with the algebra of all $k \times k$ matrices with entries in $A(H)$ ($k = 1,2,\ldots$). This identifications will play a very important role.

The proof of the following elementary algebraic lemma is left to the reader.

LEMMA 3.43. *Let* $\mathbb{M}_2(R)$ *be the ring of all* 2×2 *matrices with entries in a ring* R *with identity* 1. *Then*

(i) If $\begin{pmatrix} p & r \\ 0 & q \end{pmatrix}$ *has a right inverse* $\begin{pmatrix} a & b \\ c & d \end{pmatrix}$ *in* $\mathbb{M}_2(R)$, *then* $qd = 1$ *(in* R*). If* q *is invertible, then* $pa = 1$.

(ii) If $\begin{pmatrix} p & r \\ 0 & q \end{pmatrix}$ *has a left inverse* $\begin{pmatrix} a & b \\ c & d \end{pmatrix}$ *in* $\mathbb{M}_2(R)$, *then* $ap = 1$ *(in* R*). If* p *is invertible, then* $dq = 1$.

PROOF OF THEOREM 3.38. (i) By Corollary 1.14(vi), either $\ker(\lambda-T) \subset \operatorname{ran}(\lambda-T)$ or $\ker(\lambda-T)^* \subset \operatorname{ran}(\lambda-T)^*$ for every $\lambda \in \rho_{s-F}^{r}(T)$. If $\mu \in \rho_{s-F}^{r}(T)$ and $\mu \neq \lambda$, then $(\lambda-T)\ker(\mu-T) = \ker(\mu-T)$ and so $[\ker(\lambda-T)^*]^\perp = \operatorname{ran}(\lambda-T) \supset \ker(\mu-T)$.

By Corollary 3.42 $\|P_{\ker(\mu-T)} - P_{\ker(\lambda-T)}\| \to 0$ $(\mu \to \lambda)$, whence it readily follows that $\ker(\lambda-T) \subset [\ker(\lambda-T)^*]^{\perp}$. Hence,

$$H_r(T) = \vee\{\ker(\lambda-T): \lambda \in \rho^r_{s-F}(T)\} \perp \vee\{\ker(\lambda-T)^*: \lambda \in \rho^r_{s-F}(T)\} = H_\ell(T)$$

and, a fortiori, $H = H_r(T) \oplus H_o(T) \oplus H_\ell(T)$.

(ii) The invariance of $H_r(T)$ and $H_r(T) \oplus H_o(T) = H_\ell(T)^{\perp}$ under T is immediate. Hence, T admits a matrix representation of the form (3.8).

(iii) Let $\{\lambda_n\}_{n=1}^{\infty}$ be a denumerable dense subset of $\rho^r_{s-F}(T)$. By Corollary 3.42 $H_r(T) = \vee\{M_n\}_{n=1}^{\infty}$, where $M_n = \ker(\lambda_n - T)$. It is completely apparent that T_r has an upper triangular matrix with respect to an ONB obtained by Gram-Schmidt orthonormalization of a union of ONB's of the subspaces $\{M_n\}_{n=1}^{\infty}$ and $d(T_r) = \{\lambda_n\}_{n=1}^{\infty}$ in this representation (where $d(T_r)$ has the meaning of Corollary 3.40).

By Corollary 3.40, $\sigma(T_r) = \sigma_\ell(T_r) = \sigma_{\ell re}(T_r) \cup \sigma_p(T_r)$, every component of $\sigma(T_r)$ intersects $d(T_r)^- \subset [\rho^+_{s-F}(T_r)]^-$ (Observe that $\lambda - T_r$ is onto for each $\lambda \in \rho^r_{s-F}(T)$) and $\sigma_p(T_r^*) \subset d(T_r) = \{\lambda_n\}_{n=1}^{\infty}$. Since $\{\lambda_n\}_{n=1}^{\infty}$ can be arbitrarily chosen (under the condition $\{\lambda_n\}^- = [\rho^r_{s-F}(T)]^-$, of course), it readily follows that $\sigma_p(T_r^*) = \emptyset$ and, a fortiori, that $\rho^s_{s-F}(T_r) = \emptyset$ and $\text{min.ind}(\lambda - T_r) = 0$ for all $\lambda \in \rho_{s-F}(T_r)$.

It is also clear that $\sigma(T_r)$ does not have any isolated point, i. e., either $H_r(T) = \{0\}$ and $\sigma(T_r) = \emptyset$, or $\sigma(T_r)$ is a nonempty perfect set and $\sigma(T_r) = \sigma_{\ell re}(T) \cup \sigma_p(T_r) = \sigma_{\ell re}(T) \cup [\rho_r(T_r) \setminus \rho(T_r)]$. Since $\rho^+_{s-F}(T_r) = \rho_r(T_r) \setminus \rho(T_r)$, it follows that every component of $\sigma_{\ell re}(T_r)$ intersects $[\rho_r(T_r) \setminus \rho(T_r)]^-$.

In order to prove (iv), we only have to apply the arguments of (iii) to $T^*|H_\ell(T)$ (Indeed, observe that the equivalence of (3.9) and (3.10) and Corollary 1.14(vi) imply that $\rho^r_{s-F}(T) = \rho^r_{s-F}(T^*)^*$); (v) is the content of Corollary 3.32 and (vi) follows from (iii), (iv) and their proofs: If $\lambda \in \rho_{s-F}(T) \setminus \rho_r(T_r)$, then $\lambda \in \sigma_{\ell re}(T_r) \subset \sigma_e(T)$, a contradiction. Hence $\rho_{s-F}(T) \subset \rho_r(T_r)$. A similar argument shows that $\rho_{s-F}(T) \subset \rho_\ell(T_\ell)$.

(vii) Assume that $H_r(T) \neq \{0\}$ and $H_\ell(T) \neq \{0\}$; then $H_r(T)$, $H_\ell(T)$, $H_r(T) \oplus H_o(T)$ and $H_o(T) \oplus H_\ell(T)$ are infinite dimensional spaces.

Suppose that $\lambda \in \rho_{s-F}(T)$ and $\text{ind}(\lambda-T) > -\infty$; then $\lambda - \tilde{T}_\ell$ is invertible and $\lambda - \tilde{T}$ is right invertible. By decomposing H as $[H_r(T) \oplus H_o(T)] \oplus H_\ell(T)$, $\lambda - \tilde{T}$ can be written as the 2×2 matrix

$$\lambda - \tilde{T} = \begin{pmatrix} \widetilde{\left(\lambda - \begin{pmatrix} T_r & * \\ 0 & T_o \end{pmatrix}\right)} & \tilde{R} \\ 0 & \lambda - \tilde{T}_\ell \end{pmatrix} = \begin{pmatrix} p & r \\ 0 & q \end{pmatrix}.$$

By Lemma 3.43(i), p is right invertible. Since

$$p = \begin{pmatrix} \lambda-\tilde{T}_r & * \\ 0 & \lambda-\tilde{T}_o \end{pmatrix},$$

it follows from the same lemma that $\lambda-\tilde{T}_o$ is right invertible.

Similarly, if $\text{ind}(\lambda-T) < \infty$, we conclude from Lemma 3.43(ii) that $\lambda-\tilde{T}_o$ is left invertible.

Hence $\lambda \in \rho_{s-F}(T_o)$ and therefore either $\lambda-\tilde{T}_o$ is invertible, or $\text{nul}(\lambda-T_o) \neq 0$ or $\text{nul}(\lambda-T_o)^* \neq 0$. Thus, in order to complete the proof, it only remains to show that $\lambda \notin \sigma_p(T_o)\cup\sigma_p(T_o^*)^*$.

Assume that $(\lambda-T_o)x = 0$ or equivalently $(\lambda-T)x \perp H_o(T)$. The invariance of $H_r(T)\oplus H_o(T)$ implies that $(\lambda-T)x \in H_r(T)$. Since $\lambda-T$ maps $H_r(T)$ *onto* itself, there is some $y \in H_r(T)$ such that $(\lambda-T_r)y = (\lambda-T)x$. Since $y - x \in \ker(\lambda-T) \subset H_r(T)$, x must be equal to 0. An analogous argument shows that $\text{nul}(\lambda-T_o)^* = 0$ and so $\lambda-T_o$ is invertible, i.e. $\lambda \in \rho(T_o)$.

If either $H_r(T) = \{0\}$ or $H_\ell(T) = \{0\}$, then the result follows by an even simpler argument.

(viii) Let $\mu \in \rho^s_{s-F}(T)$. If μ is not an eigenvalue of T_o, then (vi) shows that we have $\ker(\mu-T) = \ker(\mu-T_r)$ and (by Lemma 3.12) we can find a neighborhood Ω of μ such that

$$\ker(\lambda-T) = \ker(\lambda-T_r), \text{ for all } \lambda \in \Omega.$$

This implies that $\mu \in \rho^r_{s-F}(T)$, a contradiction.

Hence, $\mu \in \sigma_p(T_o)$. On the other hand, the first part of the proof of (vii) indicates that $\mu \in \rho_{s-F}(T_o)$ and, by (vii) and Corollary 1.14 (v), μ is an isolated point of $\sigma(T_o)$. Therefore, $\mu \in \sigma_o(T_o)$.

(ix) This follows immediately from (vi) and (viii).

(x) Let Λ be a finite subset of $\rho^s_{s-F}(T)$. It follows from (viii) that $\Lambda \subset \sigma_o(T_o)$ and, a fortiori (since Λ is closed), Λ is a clopen subset of $\sigma(T_o)$, so that (by Riesz' decomposition theorem)

$$T_o = \begin{pmatrix} A & T_{23} \\ 0 & B \end{pmatrix},$$

where A acts on a finite dimensional space, $\sigma(A) = \Lambda$ and $\Lambda\cap\sigma(B) = \emptyset$.

Therefore T admits the decomposition

$$T = \begin{pmatrix} T_r & T_{12} & T_{13} & T_{14} \\ 0 & A & T_{23} & T_{24} \\ 0 & 0 & B & T_{34} \\ 0 & 0 & 0 & T_\ell \end{pmatrix} = \begin{pmatrix} T_r & T_{12} & B_{13} \\ 0 & A & B_{23} \\ 0 & 0 & C_\ell \end{pmatrix}, \quad C_\ell = \begin{pmatrix} B & T_{34} \\ 0 & T_\ell \end{pmatrix},$$

where $\Lambda = \sigma(A) \subset \rho_r(T_r)\cap\rho_\ell(T_\ell)$, i.e., $\sigma_r(T_r)\cap\sigma_\ell(A) = \sigma_r(A)\cap\sigma_\ell(T_\ell) = \sigma(A)\cap \sigma(B) = \sigma_r(A)\cap\sigma_\ell(C_\ell) = \emptyset$.

By Corollary 3.22 and its proof, there exist operators X, Y such that

$$\begin{pmatrix} 1 & X \\ 0 & 1 \end{pmatrix} \begin{pmatrix} T_r & T_{12} \\ 0 & A \end{pmatrix} \begin{pmatrix} 1 & X \\ 0 & 1 \end{pmatrix}^{-1} = \begin{pmatrix} T_r & 0 \\ 0 & A \end{pmatrix}$$

and

$$\begin{pmatrix} 1 & Y \\ 0 & 1 \end{pmatrix} \begin{pmatrix} A & B_{12} \\ 0 & C_\ell \end{pmatrix} \begin{pmatrix} 1 & Y \\ 0 & 1 \end{pmatrix}^{-1} = \begin{pmatrix} A & 0 \\ 0 & C_\ell \end{pmatrix}.$$

Hence

$$T \sim \left(\begin{pmatrix} 1 & 0 & 0 \\ 0 & 1 & Y \\ 0 & 0 & 1 \end{pmatrix} \begin{pmatrix} 1 & X & 0 \\ 0 & 1 & 0 \\ 0 & 0 & 1 \end{pmatrix} \right) \begin{pmatrix} T_r & T_{12} & B_{13} \\ 0 & A & B_{23} \\ 0 & 0 & C_\ell \end{pmatrix} \left(\begin{pmatrix} 1 & 0 & 0 \\ 0 & 1 & Y \\ 0 & 0 & 1 \end{pmatrix} \begin{pmatrix} 1 & X & 0 \\ 0 & 1 & 0 \\ 0 & 0 & 1 \end{pmatrix} \right)^{-1}$$

$$= \begin{pmatrix} T_r & 0 & * \\ 0 & A & 0 \\ 0 & 0 & C_\ell \end{pmatrix} \simeq A \oplus \begin{pmatrix} T_r & * \\ 0 & C_\ell \end{pmatrix} = A \oplus \begin{pmatrix} T_r & * & * \\ 0 & B & * \\ 0 & 0 & T_\ell \end{pmatrix},$$

$T_\Lambda = A$ has the desired properties and $\Lambda \in \rho^r_{s-F}(T'_\Lambda)$, where

$$T'_\Lambda = \begin{pmatrix} T_r & * & * \\ 0 & B & * \\ 0 & 0 & T_\ell \end{pmatrix}.$$ □

From Theorem 3,38 and Lemma 3.36, we obtain the following

COROLLARY 3.44. *Let* $T \in L(H)$ *with triangular representation* (3.8); *then* $T \underset{\text{sim}}{\rightarrow} T_r \oplus T_o \oplus T_\ell$.

3.4 Correction by compact perturbation of the singular behavior of operators

The *Weyl spectrum* of $T \in L(H)$ is defined by

$$\sigma_W(T) = \cap\{\sigma(T+K): \quad K \in K(H)\}.$$

It is easily seen that $\sigma_W(T)$ is the largest subset of the spectrum that is invariant under compact perturbations and that (by Theorem 1.13(v)) $\sigma_W(T)$ contains every complex λ such that $\lambda-T$ is not a Fredholm operator of order 0. Furthermore, it is well-known that $\sigma_W(T)$ actually coincides with this set, i.e.,

$$\sigma_W(T) = \{\lambda \in \mathbb{C}: \quad \lambda-T \text{ is not a Fredholm operator of index } 0\}. \quad (3.11)$$

(The inclusion '$\subset$' will easily follow as a corollary of Theorem 3.48 below.)

The *Browder spectrum* of T is the complement of $\sigma_o(T)$ in the spectrum, i.e.,

$$\sigma_B(T) = \sigma(T) \setminus \sigma_o(T).$$

Clearly, $\sigma(T) \supset \sigma_B(T) \supset \sigma_W(T) \supset \sigma_e(T) \supset \sigma_{\ell re}(T)$ and $\emptyset \neq \partial\sigma_B(T) \subset \partial\sigma_W(T) \subset \partial\sigma_e(T) \subset \partial\sigma_{\ell re}(T)$.

PROPOSITION 3.45. *Given* $T \in L(H)$, *there exists* $K \in K(H)$ *such that* $\rho^s_{s-F}(T+K) = \emptyset$, $\sigma(T+K) = \sigma_B(T)$ *and* $\min.\mathrm{ind}(T+K-\lambda) = \min.\mathrm{ind}(T-\lambda)$ *for all* $\lambda \in \rho^r_{s-F}(T)$; *moreover,* K *can be chosen equal to a normal compact operator such that*

$$\|K\| = \max\{\mathrm{dist}[\lambda, \partial\rho_{s-F}(T)]: \ \lambda \in \rho^s_{s-F}(T)\}.$$

PROOF. Let

$$T = \begin{pmatrix} T_r & T_{12} & T_{13} \\ 0 & T_o & T_{23} \\ 0 & 0 & T_\ell \end{pmatrix} \begin{matrix} H_r(T) \\ H_o(T) \\ H_\ell(T) \end{matrix}$$

be the triangular representation (3.8) of T and let $\{\mu_n\}_{1\le n<m}$ $(0 \le m \le \infty)$ be an enumeration of the eigenvalues of $\sigma_o(T)$. Then T_o admits the representation

$$T_o = \left(\begin{array}{cccccccc|c} \mu_1+T_1 & & & & & & & & \\ & \mu_2+T_2 & & & & & & & \\ & & \cdot & & & * & & & \\ & & & \cdot & & & & & \\ & & & & \cdot & & & & * \\ & & & & & \mu_n+T_n & & & \\ & & 0 & & & & \cdot & & \\ & & & & & & & \cdot & \\ & & & & & & & & \cdot \\ \hline & & & & 0 & & & & T_\infty \end{array}\right) \begin{matrix} H_1 \\ H_2 \\ \cdot \\ \cdot \\ \cdot \\ H_n \\ \cdot \\ \cdot \\ \cdot \\ H_\infty \end{matrix},$$

where $H_1, H_2, \ldots, H_n, \ldots$ are defined so that $\oplus_{j=1}^n H_j$ coincides with the Riesz subspace of T_o corresponding to the clopen subset $\{\mu_1, \mu_2, \ldots, \mu_n\}$ of $\sigma(T_o)$ $(1 \le n < m)$ and $H_\infty = H_o(T) \ominus \{\oplus_{1\le n<m} H_n\}$. It is not difficult to check that $\sigma(T_\infty) \subset \sigma(T_o) \setminus \sigma_o(T_o) = \sigma_B(T_o)$.

For each μ_n, let $\lambda_n \in \partial\rho_{s-F}(T)$ be any point such that $|\mu_n - \lambda_n| = \mathrm{dist}[\mu_n, \sigma(T) \setminus \rho_{s-F}(T)]$. It is easily seen that $\lambda_n \in \sigma_{\ell re}(T)$. Thus, if T_o' is the operator obtained from T_o by replacing μ_n by λ_n $(1 \le n < m)$, and

$$T' = \begin{pmatrix} T_r & T_{12} & T_{13} \\ 0 & T_o' & T_{23} \\ 0 & 0 & T_\ell \end{pmatrix},$$

then it follows from Theorem 3.38 (Observe that $(T')_r = T_r$ and $(T')_\ell = T_\ell$!) that $\rho^s_{s-F}(T') = \emptyset$, $\sigma(T') = \sigma_B(T') = \sigma_B(T)$ and $K = T' - T$ is a diagonal normal operators with eigenvalues $\lambda_1 - \mu_1$ (with multiplicity equal to $\dim H_1$), $\lambda_2 - \mu_2$ (with multiplicity equal to $\dim H_2$), ..., $\lambda_n - \mu_n$ (with multiplicity equal to dim H_n),..., so that

$$\|K\| = \max\{|\lambda_n - \mu_n| : 1 \le n < m\} = \max\{\text{dist}\,[\mu_n, \partial\rho_{s-F}(T)] : 1 \le n < m\}. \square$$

LEMMA 3.46. *Let* $T \in L(H)$, $\mu \in \rho^s_{s-F}(T) \setminus \sigma_0(T)$ *and* $\varepsilon > 0$. *Then there exists a finite rank operator* K_μ *such that*

(i) $\|K_\mu\| < \varepsilon$, $\mu \in \rho^r_{s-F}(T+K_\mu)$.

(ii) $\text{min.ind}(\lambda-(T+K_\mu))^k = \text{min.ind}(\lambda-T)^k$, $\lambda \in \rho_{s-F}(T)\setminus\{\mu\}$, $k = 1,2,\ldots$

(iii) If $\nu \in \rho^s_{s-F}(T)\setminus\{\mu\}$ *and* $T \sim T_\nu \oplus T'_\nu$, *where* T_ν *acts on a finite dimensional space*, $\sigma(T_\nu) = \{\nu\}$ *and* $\nu \in \rho^r_{s-F}(T'_\nu)$, *then* $\nu \in \rho^s_{s-F}$ $(T+K_\mu)$ *and* $T+K_\mu \sim T_\nu \oplus T''_\nu$, *where* $\nu \in \rho^r_{s-F}(T''_\nu)$.

PROOF. By Theorem 3.38, $\mu \in \sigma_p(T_r) \cup \sigma_p((T_\ell)^*)^*$. Suppose that $\mu \in \sigma_p((T_\ell)^*)^*$ and denote by M the Riesz spectral subspace of T_0 corresponding to $\sigma(T_0)\setminus\{\mu\}$. If we put $H_\mu = H_0(T) \ominus M$ then

$$T = \begin{pmatrix} T_r & T_{12} & T_{13} & T_{14} \\ 0 & T_0|M & T_{23} & T_{24} \\ 0 & 0 & T_\mu & T_{34} \\ 0 & 0 & 0 & T_\ell \end{pmatrix} \begin{matrix} H_r(T) \\ M \\ H_\mu \\ H_\ell(T) \end{matrix}.$$

Let

$$S' = \begin{pmatrix} T_\mu & T_{34} \\ 0 & T_\ell \end{pmatrix}.$$

Since $\sigma(T_\mu) = \{\mu\}$, we can find a (finite rank!) operator $A \in L(H_\mu)$ such that $\|A\| < \varepsilon/2$, $\sigma(T_\mu + A) = \{\mu\}$ and $\text{nul}(T_\mu + A - \mu) = 1$. (Thus, if $x \in \ker(T_\mu + A - \mu)^*$, $x \ne 0$, then x is a cyclic vector for $T_\mu + A$.) Define

$$S'' = \begin{pmatrix} T_\mu + A & * \\ 0 & T_\ell \end{pmatrix} \begin{matrix} H_\mu \\ H_\ell(T) \end{matrix}.$$

By Corollary 3.22, $S'' \sim (T_\mu + A) \oplus T_\ell$, i.e., S'' has an invariant subspace R complementary to H_μ such that $S''|R \in T_\ell$. Let $W \in L(H_\mu \oplus H_\ell(T))$ be an invertible operator such that

$$S_1 = WS''W^{-1} = (T_\mu + A) \oplus T_\ell.$$

Let $f \in \ker(T_\mu + A - \mu)$, $0 < \|f\| < \varepsilon/(2\|W\|.\|W^{-1}\|)$, and let $e \in \ker(T_\ell - \mu)^*$, $\|e\| = 1$. Let $S_2 = S_1 + e \otimes f \in L(H_\mu \oplus H_\ell(T))$, $\lambda \in \mathbb{C}$, $y \in H_\ell(T)$ and $z \in H_\mu$ and assume that $(S_2 - \mu)(y+z) = 0$; then $(S_1 - \lambda)y + \langle z,f\rangle e + (S_1 - \lambda)z = 0$. Hence

$$(T_\ell - \lambda)y + \langle z,f\rangle e = 0, \quad (T_\mu + A - \lambda)z = 0.$$

If $\lambda \ne \mu$, then $\lambda \notin \sigma(T_\mu + A) = \{\mu\}$. It follows that $z = 0$ and therefore, $(T_\ell - \lambda)y = 0$. Since $\sigma_p(T_\ell) = \emptyset$, we see that $y = 0$. If $\lambda = \mu$, then $z \in \ker(T_\mu + A - \mu)$ and therefore either $z = 0$, or $\langle z,f\rangle \ne 0$ and $(T_\ell - \mu)y = -\langle z,f\rangle e \ne 0$. But this is impossible because $e \perp \text{ran}(T_\ell - \mu)$. We conclude that $\sigma_p(S_2) = \emptyset$.

Observe that

$$S_2 = \begin{pmatrix} T_\mu + A & 0 \\ e\otimes f & T_\ell \end{pmatrix} \begin{matrix} H_\mu \\ H_\ell(T) \end{matrix}, \text{ so that } (S_2-\lambda)^* = \begin{pmatrix} (T_\mu+A-\lambda)^* & f\otimes e \\ 0 & (T_\mu-\lambda)^* \end{pmatrix}.$$

Since H_μ is finite dimensional, it is not difficult to check that $H_\mu = \ker(S_2-\mu)^{*d(\mu)}$ (where $d(\mu) = \dim H_\mu$) and $H_\mu \subset \bigvee \{\ker(S_2-\lambda)^*: \lambda \in \rho^r_{s-F}(T)\}$, whence we conclude that $H_\ell(S_2) = H_\mu \oplus H_\ell(T)$ and $(S_2)_\ell = S_2$.

Let $S_3 = W^{-1}S_2W$ and set

$$S = \begin{pmatrix} T_r & T_{12} & S_{13} \\ 0 & T_o|M & S_{23} \\ 0 & 0 & S_3 \end{pmatrix} \begin{matrix} H_r(T) \\ M \\ H_\mu \oplus H_\ell(T) \end{matrix},$$

where $S_{13} = (T_{13}\ T_{14})$ and $S_{23} = (T_{23}\ T_{24})$; then $K_\mu = S - T$ is a finite rank operator such that $\|K_\mu\| \le \|A\| + \|W\|.\|W^{-1}\|.\|e\otimes f\| < \varepsilon/2+\varepsilon/2 = \varepsilon$, $H_r(S) = H_r(T)$ and $H_\ell(S) = H_\mu \oplus H_\ell(T)$ and $S_\ell = S_3$, and $H_o(S) = M$ and $S_o = T_o|M$. Now (i), (ii) and (iii) can be easily verified by using Theorem 3.38.

If $\mu \in \sigma_p(T_r)$, then we decompose T as

$$T = \begin{pmatrix} T_r & T_{12} & T_{13} & T_{14} \\ 0 & T_\mu & T_{23} & T_{24} \\ 0 & 0 & (T_o)_M & T_{34} \\ 0 & 0 & 0 & T_\ell \end{pmatrix} \begin{matrix} H_r(T) \\ H_\mu \\ M \\ H_\ell(T) \end{matrix},$$

where H_μ is the (finite dimensional) Riesz subspace of T_o corresponding to $\{\mu\}$ and $M = H_o(T)\ominus H_\mu$. Applying the above arguments to

$$\begin{pmatrix} T_r & T_{12} \\ 0 & T_\mu \end{pmatrix}^*$$

instead of S', we can find a finite rank operator $K'_\mu \in L(H_r(T)\oplus H_\mu)$ such that $\|K'\| < \varepsilon$, and

$$S_4 = \begin{pmatrix} T_r & T_{12} \\ 0 & T_\mu \end{pmatrix} + K'_\mu$$

satisfies $H_r(S_4) = H_r(T)\oplus H_\mu$ and $(S_4)_r = S_4$.

In this case we define

$$S = \begin{pmatrix} S_4 & R_{13} & R_{14} \\ 0 & (T_o)_M & T_{34} \\ 0 & 0 & T_\ell \end{pmatrix} \begin{matrix} H_r(T)\oplus H_\mu \\ M \\ H_\ell(T) \end{matrix},$$

where $R_{13} = \begin{pmatrix} T_{13} \\ T_{23} \end{pmatrix}$ and $R_{14} = \begin{pmatrix} T_{14} \\ T_{24} \end{pmatrix}$. It is easily seen that $K_\mu = S - T \simeq K'_\mu \oplus 0$ has the desired properties. □

PROPOSITION 3.47. *Let* $T \in L(H)$ *and* $\varepsilon > 0$; *then there exists* $K \in K(H)$, $\|K\| < \varepsilon$ *such that (i)* $\sigma_o(T+K) = \sigma_o(T)$ *and* $(T+K)|H(\lambda;T+K) \sim T|H(\lambda;T)$ *for all* $\lambda \in \sigma_o(T)$;

(ii) $\rho^r_{s-F}(T+K) = \rho_{s-F}(T)\setminus\sigma_o(T)$;

(iii) $\min.\text{ind}(T+K-\lambda) = \min.\text{ind}(T-\lambda)$ *for all* $\lambda \in \rho^r_{s-F}(T)$.

PROOF. Let $\Delta = \{\lambda \in \rho_{s-F}(T)\setminus[\sigma_o(T)\cup\rho(T)]: \ \text{dist}[\lambda,\partial\rho_{s-F}(T)] \geq \varepsilon/2\}$. Clearly, Δ is a compact set and $\Lambda = \Delta\cap[\rho^s_{s-F}(T)\setminus\sigma_o(T)]$ is finite. Applying Lemma 3.46 a finite number of times (once for each point in Λ), we can find a finite rank operator K_1 such that $\|K_1\| < \varepsilon/2$ and $T+K_1$ satisfies (i), (iii) and

(ii') $\rho^s_{s-F}(T+K_1)\cap\Delta = \emptyset$.

Now we can repeat the argument of the proof of Proposition 3.45 in order to "push" the points of $\rho^s_{s-F}(T+K_1)\setminus\sigma_o(T)$ to $\rho_{s-F}(T+K_1)$ by means of a compact perturbation K_2, $\|K_2\| < \varepsilon/2$, which only affects the action of $(T+K_1)_o$.

It is easily seen that $K = K_1+K_2$ satisfies all our requirements.

□

THEOREM 3.48. *Let* $T \in L(H)$ *and* $\varepsilon > 0$; *then there exists* $K \in K(H)$ *such that*

$$\|K\| < \varepsilon+\max\{\text{dist}[\lambda,\partial\rho_{s-F}(T)]: \ \lambda \in \sigma_o(T)\}$$

and $\min.\text{ind}(T+K-\lambda) = 0$ *for all* $\lambda \in \rho_{s-F}(T)$.

In particular,

$$\sigma(T+K) = \{\lambda \in \mathbb{C}: \ \lambda-T \textit{ is not a Fredholm operator of index } 0\} = \sigma_W(T).$$

PROOF. By Propositions 3.45 and 3.47, we can find a compact operator K_1 such that

$$\|K_1\| < \varepsilon/2+\max\{\text{dist}[\lambda,\partial\rho_{s-F}(T)]: \ \lambda \in \sigma_o(T)\}$$

and the operator $A = T+K_1$ satisfies the following properties: $\rho^r_{s-F}(A) = \rho_{s-F}(A)$ and $\min.\text{ind}(\lambda-A) = \min.\text{ind}(\lambda-T)$ for all $\lambda \in \rho^r_{s-F}(T)$.

Thus, in order to complete the proof it suffices to show that there exists $K_2 \in K(H)$, $\|K_2\| < \varepsilon/2$, such that $\min.\text{ind}(A+K_2-\lambda) = 0$ for all $\lambda \in \rho_{s-F}(A) = \rho_{s-F}(T)$. (Indeed, $K = K_1+K_2$ will obviously satisfy all our requirements.)

Let $\Delta = \{\lambda \in \rho_{s-F}(A): \ \text{dist}[\lambda,\partial\rho_{s-F}(A)] \geq \varepsilon/6\}$ and $\Delta_1 = \{\lambda \in \Delta: \min.\text{ind}(\lambda-A) \neq 0\}$. By Proposition 3.16 we can find $y,z \in H$ such that

$P_{\ker(\lambda-A)}y \neq 0$, $P_{\ker(\lambda-A)^*}z \neq 0$, for all $\lambda \in \Delta_1$.

Let $C_\alpha = \alpha y\otimes z$, $\alpha > 0$. We have

$$\min.\text{ind}(A+C_\alpha-\lambda) = \min.\text{ind}(A-\lambda)-1, \text{ for all } \lambda \in \Delta_1,\ \alpha > 0.$$

Since $\min.\text{ind}(\lambda-A) = 0$ for all $\lambda \in \Delta\setminus\Delta_1$, it readily follows that

$\Delta\backslash\Delta_1 \subset \rho_\ell(A)\ \rho_r(A)$. Therefore, there is a positive constant η_1 such that

$$\|(\lambda-A)x\| \geq \eta_1 \text{ for all } \lambda \in \rho_\ell(A)\cap(\Delta\backslash\Delta_1) \text{ and}$$
$$\|(\lambda-A)^*x\| \geq \eta_1 \text{ for all } \lambda \in \rho_r(A)\cap(\Delta\backslash\Delta_1),$$

for all $x \in H$, $\|x\| = 1$.

Thus, if $0 < \alpha < \min\{\varepsilon/6, \eta_1/2\}$, then $\|C_\alpha\| < \varepsilon/6$ and $\min.\text{ind}(A+C_\alpha-\lambda) = 0$ for all $\lambda \in \Delta\backslash\Delta_1$.

Now we can apply the same arguement to $A+C_\alpha$, etc. After finitely many steps we obtain a finite rank operator C_1 such that $\|C_1\| < \varepsilon/6$,

$$\min.\text{ind}(A+C_1-\lambda) = 0 \text{ for all } \lambda \in \Delta \text{ and } \sigma(A+C_1) \subset \text{interior}\ \sigma(A)_{\varepsilon/6}. \tag{3.12}$$

Let $\{\Omega_j\}_{1\leq n<m}$ $(0 \leq m \leq \infty)$ be an enumeration of the components of $\sigma(A+C_1)\cap\rho_{s-F}(A)$ such that $\min.\text{ind}(\lambda-A) \neq 0$ for all $\lambda \in \Omega_j$. (Clearly, $\cup_j \Omega_j \subset [\partial\rho_{s-F}(A)]_{\varepsilon/3}$.) A repetition of our previous argument shows that there exists a positive constant $\eta < \min\{\varepsilon/6, \eta_1/2\}$ such that the relations (3.12) remain true if $A+C_1$ is replaced by $A+C_1+B$ for any B in $L(H)$ such that $\|B\| \leq \eta$.

If $\lambda_1 \in \rho^r_{s-F}(A+C_1)\cap\Omega_1$, then the arguments used for the construction of C_α can be repeated here in order to find a finite rank operator $C_{2,1}$, $\|C_{2,1}\| < \eta/2$, such that $\min.\text{ind}(A+C_1+C_{2,1}-\lambda_1) = 0$. By induction, we can construct finite rank operators $C_{2,1}, C_{2,2}, \ldots, C_{2,n}, \ldots$, $\|C_{2,n}\| < \eta/2^n$ $(n = 1,2,\ldots)$ such that $C_2 = \sum_{n=1}^\infty C_{2,n} \in K(H)$, $\|C_2\| < \eta$, $\min.\text{ind}(A+C_1+C_2-\lambda) = 0$ for all $\lambda \in \rho^r_{s-F}(A+C_1+C_2)$,

$$\Delta \subset \rho^r_{s-F}(A+C_1+C_2) \quad\text{and}\quad \sigma(A+C_1+C_2) \subset \text{interior}\ \sigma(A)_{\varepsilon/6}.$$

Since

$$\max\{\text{dist}[\lambda, \partial\rho_{s-F}(A)]:\ \lambda \in \rho^s_{s-F}(A+C_1+C_2)\} < \varepsilon/6,$$

we can apply the arguments of Proposition 3.45 in order to find a compact normal operator C_3, $\|C_3\| < \varepsilon/6$, such that

$$\min.\text{ind}(A+C_1+C_2+C_3-\lambda) = 0 \text{ for all } \lambda \in \rho_{s-F}(A).$$

Now take $K_2 = C_1+C_2+C_3$. □

An operator $T \in L(H)$ such that

$$\min.\text{ind}(\lambda-T) = 0 \text{ for all } \lambda \in \rho_{s-F}(T) \tag{3.13}$$

will be called a *smooth operator*.

3.5 Apostol-Foiaş-Voiculescu's theorem on normal restrictions of compact perturbations of operators

Given T in $L(H)$ we cannot expect, a priori, to find an infinite dimensional invariant subspace M such that $T|M$ is normal. An important approximation argument (due to C. Apostol, C. Foiaş and D. Voiculescu) asserts that some arbitrarily small compact perturbation of T always has that property. We shall need some extra notation to make it more clear.

3.5.1 Schatten p-classes

Recall that if K is compact, then $(K^*K)^{\frac{1}{2}}$ can be written as $(K^*K)^{\frac{1}{2}} = \sum_{n=1}^{\infty} \lambda_n e_n \otimes e_n$ with respect to some ONB $\{e_n\}_{n=1}^{\infty}$, where $\lambda_1 = \|K\| \geq \lambda_2 \geq \lambda_3 \geq \ldots \geq \lambda_n \geq \lambda_{n+1} \geq \ldots \geq 0$ and $\lambda_n \to 0$ $(n \to \infty)$.

The *Schatten p-class* $C^p(H)$ of compact operators is defined by

$$C^p(H) = \{K \in K(H): \sum_{n=1}^{\infty} \lambda_n^p < \infty\},\ 0 < p < \infty,$$

and $C^{\infty}(H) = K(H)$. The reader is referred to [77], [107], [183] for the properties of these ideals of operators. In particular, we have:

(a) If $K \in C^p(H)$, then $K^* \in C^p(H)$;

(b) $C^p(H)$ $(C^{\infty}(H))$ is a Banach space under the norm $|K|_p = |K^*|_p = (\sum_{n=1}^{\infty} \lambda_n^p)^{1/p}$ for $1 < p < \infty$ ($|K|_{\infty} = |K^*|_{\infty} = \|K\|$, resp.);

(c) If $0 < p < q < \infty$, then $C^p(H) \subset C^q(H) \subset C^{\infty}(H)$ (The inclusions are proper); moreover, if $p \geq 1$ and $K \in C^p(H)$, then $|K|_p \geq |K|_q \geq \|K\|$.

(d) If $K \in C^1(H)$, then it is possible to define the *trace* of K by

$$\operatorname{tr}(K) = \sum_{n=1}^{\infty} \langle Kf_n, f_n \rangle$$

(with respect to some ONB $\{f_n\}_{n=1}^{\infty}$ of H. The result is independent of the particular ONB.);

(e) $C^1(H)$ = the ideal of trace class operators is isometrically isomorphic to the dual $K(H)^{\dagger}$ of $K(H)$. This isomorphism is defined by $C^1(H) \ni K \leftrightarrow \phi_K \in K(H)^{\dagger}$, where

$$\phi_K(X) = \operatorname{tr}(KX),\ X \in K(H);$$

(f) $C^1(H)^{\dagger} = K(H)^{\dagger\dagger}$ is isometrically isomorphic to $L(H)$. This isomorphism is defined by $L(H) \ni A \leftrightarrow \Phi_A \in C^1(H)^{\dagger}$, where

$$\Phi_A(K) = \operatorname{tr}(AK),\ K \in C^1(H);$$

(g) If $A \in L(H)$ is a non-negative hermitian operator, then

$$\operatorname{tr}(A) = \sum_{n=1}^{\infty} \langle Af_n, f_n \rangle \quad (0 \leq \operatorname{tr}(A) \leq \infty)$$

is well-defined and the result is independent of the particular ONB $\{f_n\}_{n=1}^{\infty}$ of H.

(h) $(\mathcal{C}^2(H), |\cdot|_2)$ (= the ideal of Hilbert-Schmidt operators) is a Hilbert space. If $A, B \in \mathcal{C}^2(H)$, then $AB^* \in \mathcal{C}^1(H)$ and the inner product of $\mathcal{C}^2(H)$ is given by $\langle A,B\rangle = \mathrm{tr}(AB^*)$.

3.5.2 Normal restrictions

With the above notation in mind, we have the following

THEOREM 3.49. *Given* $T \in L(H)$, *a nonempty compact subset* Γ_ℓ *of* $\sigma_{\ell e}(T)$ *and* $\varepsilon > 0$, *there exists an infinite dimensional subspace* $H_\varepsilon \subset H$ *and* $K_\varepsilon \in \mathcal{C}^1(H)$ *such that* $|K_\varepsilon|_1 < \varepsilon$ *(so that* $\|K_\varepsilon\| < \varepsilon$*)*, $(T-K_\varepsilon)H_\varepsilon \subset H_\varepsilon$, $T_\varepsilon = (T-K_\varepsilon)|H_\varepsilon$ *is a diagonal normal operator of uniform infinite multiplicity and* $\sigma(T_\varepsilon) = \sigma_e(T_\varepsilon) = \Gamma_\ell$.

PROOF. Let $\{\lambda_n\}_{n=1}^\infty$ be a sequence of complex numbers such that $\{\lambda_n\}^- = \Gamma_\ell$ and

$$\mathrm{card}\{m:\ \lambda_m = \lambda_n\} = \infty \tag{3.14}$$

for each $n = 1,2,\ldots$. It is obvious that $\Gamma_\ell \subset \sigma_{\ell e}(T) \subset \sigma_\ell(T)$. Therefore, we can find a unit vector $e_1 \in H$ such that $\|(\lambda_1 - T)e_1\| < \varepsilon/4$. Consider the decomposition

$$T = \begin{pmatrix} \lambda_1 + t_{11} & T_{12}^1 \\ T_{21}^1 & T_{22}^1 \end{pmatrix} \begin{matrix} e_1 \\ \{e_1\}^\perp \end{matrix};$$

then $\begin{pmatrix} t_{11} \\ T_{21}^1 \end{pmatrix} : \vee\{e_1\} \to \{e_1\}^\perp$ is a rank-one operator, $\left\|\begin{pmatrix} t_{11} \\ T_{21}^1 \end{pmatrix}\right\| < \varepsilon/4$ and $T_{12}^1 : \{e_1\}^\perp \to \vee\{e_1\}$ is also a rank-one operator. Clearly, we can find a finite dimensional subspace $M_1 \supset \vee\{e_1\}$ such that $\|T_{21}^1|M_1^\perp\| < \varepsilon/4$. Let R_1 be the orthogonal projection of H onto $M_1^\perp$ and define C_1 by

$$C_1 = \begin{pmatrix} t_{11} & 0 \\ T_{21}^1 & 0 \end{pmatrix} + \begin{pmatrix} 0 & T_{12}^1 \\ 0 & 0 \end{pmatrix} R_1.$$

It is easily seen that $|C_1|_1 < \left|\begin{pmatrix} t_{11} \\ T_{21}^1 \end{pmatrix}\right|_1 + |T_{12}^1|M_1^\perp|_1 < \varepsilon/2$ and

$$T - C_1 = \begin{pmatrix} \lambda_1 & T_{12}^1|N_1 & 0 \\ 0 & A_1 & B_1 \\ 0 & D_1 & T_1 \end{pmatrix} \begin{matrix} e_1 \\ N_1 \\ M_1^\perp \end{matrix} = \begin{pmatrix} \lambda_1 & 0 & T_{12}^1|N_1 \\ 0 & T_1 & D_1 \\ 0 & B_1 & A_1 \end{pmatrix} \begin{matrix} e_1 \\ M_1^\perp, \\ N_1 \end{matrix}$$

where $N_1 = M_1 \cap \{e_1\}^\perp$.

Clearly, $\sigma_{\ell e}(T_1) = \sigma_{\ell e}(T)$. Thus, we can apply the same argument to T_1 in order to obtain a unit vector $e_2 \in M_1$, a trace class operator C_2 such that $|C_2|_1 < \varepsilon/4$ and a finite dimensional subspace $N_2 \supset \vee\{e_1,$

$e_2\}$ such that

$$T-(C_1+C_2)=\begin{pmatrix}\begin{pmatrix}\lambda_1 & 0\\ 0 & \lambda_2\end{pmatrix} & 0 & E_2\\ 0 & T_2 & D_2\\ 0 & B_2 & A_2\end{pmatrix}\begin{matrix}\vee\{e_1,e_2\}\\ N_2\ominus\vee\{e_1,e_2\}.\\ N_2\end{matrix}$$

By induction, we can find orthonormal vectors $e_1,e_2,\ldots e_n,\ldots$, and trace class operators $C_1,C_2,\ldots,C_n,\ldots$ such that $K_\varepsilon=\sum_{n=1}^{\infty}C_n\in C^1(H)$, $|K_\varepsilon|_1\le\sum_{n=1}^{\infty}|C_n|_1<\varepsilon$ and

$$T-K_\varepsilon=\begin{pmatrix}T_\varepsilon & E\\ 0 & A\end{pmatrix}\begin{matrix}H_\varepsilon\\ H_\varepsilon^{\perp}\end{matrix},$$

where $H_\varepsilon=\vee\{e_n\}_{n=1}^{\infty}$ and $T_\varepsilon=\sum_{n=1}^{\infty}\lambda_n e_n\otimes e_n$ (strong limit of the partial sums).

The condition (3.14) guarantees that T has the desired properties. □

COROLLARY 3.50. *Given* $T\in L(H)$, *nonempty compact subsets* $\Gamma_\ell\subset\sigma_{\ell e}(T)$ *and* $\Gamma_r\subset\sigma_{re}(T)$, $\varepsilon>0$ *and diagonal normal operators* N_ℓ *and* N_r *of uniform infinite multiplicity such that* $\Gamma_\ell\subset\sigma(N_\ell)=\sigma_e(N_\ell)\subset(\Gamma_\ell)_\varepsilon$ *and* $\Gamma_r\subset\sigma(N_r)=\sigma_r(N_r)\subset(\Gamma_r)_\varepsilon$, *there exists an operator* $L\in L(H)$,

$$L\simeq\begin{pmatrix}N_\ell & * & *\\ 0 & A & *\\ 0 & 0 & N_r\end{pmatrix}$$

such that $\|T-L\|<2\varepsilon$.

PROOF. Applying the inductive argument of the proof of Theorem 3.49 alternatively to T and to T*, we can find $K\in C^1(H)$, $\|K\|\le|K|_1<\varepsilon/2$ such that

$$T-K=\begin{pmatrix}M_\ell & * & *\\ 0 & A & *\\ 0 & 0 & M_r\end{pmatrix},$$

where M_ℓ and M_r are diagonal normal operators of uniform infinite multiplicity such that $\sigma(M_\ell)=\Gamma_\ell$ and $\sigma(M_r)=\Gamma_r$, resp..

Since $\Gamma_\ell\subset\sigma(N_\ell)=\sigma_e(N_\ell)\subset(\Gamma_\ell)_\varepsilon$ and $\Gamma_r\subset\sigma(N_r)=\sigma_e(N_r)\subset(\Gamma_r)_\varepsilon$, and N_ℓ and N_r are diagonal normal operators of uniform infinite multiplicity, we can easily "spread" the spectral measures of M_ℓ and M_r (concentrated in $\sigma_p(M_\ell)$ and $\sigma_p(M_r)$, resp.) in order to obtain normal operators N_ℓ' and N_r' such that $N_\ell'\simeq N_\ell$, $N_r'\simeq N_r$, $\|M_\ell-N_\ell'\|<3\varepsilon/2$ and $\|M_r-N_r'\|<3\varepsilon/2$.

The operator L obtained by replacing M_ℓ and M_r by the normal operators N_ℓ' and N_r', respectively, satisfies all our requirements. □

3.5.3 Density of sets of operators with bad properties

We shall say that a certain property (P) (of operators acting on a Hilbert space H) is a *bad* property if:

(I) If A has the property (P), then $\alpha+\beta A$ has the property (P) for all $\alpha \in \mathbb{C}$ and all $\beta \neq 0$,

(II) If A has the property (P) and $T \sim A$, then T has the property (P), and

(III) If A has the property (P) and $\sigma(A)\cap\sigma(B) = \emptyset$, then $A\oplus B$ has the property (P).

Examples of "bad properties" are frequent in the literature; namely, (1) T is not cyclic, (2) The spectrum of T is disconnected (or $\sigma(T)$ has infinitely many components, or c components, where c is the power of the continuum), (3) $\sigma_p(T)$ has nonempty interior, (4) The commutant of T is not abelian, (5) T is not similar to a normal operator, (6) T is not algebraic, etc, are examples of properties satisfying (I), (II) and (III).

THEOREM 3.51. *If* (P) *is a bad property and there exists some operator* A *with the property* (P), *then the set*

$$\{T \in L(H): \ T \textit{ satisfies } (P)\}$$

is dense in $L(H)$.

PROOF. Let $\mu \in \partial\sigma_e(T)\cap\partial\rho(T)$. Given $\varepsilon > 0$, we can find $\lambda \in \rho(T)$ such that $|\lambda-\mu| < \varepsilon/2$. By Corollary 3.50, we can find $L \in L(H)$ such that $||T-L|| < \varepsilon/2$ and

$$L = \begin{pmatrix} \lambda & R \\ 0 & B \end{pmatrix}$$

with respect to some decomposition $H = M\oplus M^{\perp}$, where M and $M^{\perp}$ are infinite dimensional subspaces and $\lambda \notin \sigma(B)$. (Observe that $\mu \in \partial\sigma_e(T) \subset \sigma_{\ell re}(T)$ and take $\Gamma_\ell = \{\lambda\}\cup\sigma_{\ell e}(T)$, $\Gamma_r = \sigma_{re}(T)$.)

Let $\eta = \text{dist}[\lambda,\sigma(B)]$. Clearly, $0 < \eta < \varepsilon/2$; thus, if $0 < \delta < \eta/(1+||A||)$, then it follows from Corollary 3.22 that

$$M = \begin{pmatrix} \lambda+\delta A & R \\ 0 & B \end{pmatrix} \sim (\lambda+\delta A)\oplus B.$$

Now (I), (II) and (III) imply that M has the property (P). Since $||T-M|| \le ||T-L||+\delta||A|| < \varepsilon$, and ε can be chosen arbitrarily small, it readily follows that $T \in \{S \in L(H): \ S \text{ satisfies } (P)\}^{-}$. □

An inductive repetition of the same argument yields the following

COROLLARY 3.52. *If* $\{(P_n)\}_{n=1}^{\infty}$ *is a denumerable set of bad properties, then* $(P) = \cup_{n=1}^{\infty} (P_n)$ *is a bad property. Furthermore, if there exist operators* $A_1, A_2, \ldots, A_n, \ldots$ *in* $L(H)$ *such that* A_n *satisfies* (P_n) *(for each* $n = 1,2,\ldots$*), then* $\{T \in L(H): T$ *satisfies* $(P)\}$ *is dense in* $L(H)$.

3.6 Notes and remarks

The operators τ_{AB} were first systematically studied by M. Rosenblum in [174]. Rosenblum's theorem has been proved there, except that the equality $\sigma(\tau_{AB}) = \sigma(A) - \sigma(B)$ is replaced by the weaker statement $\sigma(\tau_{AB}) \subset \sigma(A) - \sigma(B)$. The fact that this inclusion is actually an equality is due to D. C. Kleinecke (see [157,Introduction]). Corollary 3.22, for the special case when $\sigma(A) \cap \sigma(B) = \emptyset$, is usually called Rosenblum's corollary (see, e.g., [171,Chapter 0]). In [69], C. Davis and P. Rosenthal re-analyzed Rosenblum's results and proved that $\sigma_\delta(\tau_{AB}) = \sigma_r(A) - \sigma_\ell(B)$ and $\sigma_\pi(\tau_{AB}) = \sigma_\ell(A) - \sigma_r(B)$ and extended Rosenblum's corollary to Corollary 3.22. Their arguments include the use of the Berberian-Quigley construction (Lemma 3.3; see [45], [172,p.25]) in order to prove Theorem 3.4 as given here. Finally, L. A. Fialkow modified Rosenblum's original argument (by using analytic one-sided inverses instead of resolvents) in order to show that $\sigma_\delta(\tau_{AB}) = \sigma_r(\tau_{AB})$ and $\sigma_\pi(\tau_{AB}) = \sigma_\ell(\tau_{AB})$ [87]. More precisely, Fialkow's results read as follows

THEOREM 3.53. *Let* $A, B \in L(H)$*; then the following are equivalent for* τ_{AB}*:*

(i) τ_{AB} *is surjective;*

(ii) $\sigma_r(A) \cap \sigma_\ell(B) = \emptyset$*;*

(iii) ran τ_{AB} *contains the minimal ideal* $F(H)$*;*

(iv) τ_{AB} *is right invertible in* $L(L(H))$.

Clearly, τ_{AB} *maps every norm ideal (in the sense of R. Schatten* [183]*)* $J \subset K(H)$ *into itself, and each of the above conditions is also equivalent to each of the following ones*

(v) $\tau_{AB}|J$ *is surjective for some norm ideal* J*;*

(vi) $\tau_{AB}|J$ *is surjective for every norm ideal* J*;*

(vii) $\tau_{AB}|J$ *is right invertible in* $L(J)$ *for some norm ideal* J*;*

(viii) $\tau_{AB}|J$ *is right invertible for every norm ideal* J.

Moreover, in this case ker τ_{AB} *and* ran τ_{AB} *are complementary subspaces of* $L(H)$ *and* ker $\tau_{AB}|J$ *and* ran $\tau_{AB}|J$ *are complementary subspaces of* J *for each norm ideal* J.

THEOREM 3.54. *The following are equivalent for* τ_{AB}:

(i) τ_{AB} *is bounded below;*

(ii) $\sigma_\ell(A)\cap\sigma_r(B) = \emptyset$;

(iii) $\tau_{AB}|F_1(H)$ *is bounded below (where* $F_1(H)$ *is the set of all rank-one operators);*

(iv) τ_{AB} *is left invertible in* $L(L(H))$;

(v) $\tau_{AB}|J$ *is bounded below for some norm ideal* J;

(vi) $\tau_{AB}|J$ *is bounded below for every norm ideal* J;

(vii) $\tau_{AB}|J$ *is left invertible for some norm ideal* J;

(viii) $\tau_{AB}|J$ *is left invertible for every norm ideal* J.

Moreover, in this case ran τ_{AB} *and* ker τ_{AB} *(*ran $\tau_{AB}|J$ *and* ker $\tau_{AB}|J$*) are complementary subspaces of* $L(H)$ *(of* J *for every norm ideal* J*, respectively).*

Furthermore, Fialkow also considered the operator $\tilde{\tau}_{\tilde{A}\tilde{B}}$ induced by τ_{AB} in the Calkin algebra, defined by $\tilde{\tau}_{\tilde{A}\tilde{B}}(\tilde{X}) = \tilde{A}\tilde{X}-\tilde{X}\tilde{B}$. His results for these operators can be summed up as follows (see [59], [86], [87], [88], [89], [90]).

THEOREM 3.55. *The following are equivalent for* $\tilde{\tau}_{\tilde{A}\tilde{B}}$:

(i) $\tilde{\tau}_{\tilde{A}\tilde{B}}$ *is surjective;*

(ii) $\tilde{\tau}_{\tilde{A}\tilde{B}}$ *has dense range;*

(iii) $\sigma_{re}(A)\cap\sigma_{\ell e}(B) = \emptyset$;

(iv) $\tilde{\tau}_{\tilde{A}\tilde{B}}$ *is right invertible in* $L(A(H))$. *In this case, a right inverse for* $\tilde{\tau}_{\tilde{A}\tilde{B}}$ *is given by*

$$\tilde{\phi}(\tilde{Y}) = \frac{-1}{2\pi i}\int_{\partial\Omega} R_{\tilde{A}}(\lambda)\tilde{Y}L_{\tilde{B}}(\lambda)\, d\lambda,$$

where Ω *is a Cauchy domain such that* $\sigma_{re}(A) \subset \Omega$ *and* $\sigma_{\ell e}(B)\cap\Omega^- = \emptyset$, $R_{\tilde{A}}(\lambda)$ *is an analytic right inverse of* $(\lambda-\tilde{A})$ *defined on a neighborhood of* $\mathbb{C}\setminus\Omega$ *and* $L_{\tilde{B}}(\lambda)$ *is an analytic left inverse of* $(\lambda-\tilde{B})$ *defined on a neighborhood of* Ω^-. *Moreover,* ker $\tilde{\tau}_{\tilde{A}\tilde{B}}$ *and* ran $\tilde{\tau}_{\tilde{A}\tilde{B}}$ *are complementary subspaces of* $A(H)$.

THEOREM 3.56. *The following are equivalent for* $\tilde{\tau}_{\tilde{A}\tilde{B}}$:

(i) $\tilde{\tau}_{\tilde{A}\tilde{B}}$ *is bounded below;*

(ii) $\sigma_{\ell e}(A)\cap\sigma_{re}(B) = \emptyset$;

(iii) $\tilde{\tau}_{\tilde{A}\tilde{B}}$ *has a left inverse in* $L(A(H))$. *In this case, a left inverse for* $\tilde{\tau}_{\tilde{A}\tilde{B}}$ *is given by*

$$\tilde{\psi}(\tilde{Y}) = \frac{1}{2\pi i}\int_{\partial\Omega} L_{\tilde{A}}(\lambda)\tilde{Y}R_{\tilde{B}}(\lambda)\, d\lambda,$$

where Ω *is a Cauchy domain such that* $\sigma_{\ell e}(A) \subset \Omega$ *and* $\sigma_{re}(B)\cap\Omega^- = \emptyset$, $L_{\tilde{A}}(\lambda)$ *is an analytic left inverse of* $(\lambda-\tilde{A})$ *defined on a neighborhood of*

$\mathbb{C}\setminus\Omega$ *and* $R_{\tilde{B}}(\lambda)$ *is an analytic right inverse of* $(\lambda-\tilde{B})$ *defined on a neighborhood of* Ω^-. *Moreover,* ker $\tilde{\tau}_{\tilde{A}\tilde{B}}$ *and* ran $\tilde{\tau}_{\tilde{A}\tilde{B}}$ *are complementary subspaces of* $A(H)$.

THEOREM 3.57. *(i)* τ_{AB} *is a Fredholm operator in* $L(L(H))$ *if and only if* $\sigma_e(A)\cap\sigma(B)=\sigma(A)\cap\sigma_e(B)=\emptyset$*;* $\sigma_e(\tau_{AB})=[\sigma(A)-\sigma_e(B)]\cup[\sigma_e(A)-\sigma(B)]$.

(ii) τ_{AB} *is a semi-Fredholm operator if and only if* $\sigma_{\ell e}(A)\cap\sigma_r(B) = \sigma_\ell(A)\cap\sigma_{re}(B)=\emptyset$ *or* $\sigma_{re}(A)\cap\sigma_\ell(B)=\sigma_r(A)\cap\sigma_{\ell e}(B)=\emptyset$.

Moreover, Fialkow also proved the following asymptotic version of Theorem 3.19(i) [90,Theorem 1.1].

THEOREM 3.58. *The following are equivalent*

(i) ran τ_{AB} *is dense in* $L(H)$*;*

(ii) *1)* $\sigma_{re}(A)\cap\sigma_{\ell e}(B)=\emptyset$, *and*

2) There exists no nonzero $X\in C^1(H)$ *such that* $BX = XA$*;*

(iii) Given $Y\in L(H)$ *and* $\varepsilon>0$*, there exists* $X\in L(H)$ *such that* $\tau_{AB}(X)-Y\in K(H)$ *and* $\|\tau_{AB}(X)-Y\|<\varepsilon$.

Combining this result with the proof of Corollary 3.22 and Lemma 3.36, we obtain the following asymptotic version of Corollary 3.22.

COROLLARY 3.59 ([90]). *Let* H_1 *and* H_2 *be two Hilbert spaces, let* $A\in L(H_1)$, $B\in L(H_2)$ *and* $C\in L(H_2,H_1)$ *and assume that* ran τ_{AB} *is dense. Then the operators*

$$\begin{pmatrix} A & C \\ 0 & B \end{pmatrix} \quad \text{and} \quad A\oplus B$$

(acting on $H=H_1\oplus H_2$*) are asymptotically similar.*

The construction of one-sided resolvents (Theorem 3.8 and Corollary 3.9) is due to C. Apostol and K. Clancey [15] (see also [14]). The proof strongly depends on a previous deep result of C. Apostol (Proposition 3.16 [10]). This result extend to the case of an arbitrary compact set σ the one obtained in [26] for the case when σ is simply connected.

The results of Section 3.1.6 and 3.1.7 have been taken from J. G. Stampfli's article [189]. Indeed, Stampfli proved that if C is an irreducible C*-subalgebra with identity of $L(H)$ (in particular, if $C = L(H)$), then $\|\delta_T\| = 2\min\{\|T-\lambda\|: \lambda\in\mathbb{C}\}$ for each inner derivation δ_T in C (and the analogous result for the operators τ_{AB}, $A, B\in C$). Extensions of Stampfli's results to derivations on certain C*-algebras and

W*-algebras were given by C. Apostol and L. Szidó [31]. For the case of an inner derivation in the Calkin algebra, it is known that

THEOREM 3.60 (C.-K. Fong [99]). *If* $T \in L(H)$, *then* $\|\delta_{\tilde{T}}\| = 2 \min \{\|\tilde{T} - \lambda\|: \lambda \in \mathbb{C}\}$. *Moreover, there exists an orthogonal projection* $X \in L(H)$ *such that* $\|\delta_{\tilde{T}}\| = \|\delta_{\tilde{T}}(2X-1)\|$.

The interested reader will find a very detailed analysis of the spectra, left spectra and right spectra of large families of operators "somehow related" with the operators τ_{AB} (including tensor products of operators; the operator $X \to \sum_{j=1}^{n} A_j X B_j$, where $\{A_1, A_2, \dots, A_n\}$ and $\{B_1, B_2, \dots, B_n\}$ are separately commuting families of operators in $L(H)$, etc), in [55], [79], [91], [92], [100], [187] and, very especially, in several papers by R. E. Harte [127], [128]. Very recently, the spectrum of the operator $X \to \sum_{j=1}^{n} A_j X B_j$ has been completely determined by R. Curto [68].

In [175], Gian-Carlo Rota proved that the backward shift of infinite multiplicity is a "universal model" for all operators T in $L(H)$ such that sp(T) < 1 (see also [130, Lecture X]). Two (very close) universal models for an operator T with arbitrary spectrum and the spectrum of the model equal to a suitable neighborhood of $\sigma(T)$ were independently obtained by D. A. Herrero [142] and D. Voiculescu [194, Proposition 2.1]. The model given here (Theorem 3.33) is the one due to Voiculescu. J. A. Ball obtained analogous models for commuting n-tuples of operators [41] and F. Gilfeather used Rota's original construction and an iterative argument to give a different (and independent) proof of Herrero's corollary (Corollary 3.35); moreover, Gilfeather [105] actually showed that the invertible operator W such that $T' = WTW^{-1}$ satisfies the conditions of that corollary can be constructed in the C*-algebra C*(T) generated by T and 1.

Theorem 3.38 (except for those details concerning the structures of T_r and T_ℓ given by (iii) and (iv)), Proposition 3.47 and Theorem 3.48 are due to C. Apostol [10]. A simplified version of Apostol's triangular representation has been obtained by R. Bouldin in [52]. The version given here (Theorem 3.38) combines the arguments of [10], [52] and some improvements of D. A. Herrero (items (iii) and (iv)), based on Corollary 3.40 which, in turn, is a particular case of [144] (Theorem 1 or Theorem 2). However, the proof of Corollary 3.40(i) given here is not the same as the one given in [144]. This new proof has been based on Lemma 3.39 (due to R. G. Douglas and C. M. Pearcy [75]) which will play an important role in the characterization of quasitriangularity (see Chapter VI below).

In [190], J. G. Stampfli proved that, given $T \in L(H)$ there exists $K \in K(H)$ such that $\sigma(T+K) = \sigma_W(T)$ (without any estimate on $||K||$); moreover (as observed by N. Salinas; see Remark on p.174 of [190]), the fact that $\sigma(T+K) = \sigma_B(T)$ for some K in $L(H)$ is implicitly contained in that paper. A more general theorem, containing both results, has been obtained by C. Apostol, C. M. Pearcy and N. Salinas in [28] (see also Section 4.3 of Chapter IV below.)

Theorem 3.49 is Theorem 2.2 of [23]. This clever approximation argument is one of the most important ingredients for the theorem on spectral characterization of quasitriangularity of C. Apostol, C. Foiaş and D. Voiculescu (obtained in that paper; see Chapter VI below). The result can be regarded as an asymmetric version of the analogous result of C. M. Pearcy and N. Salinas for semi-normal operators [162]. These results (and many other related ones scattered thorugh the literature; see, e.g., [3], [139], [163], [179] and, very especially, [56] and [57]) are based on an argument due to F. Wolf [199]. (See also the partially related articles [71],[95] and [96].) Corollary 3.50 appears in [27]. Theorem 3.51 is a mild generalization of Herrero's theorem for non-cyclic operators. Indeed, the same result is actually true for the Banach spaces ℓ^p $(1 \leq p < \infty)$ and c_o (see [146,Theorem 4]). The underlying idea is contained in a paper of N. Salinas [178].

4 Two results borrowed from the theory of C*-algebras

The similarity orbit of $T \in L(H)$ contains the *unitary orbit* of T

$$U(T) = \{UTU^*: \quad U \in U(H)\}$$

(where $U(H)$ is the unitary group of $L(H)$), which plays a very important role in approximation problems.

We shall see later that, since $U(T)^{-} \subset S(T)^{-}$, a large amount of information about $S(T)^{-}$ can be obtained from the structure of $U(T)^{-}$. This structure is closely related with the C*-algebra C*(T) generated by T and the identity $1 \in L(H)$.

Most of those results related with closure of unitary orbits (and many others) connected with approximation problems can be derived from two deep results of the theory of C*-algebras. These results (the Brown-Douglas-Fillmore theorem and Voiculescu's theorem) will be stated here without proofs (suitable references will be given). Several consequences of these two results will be completely developed here.

4.1 Essentially normal operators

An operator $T \in L(H)$ will be called *essentially normal* if T is a normal element of the Calkin algebra or, equivalently, if the self-commutator $[T^*,T] = T^*T - TT^*$ is compact. One way in which these operators arise is as compact perturbations of normal operators. The interest in essentially normal operators is due to two facts: a) Not all of them arise in that fashion; b) There is a rich family of operators of the form "normal+compact" whose structure is not obvious from their analytic expression. (This is, by far, the most important fact from our viewpoint.)

4.1.1 Brown-Douglas-Fillmore theorem

Essentially normal operators have been completely classified in the famous article [60] of L. G. Brown, R. G. Douglas and P. A. Fillmore. Their main results will be quoted here in the most convenient form for our purposes. (The reader is referred to the above cited article for proofs.)

It is completely apparent from the definition that if T is essentially normal, then $\sigma_{\ell e}(T) = \sigma_{re}(T)$ and therefore $-\infty < \text{ind}(\lambda - T) < \infty$ for all $\lambda \in \rho_{s-F}(T)$, $\rho_{s-F}(T) = \rho_F(T)$ and $\sigma_{\ell re}(T) = \sigma_e(T)$.

THEOREM 4.1 (BDF theorem). *If* T_1 *and* T_2 *are essentially normal operators on* H, *then a necessary and sufficient condition that* T_1 *be unitarily equivalent to some compact perturbation of* T_2 *is that* $\sigma_e(T_1) = \sigma_e(T_2)$ *and* $\text{ind}(\lambda - T_1) = \text{ind}(\lambda - T_2)$ *for all* $\lambda \notin \sigma_e(T_1)$.

There is a special case of particular interest.

COROLLARY 4.2 (BDF corollary). *(i) If* T *is an essentially normal operator on* H *such that* $\text{ind}(\lambda - T) = 0$ *for all* $\lambda \in \rho_{s-F}(T)$, *then* T *is in*

$$Nor(H) + K(H) = \{N+K: \ N \text{ is normal, } K \text{ is compact}\}$$

(ii) In particular, every essentially normal operator T *such that* $\sigma_e(T)$ *does not disconnect the plane is in* $Nor(H) + K(H)$.

(iii) $Nor(H) + K(H)$ *is a closed subset of* $L(H)$.

4.1.2 Berger-Shaw trace inequality

In order to make the above results useful for approximation purposes, a large family of essentially normal operators will be exhibited. We shall need some extra notation.

Given $T \in L(H)$, let $A(T)$ (= the weak closure of the polynomials in 1 and T), $A^a(T)$ (= the weak closure of the rational functions of T with poles outside $\sigma(T)$), $A'(T) = \{A \in L(H): \ AT = TA\}$ (= the commutant of T) and $A''(T) = \{B \in L(H): \ BA = AB \text{ for all } A \text{ in } A'(T)\}$ (= the double commutant of T) be the four (weakly closed) sublagebras of $L(H)$ naturally associated with T. It is apparent that $A(T) \subset A^a(T) \subset A''(T) \subset A'(T)$.

The *multiplicity* $\mu(A)$ of a (weakly closed, identity containing) subalgebra A of $L(H)$ is defined by

$$\mu(A) = \inf\{\text{card}(\Gamma): \ H = \bigvee[Ax: A \in A, \ x \in \Gamma], \ \Gamma \subset H\} \qquad (4.1)$$

$(1 \le \mu(A) \le \infty)$. T is n-*multicyclic* (n-*rationally multicyclic*) if $\mu(A(T)) = n$ $(\mu(A^a(T)) = n$, resp.).

An operator $T \in L(H)$ is called *hyponormal* if $[T^*,T]$ is a positive semi-definite hermitian operator; T is called *subnormal* if there exists a normal operator N acting on a Hilbert space $\mathcal{R}$ containing $\mathcal{H}$ such that $\mathcal{H}$ is invariant under N and $T = N|\mathcal{H}$. It is not difficult to see that

(Normal) => (Subnormal) => (Hyponormal).

None of these implications can be reversed; **e. g.**, the operators $M_+(\Gamma)$ of (3.2) are subnormal, but not normal.

It readily follows from (g) of Section 3.5.1 that

$$\mathrm{tr}[T^*,T] = \sum_{n=1}^{\infty} \langle (T^*T - TT^*)f_n, f_n\rangle = \sum_{n=1}^{\infty} (\|Tf_n\|^2 - \|T^*f_n\|^2) \quad (4.2)$$

$(0 \le \mathrm{tr}[T^*,T] \le \infty)$ is well-defined for all hyponormal T. The main result of [50] asserts that if T is a hyponormal n-rationally multicyclic operator, then $[T^*,T] \in C^1(\mathcal{H})$; moreover,

$$|[T^*,T]|_1 = \mathrm{tr}[T^*,T] \le (n/\pi)\ \mathrm{meas}_2(\sigma(T)), \quad (4.3)$$

where meas_2 denotes the planar Lebesgue measure.

In order to prove this inequality, we shall need an auxiliary result.

LEMMA 4.3. *If* $\mu(A^a(T)) = n$, *then* $n \le \mu(A(T)) \le n+1$.

PROOF. It is trivial that $\mu(A(T)) \ge \mu(A^a(T)) = n$.

Assume that $\mathcal{H} = \bigvee\{Ax_j\colon\ A \in A^a(T),\ j = 1,2,\ldots,n\}$ for a suitable finite family $\{x_1,x_2,\ldots,x_n\} \subset \mathcal{H}$ and let $\{\lambda_{kj}\}_{k=1}^{\infty}$ $(j = 1,2,\ldots,n)$ be n pairwise disjoint dense subsets of $\rho(T)$. If the positive constants $\{c_{kj}\}_{k=1,j=1,2,\ldots,n}^{\infty}$ are suitably chosen, then $\sum_{j=1}^{n}\sum_{k=1}^{\infty} \|c_{kj}(\lambda_{kj}-T)^{-1} x_j\| < \infty$ and therefore

$$x_o = \sum_{j=1}^{n}\sum_{k=1}^{\infty} c_{kj}(\lambda_{kj}-T)^{-1}x_j$$

is a well-defined element of $\mathcal{H}$.

Now it is easily seen that, for suitably chosen polynomials $p_{11,i}$, $p_{11,i}^1, p_{11,i}^2, \ldots, p_{11,i}^n$, we have

$$\|(\lambda_{11}-T)^{-1}x_1 - \{p_{11,i}(T)x_o + \sum_{j=1}^{n} p_{11,i}^j(T)x_j\}\| \to 0 \ (i \to \infty),$$

so that $(\lambda_{11}-T)^{-1}x_1 \in \mathcal{M} = \bigvee\{Ax_j\colon\ A \in A(T),\ j = 0,1,2,\ldots,n\}$; moreover, the same argument shows that

$$(\lambda_{kj}-T)^{-1}x_j \in \mathcal{M} \text{ for all } j = 1,2,\ldots,n, \text{ and all } k = 1,2,\ldots .$$

Hence (by Runge's approximation theorem), $r(T)x_j \in \mathcal{M}$ for all $j = 1,2,\ldots,n$ and for all rational functions r with poles outside $\sigma(T)$.

It readily follows that $\mathcal{M}$ is invariant under $A^a(T)$. Since $\mathcal{M}$ contains the vectors $x_1,x_2,\ldots,x_n$, we conclude that $\mathcal{M} = \mathcal{H}$. Hence, $\mu(A(T)) \le \mathrm{card}\{x_o,x_1,x_2,\ldots,x_n\} = n+1$. □

THEOREM 4.4 (Berger-Shaw trace inequality). *If* $T \in L(\mathcal{H})$ *is a hypo normal* n-*rationally multicyclic operator, then* $[T^*,T] \in C^1(H)$ *and* $|[T^*,T]|_1$ *satisfies the inequality* (4.3).

In particular, T *is essentially normal.*

PROOF. It will be convenient to split the proof in several steps.

(I) Assume that A is hyponormal and p-multicylic, i.e.,

$$H = \vee\{A^k x_1, A^k x_2, \ldots, A^k x_p\}_{k=0}^{\infty}$$

for suitably chosen p vectors $x_1, x_2, \ldots, x_p$ in H.

Let $\{f_k\}_{k=1}^{\infty}$ be the ONB of H obtained by means of the Gram-Schmidt process of orthonormalization applied to the sequence $x_1, x_2, \ldots, x_p, Ax_1, Ax_2, \ldots, Ax_p, A^2x_1, A^2x_2, \ldots, A^2x_p, \ldots, A^kx_1, A^kx_2, \ldots, A^kx_p, \ldots$. It is easily seen that A admits the matrix representation

$$A = \begin{pmatrix} a_{11} & a_{12} & a_{13} & a_{14} & \cdot & \cdot \\ a_{21} & a_{22} & a_{23} & a_{24} & \cdot & \cdot \\ a_{31} & a_{32} & a_{33} & a_{34} & \cdot & \cdot \\ a_{41} & a_{42} & a_{43} & a_{44} & \cdot & \cdot \\ \cdot & \cdot & \cdot & \cdot & \cdot & \cdot \\ \cdot & \cdot & \cdot & \cdot & \cdot & \cdot \\ \cdot & \cdot & \cdot & \cdot & \cdot & \cdot \\ a_{p1} & a_{p2} & a_{p3} & a_{p4} & \cdot & \cdot \\ a_{p+1,1} & a_{p+1,2} & a_{p+1,3} & a_{p+1,4} & \cdot & \cdot \\ & a_{p+2,2} & a_{p+2,3} & a_{p+2,4} & \cdot & \cdot \\ & & a_{p+3,3} & a_{p+3,4} & \cdot & \cdot \\ & & & a_{p+4,4} & \cdot & \cdot \\ & 0 & & & \cdot & \\ & & & & & \cdot \end{pmatrix} \begin{matrix} f_1 \\ f_2 \\ f_3 \\ f_4 \\ \cdot \\ \cdot \\ \cdot \\ f_p \\ f_{p+1} \\ f_{p+2} \\ f_{p+3} \\ f_{p+4} \\ \cdot \\ \cdot \end{matrix} .$$

Hence, given $m > p+2$, we have (a_{kr} is defined equal to 0 for all $r \le 0$)

$$0 \le \sum_{k=1}^{m}(\|Af_k\|^2 - \|A^*f_k\|^2) = \sum_{k=1}^{m}(\sum_{i=1}^{k+p}|a_{ik}|^2 - \sum_{i=k-p}^{\infty}|a_{ki}|^2)$$

$$\le \sum_{k=m-p+1}^{m}(\sum_{i=m+1}^{k+p}|a_{ik}|^2 - \sum_{i=m+1}^{\infty}|a_{ki}|^2) \le \sum_{k=m-p+1}^{m}\sum_{i=m+1}^{k+p}|a_{ik}|^2$$

$$\le p\|A\|^2.$$

Since $[A^*,A] = A^*A - AA^* \ge 0$, it readily follows from (g) of Section 3.5.1 and (4.2) that

$$|[A^*,A]|_1 = \mathrm{tr}[A^*,A] = \lim(m \to \infty) \sum_{k=1}^{m}(\|Af_k\|^2 - \|A^*f_k\|^2) \le p\|A\|^2.$$

(II) If Ω is a simply connected analytic Cauchy domain, $\phi: D \to \Omega$ (where $D = \{\lambda: |\lambda| < 1\}$) is a conformal mapping with Taylor series $\phi(\lambda) = \sum_{m=0}^{\infty} c_m\lambda^m$ and $T_\phi \in L(H^2(\partial D))$ is the analytic Toeplitz operator with symbol ϕ (defined by $T_\phi f = \phi f$, $f \in H^2(\partial D)$), then T_ϕ is a cyclic subnormal operator ($e_o(\lambda) \equiv 1$ is a cyclic vector for T_ϕ) and $\sigma(T_\phi) = \phi(D)^- = \Omega^-$ [73]. If $\{e_n\}_{n=0}^{\infty}$ is the canonical ONB of $H^2(\partial D)$ (defined by $e_n(\lambda) \equiv \lambda^n$, $n = 0,1,2,\ldots$), then

$$|[T_\phi^*,T_\phi]|_1 = \mathrm{tr}[T_\phi^*,T_\phi] = \sum_{n=0}^{\infty}(\|T_\phi e_n\|^2 - \|T_\phi^* e_n\|^2)$$

$$= \sum_{n=0}^{\infty}(\|\sum_{m=0}^{\infty} c_m\lambda^{n+m}\|^2 - \|\sum_{m=0}^{n} \overline{c_m}\, \lambda^{n-m}\|^2)$$

$$= \sum_{n=0}^{\infty}(\sum_{m=0}^{\infty} |c_m|^2 - \sum_{m=0}^{n}|c_m|^2) = \sum_{n=1}^{\infty}\sum_{m=n+1}^{\infty}|c_m|^2$$

$$= \sum_{m=1}^{\infty} m|c_m|^2 = (1/\pi)\int_D |\phi'(\lambda)|^2\, dA = (1/\pi)\mathrm{meas}_2(\Omega)$$

(where dA denotes the planar Lebesgue measure).

Hence, $|[T_\phi^*,T_\phi]|_1 = (1/\pi)\mathrm{meas}_2(\Omega) = (1/\pi)\mathrm{meas}_2[\sigma(T_\phi)]$.

(III) Given $\varepsilon > 0$, we can find finitely many simply connected analytic Cauchy domains $\Omega_1, \Omega_2, \ldots, \Omega_q$ such that

1) $\sigma(T)\cup(\cup_{r=1}^{q} \Omega_r)^- \subset \Delta = \{\lambda: |\lambda| \le \|T\|\}$;
2) $\Omega_r^- \cap \Omega_s^- = \emptyset$, if $r \neq s$;
3) $\sigma(T)\cap(\cup_{r=1}^{q} \Omega_r)^- = \emptyset$; and
4) $\mathrm{meas}_2(\Delta\backslash[\sigma(T)\cup(\cup_{r=1}^{q} \Omega_r)^-]) < \varepsilon$.

If $\phi_r : D \to \Omega_r$ is a conformal mapping ($r = 1,2,\ldots,q$), $k \ge 1$ and

$$A_k = [T\oplus(\oplus_{r=1}^{q} T_{\phi_r})^{(n)}]^{(k)},$$

then A_k is a hyponormal operator with $\|A\| = \max\{\|T\|, \max_{1\le r\le q} \|T_{\phi_r}\|\} = \|T\|$ and $\sigma(A_k) = \sigma(T)\cup(\cup_{r=1}^{q} \Omega_r)^-$. Moreover, if $e_{o,r}^{(i)}$ denotes the function identically 1 in the (i,r)-th copy of $H^2(\partial D)$ ($i = 1,2,\ldots,nk$; $r = 1,2,\ldots,q$) and $H = \vee\{Ly_j : L \in A^a(T),\ j = 1,2,\ldots,n\}$, it is not difficult to check that

$$(H\oplus[\oplus_{r=1}^{q} H^2(\partial D)^{(n)}])^{(k)} = \vee\{B_k[y_j\oplus(\oplus_{r=1}^{q} e_{o,r}^{(i)})]:\ B_k \in A^a(A_k),$$
$$i = 1,2,\ldots,nk,\ 1 \le j \le n \text{ and } j \equiv i \pmod n\},$$

so that A_k is nk-rationally multicyclic.

(IV) By Lemma 4.3, A_k is (nk+1)-multicyclic. Thus, applying (I) to $A = A_k$, we obtain

$$|[A_k^*,A_k]|_1 = k|[T^*,T]|_1 + nk\sum_{r=1}^{q}|[T_{\phi_r}^*,T_{\phi_r}]|_1$$

$$\le (nk+1)\|A_k\|^2 = (nk+1)\|T\|^2 = [(nk+1)/\pi]\mathrm{meas}_2(\Delta).$$

From this inequality, (II) and 1) - 4) of (III), it follows that

$$k|[T^*,T]|_1 < [(nk+1)/\pi]\{\mathrm{meas}_2[\sigma(T)] + \sum_{r=1}^{q} \mathrm{meas}_2(\Omega_r) + \varepsilon\}$$

$$-(nk/\pi)\sum_{r=1}^{q} \mathrm{meas}_2(\Omega_r) = [(nk+1)/\pi]\mathrm{meas}_2[\sigma(T)]$$

$$+(1/\pi)\sum_{r=1}^{q}\operatorname{meas}_2(\Omega_r)+(nk+1)(\varepsilon/\pi).$$

Hence,

$$|[T^*,T]|_1 < [(n+1/k)/\pi]\operatorname{meas}_2[\sigma(T)]+\|T\|^2/k+n\varepsilon.$$

Since k can be taken arbitrarily large and ε can be chosen arbitrarily small, we conclude that

$$|[T^*,T]|_1 \le (n/\pi)\operatorname{meas}_2[\sigma(T)]. \qquad \square$$

4.1.3 Examples of essentially normal operators

The operator $M_+(\Gamma)$ of (3.2) is subnormal and rationally cyclic (the function $e_0(\lambda) \equiv 1$ being a cyclic vector for $A^a(M_+(\Gamma))$.) Thus, by Theorem 4.4, $M_+(\Gamma)$ is an essentially normal operator. On the other hand, since

$$0=[M(\Gamma)^*,M(\Gamma)]=\begin{pmatrix}[M_+(\Gamma)^*,M_+(\Gamma)]-ZZ^* & M_+(\Gamma)^*Z-ZM_-(\Gamma)^* \\ Z^*M_+(\Gamma)-M_-(\Gamma)Z^* & Z^*Z+[M_-(\Gamma)^*,M_-(\Gamma)]\end{pmatrix}\begin{matrix}H^2(\Gamma)\\ H^2(\Gamma)^\perp\end{matrix},$$

it readily follows from the same proposition that $[M_+(\Gamma)^*,M_+(\Gamma)]=ZZ^* \in C^1(H^2(\Gamma))$. A fortiori, $[M_-(\Gamma)^*,M_-(\Gamma)]=-Z^*Z \in C^1(H^2(\Gamma)^\perp)$, so that $M_-(\Gamma)$ is also essentially normal and Z is a Hilbert-Schmidt operator. (Indeed, Z is a finite rank operator.)

From Corollary 4.2, we obtain the following

COROLLARY 4.5. *Let Ω be an analytic Cauchy domain with boundary $\Gamma = \partial\Omega$ and let* **M** *be a normal operator such that $\sigma(\mathbf{M})=\sigma_e(\mathbf{M})=\Omega^-$. Then* $\mathbf{M}\oplus M_+(\Gamma)$ *and* $\mathbf{M}\oplus M_-(\Gamma)$ *are unitarily equivalent to compact perturbations of* **M**.

A different kind of examples can be constructed as follows: Let Ω be a nonempty bounded open set such that $\Omega=$ interior Ω^- and let $N(\Omega)$ be the "multiplication by λ" operator acting on $L^2(\Omega,dA)$. The subspace $A^2(\Omega)$ spanned by the rational functions with poles outside Ω^- is invariant under $N(\Omega)$. By $N_+(\Omega)$ and $N_-(\Omega)$ we shall denote the restriction of $N(\Omega)$ to $A^2(\Omega)$ and its compression to $L^2(\Omega)\ominus A^2(\Omega)$, respectively, i.e.,

$$N(\Omega)=\begin{pmatrix}N_+(\Omega) & G\\ 0 & N_-(\Omega)\end{pmatrix}\begin{matrix}A^2(\Omega)\\ A^2(\Omega)^\perp\end{matrix}. \qquad (4.4)$$

As in the case when $M_+(\Omega)$ (Section 3.2), $A^2(\Omega)$ is a reproducing kernel space and it can be realized as a space of analytic functions defined of Ω; furthermore,

$$\begin{cases} \sigma(N(\Omega)) = \sigma_e(N(\Omega)) = \sigma(N_+(\Omega)) = \sigma(N_-(\Omega)) = \sigma_e(N_-(\Omega)) = \Omega^-, \\ \sigma_e(N_+(\Omega)) = \partial\Omega,\ \mathrm{nul}(\mu - N_+(\Omega)) = 0 \text{ and } \mathrm{ind}(\mu - N_+(\Omega)) = -1 \\ \text{for all } \mu \in \Omega. \\ N(\Omega) \text{ is a normal operator, } N_+(\Omega) \text{ is a rationally cyclic} \\ \text{subnormal operator (Hence, } N_+(\Omega) \text{ is an essentially normal} \\ \text{operator, by Theorem 4.4) and } \|r(N(\Omega))\| = \|r(N_+(\Omega))\| = \max \\ \{|r(\lambda)|:\ \lambda \in \partial\Omega\} \text{ for each rational function } r \text{ with poles} \\ \text{outside } \Omega^-. \end{cases} \tag{4.5}$$

Now it is clear that $N_+(\Omega^*)^*$ is also an essentially normal operator such that

$$\begin{cases} \mathrm{nul}(\mu - N_+(\Omega^*)^*) = \mathrm{ind}(\mu - N_+(\Omega^*)^*) = 1 \text{ for all } \mu \in \Omega \text{ and} \\ \|r(N_+(\Omega^*)^*)\| = \max\{|r(\lambda)|:\ \lambda \in \partial\Omega\} \text{ for each rational function} \\ r \text{ with poles outside } \Omega^-. \end{cases}$$

Indeed, it can be actually shown that (see [50], [51])

$$\mathrm{tr}[N_+(\Omega)^*, N_+(\Omega)] \le (1/\pi)\mathrm{meas}_2(\Omega) \quad (\text{not } \Omega^- \ !).$$

Now it follows from the BDF theorem that

COROLLARY 4.6. *If* N *is a normal operator such that* $\sigma(N)$ *(=* $\sigma_e(N)$*)* $= \partial\Omega$*, then* $N_+(\Omega) \oplus N_+(\Omega^*)^*$ *is unitarily equivalent to a compact perturbation of* N.

COROLLARY 4.7. *If* T *is an essentially normal operator,* $\Omega_n = \{\lambda \in \mathbb{C}: \mathrm{ind}(\lambda - T) = n\}$ *(*$1 \le |n| < \infty$*) and* N *is an arbitrary normal operator such that* $\sigma(N) = \sigma_e(N) = \sigma_e(T)$*, then* T *is unitarily equivalent to a compact pertubation of*

$$L = N \oplus \{\oplus_{n=1}^{\infty} [N_+(\Omega_n^*)^*]^{(n)}\} \oplus \{\oplus_{n=1}^{\infty} N_+(\Omega_{-n})^{(n)}\}$$

(where $N_+(\Omega)$ *must be interpreted as the* 0 *operator acting on* $\mathbb{C}^0$ *in the case when* $\Omega = \emptyset$*.)*

Indeed, the conditions $\|r(N_+(\Omega))\| = \|r(N_+(\Omega^*)^*)\| = \max\{|r(\lambda)|:\ \lambda \in \partial\Omega\}$ imply, in particular, that $\|[\mu - N_+(\Omega)]^{-1}\| = \|[\mu - N_+(\Omega^*)^*]^{-1}\| = 1/\mathrm{dist}[\mu, \Omega^-]$ and this suffices to guarantee that L is well-defined. Since $\mathrm{tr}[N_+(\Omega_n)^*, N_+(\Omega_n)] \le (1/\pi)\mathrm{meas}_2(\Omega_n)$, it is not difficult to check that L is essentially normal.

4.1.4 An application to approximation problems

It is easily seen that if N is a normal operator, then so is every operator M in the closure $\mathcal{U}(N)^-$ of $\mathcal{U}(N)$. In this case, $\mathcal{U}(N)^-$ admits a very simple characterization. The following result is straightforward.

LEMMA 4.8. *If* $N \in \mathcal{L}(\mathcal{H})$ *is normal, then*

$$\mathcal{U}(N)^- = \{M \in \mathcal{L}(\mathcal{H}):\ M \textit{ is normal, } \sigma(M) = \sigma(N), \sigma_e(M) = \sigma_e(N) \textit{ and } \mathrm{nul}(\lambda - M) = \mathrm{nul}(\lambda - N) \textit{ for all } \lambda \in \sigma(N)\setminus\sigma_e(N)\}.$$

We have the following

PROPOSITION 4.9. *If* M *is a normal operator such that* $\sigma(M)$ *is a perfect set,* K *is compact and* $\sigma(M+K) = \sigma(M)$, *then* $M \underset{\text{sim}}{\to} M+K$.

PROOF. It is easily seen that $\sigma_e(M) = \sigma_e(M+K) = \sigma(M)$. Let N be a diagonal normal operator such that $\sigma(N) = \sigma(M)$, i.e., $Ne_n = \lambda_n e_n$ with respect to some ONB $\{e_n\}_{n=1}^{\infty}$ and $\{\lambda_n\}_{n=1}^{\infty}$ is dense in $\sigma(M)$. Assume, moreover, that all the eigenvalues of N have infinite multiplicity. By Lemma 4.8, $\mathcal{U}(M)^- = \mathcal{U}(N)^-$; furthermore, the BDF theorem implies that M is unitarily equivalent to a compact perturbation of N. In other words, N can be chosen so that $M - N$ is compact, and therefore $M + K = N+C$, where $C = (M - N)+K \in \mathcal{K}(\mathcal{H})$. Clearly, $\sigma(N+C) = \sigma(M+K) = \sigma(M) = \sigma(N)$.

Since $N \in \mathcal{U}(M)^- \subset \mathcal{S}(M)^-$, it follows that $M \underset{\text{sim}}{\to} N$. Hence, it is sufficient to show that $N \underset{\text{sim}}{\to} N+C$. Let P_n be the orthogonal projection of $\mathcal{H}$ onto $\vee\{e_1,e_2,\ldots,e_n\}$; then $\|C - P_nCP_n\| \to 0$ $(n \to \infty)$ (recall that C is compact), and therefore $\|(N+C) - [(1-P_n)N(1-P_n)+P_n(N+C)P_n]\| = \|C - P_nCP_n\| \to 0$ $(n \to \infty)$.

By using this observation and the upper semicontinuity of the spectrum (Corollary 1.2(ii)), given $\varepsilon > 0$ we can find $m = m(\varepsilon)$ such that $\|C - P_mCP_m\| < \varepsilon/2$ and $\max\{\mathrm{dist}[\lambda,\sigma(N)]:\ \lambda \in \sigma[P_m(N+C)P_m]\} < \varepsilon/2$. Thus, (since $\sigma(N)$ is a perfect set) we can find an operator $F_m \in \mathcal{L}(\mathrm{ran}\, P_m)$ such that $\|F_m - P_m(N+C)P_m|\mathrm{ran}\, P_m\| < \varepsilon/2$ and $\sigma(F_m)$ is a set of m distinct points $\mu_1^m, \mu_2^m, \ldots, \mu_m^m$ of the point spectrum of N.

Let $F \in \mathcal{F}(\mathcal{H})$ be the operator defined by: $F|\mathrm{ran}\, P_m = F_m$ and $F|\mathrm{ran}\,(1-P_m) = 0$. Then $\|(N+C) - (N+F)\| = \|C - F\| \le \|C - P_mCP_m\| + \|F_m - P_m(N+C)P_m\| + \|F_m - (F+N)|\mathrm{ran}\, P_m\| < \varepsilon$, $N+F \simeq N\oplus F_m$ and $N\oplus F_m \sim N$, whence the result follows. □

COROLLARY 4.10. *If* $T \in \mathcal{L}(\mathcal{H})$, Ω *is an analytic Cauchy domain with boundary* $\Gamma = \partial\Omega$ *containinig* $\sigma(T)$ *and* M *is a normal operator such that* $\sigma(M) = \Omega^-$, *then*

$$M \underset{\text{sim}}{\to} M\oplus M_+(\Gamma)^{(\infty)}\oplus T,\quad M \underset{\text{sim}}{\to} M\oplus M_+(\Gamma)^{(\infty)},\quad M \underset{\text{sim}}{\to} M\oplus M_-(\Gamma)^{(\infty)}.$$

PROOF. By Lemma 4.8, Corollary 4.5 and Proposition 4.9,

$$M \underset{\text{sim}}{\to} M^{(\infty)} \underset{\text{sim}}{\to} (M\oplus M_+(\Gamma))^{(\infty)} = M^{(\infty)} \oplus M_+(\Gamma)^{(\infty)} \underset{\text{sim}}{\to} M\oplus M_+(\Gamma)^{(\infty)}$$

and (by a similar argument) $M \underset{\text{sim}}{\to} M\oplus M_-(\Gamma)^{(\infty)}$.

On the other hand, by Corollary 3.37, $M_+(\Gamma)^{(\infty)} \underset{\text{sim}}{\to} M_+(\Gamma)^{(\infty)} \oplus T$, whence we readily obtain $M \underset{\text{sim}}{\to} M\oplus M_+(\Gamma)^{(\infty)} \oplus T$. □

4.2 Matrix models for operators

Let $T \in L(H)$ and let $\Phi_n = \{\lambda \in \rho_F(T): \text{ind}(\lambda - T) = n\}$. Define $\Omega_n = \text{interior}(\Phi_n)^-$ $(1 \le |n| < \infty)$ (so that $\Omega_n = \text{interior}(\Omega_n)^-$), and

$$N_+(T) = \oplus_{n=1}^{\infty} N_+(\Omega_{-n})^{(n)}, \quad N_-(T) = \oplus_{n=1}^{\infty} [N_+(\Omega_n{}^*)^*]^{(n)},$$

where $N_+(\Omega)$ is defined by (4.4).

THEOREM 4.11. *Given* $T \in L(H)$, *there exist* $K \in K(H)$ *and a decomposition* $H = H^+ \oplus H^0 \oplus H^-$ *such that*

$$T+K \simeq \begin{pmatrix} N_+(T) & * & * \\ 0 & B & * \\ 0 & 0 & N_-(T) \end{pmatrix} \tag{4.6}$$

with respect to that decomposition, where B *is a smooth operator and* $\sigma(B) = \sigma_e(B) = \sigma_e(T)$.

PROOF. By Theorem 3.48, we can find $K_1 \in K(H)$ such that $T+K_1$ is smooth. Since

$$\partial\sigma_e(N_+(T)) \cup \partial\sigma_e(N_-(T)) = \partial(\cup_{1\le|n|<\infty} \Omega_n) \subset \partial(\cup_{1\le|n|<\infty} \Phi_n) \subset \partial\sigma_e(T),$$

which is contained in $\sigma_{\ell re}(T)$, it follows from Theorem 3.49 and Corollary 3.50 that

$$(T+K_1)+K_2 \simeq \begin{pmatrix} N^+ & * & * \\ 0 & B_1 & * \\ 0 & 0 & N^- \end{pmatrix}$$

(with respect to a suitable decomposition of the space), where N^+ and N^- are normal operators such that $\sigma(N^+) = \sigma_e(N^+) = \sigma_e(N_+(T))$ and $\sigma(N^-) = \sigma_e(N^-) = \sigma_e(N_-(T))$, respectively.

Observe that $\lim(n \to \infty)\{\text{meas}_2(\Omega_n)+\text{meas}_2(\Omega_{-n})\} = 0$ and that the stronger form of the Berger-Shaw trace inequality for the operators $N_+(\Omega_{-n})$ and $N_+(\Omega_n{}^*)^*$ says that

$$\text{tr}[N_+(\Omega_n{}^*)^*, N_+(\Omega_n{}^*)] \le (1/\pi)\text{meas}_2(\Omega_n)$$

and

$$\mathrm{tr}[N_+(\Omega_{-n})^*, N_+(\Omega_{-n})] \le (1/\pi)\,\mathrm{meas}_2(\Omega_{-n}),\ 1 \le |n| < \infty.$$

It is not difficult to conclude that $[N_+(T)^*, N_+(T)]$, $[N_-(T)^*, N_-(T)]$, $[N_+'(T)^*, N_+'(T)]$ and $[N_-'(T)^*, N_-'(T)]$, where

$$N_+'(T) = \oplus_{n=1}^{\infty} N_+(\Omega_{-n})^{(n)} \quad \text{and} \quad N_-'(T) = \oplus_{n=1}^{\infty} [N_+(\Omega_n^*)^*]^{(n)},$$

are compact operators.

Hence, by the BDF theorem (Theorem 4.1; see also Corollary 4.6), N^+ (N^-, resp.) is unitarily equivalent to a compact perturbation of $N_+(T)\oplus N_-'(T)$ ($N_+'(T)\oplus N_-(T)$, resp.), i.e., there exists a compact operator K_3 such that

$$(T+K_1+K_2)+K_3 \simeq \begin{pmatrix} N_+(T) & 0 & * & * & * \\ 0 & N_-'(T) & * & * & * \\ 0 & 0 & B_1 & * & * \\ 0 & 0 & 0 & N_+'(T) & 0 \\ 0 & 0 & 0 & 0 & N_-(T) \end{pmatrix} = \begin{pmatrix} N_+(T) & * & * \\ 0 & B_2 & * \\ 0 & 0 & N_-(T) \end{pmatrix},$$

where

$$B_2 = \begin{pmatrix} N_-'(T) & * & * \\ 0 & B_1 & * \\ 0 & 0 & N_+'(T) \end{pmatrix}.$$

Finally, by Theorem 3.49, there exists a compact operator $K_4 \simeq 0\oplus K_4'\oplus 0$ (K_4' acting on the subspace corresponding to B_2) such that

$$(T+K_1+K_2+K_3)+K_4 \simeq \begin{pmatrix} N_+(T) & * & * \\ 0 & B & * \\ 0 & 0 & N_-(T) \end{pmatrix},$$

where $B = B_2+K_4'$ is a smooth operator. In particular,

$$\sigma(B) = \sigma_W(B) = \sigma_e(B) \cup \{\lambda \in \rho_F(B):\ \mathrm{ind}(\lambda - B) \neq 0\} = \sigma_e(B) \cup \{\rho_F(B) \cap \sigma(B)\}.$$

Let $K = K_1+K_2+K_3+K_4$; then (see observations after Corollary 3.41)

$$\tilde{T} = (\tilde{T}+\tilde{K}) \simeq \begin{pmatrix} \widetilde{N_+(T)} & * & * \\ 0 & \tilde{B} & * \\ 0 & 0 & \widetilde{N_-(T)} \end{pmatrix}.$$

Since $\sigma_e(N_+(T)) \cup \sigma_e(N_-(T)) \subset \sigma_e(T)$, we readily conclude that $\sigma_e(B)$ must also be contained in $\sigma_e(T)$. Moreover, since $\sigma_e(N_+(T)) \cup \sigma_e(N_-(T)) = \sigma_e(N_+'(T)) \cup \sigma_e(N_-'(T)) \subset \sigma_e(B)$, it follows that $\sigma_e(B) = \sigma_e(T)$. Thus, in order to complete the proof it only remains to show that $\rho_F(B) \cap \sigma(B) = \emptyset$.

Assume it is not true. Then $\lambda - B$ is a non-invertible Fredholm operator for some complex λ; furthermore, by stability properties of the index (Corollary 1.14(i)), we can directly assume that λ does not belong to the nowhere dense set $\sigma_e(N_+(T)) \cup \sigma_e(N_-(T))$. By smoothness, ind

$(\lambda - B) = m \neq 0$. Now, if $\lambda \in \rho_F^-(T) = \rho_F^-(N_+(T))$, then $\lambda \notin \rho_F(N_-(T)) \cap \sigma(N_-(T))$ and it is straightforward that

$$\mathrm{ind}(\lambda - T) = \mathrm{ind}(\lambda - N_+(T)) + \mathrm{ind}(\lambda - B) = \mathrm{ind}(\lambda - T) + m,$$

a contradiction.

A similar argument proves that λ cannot belong to $\rho_F^+(T) = \rho_F^+(N_-(T))$. Since $\lambda \notin \rho_F^-(T) \cup \rho_F^+(T)$, it follows that

$$0 = \mathrm{ind}(\lambda - T) = \mathrm{ind}(\lambda - B) = m,$$

a new contradiction.

We conclude that $\rho_F(B) \cap \sigma(B) = \emptyset$. Therefore, the compact operator K satisfies all our requirements. □

REMARK 4.12. In the above theorem (as in Corollary 4.7), we had to use the stronger inequality

$$\mathrm{tr}[N_+(\Omega)^*, N_+(\Omega)] \leq (1/\pi)\,\mathrm{meas}_2(\Omega) \quad (\text{not } \Omega^- \,!)$$

in order to guarantee that $N_+(T)$ and $N_-(T)$ have compact self-commutators. Given ε, $0 < \varepsilon < 1$, it is not difficult to construct denumerably many pairwise disjoint open connected subsets $\Omega_1, \Omega_2, \ldots, \Omega_n, \ldots$ of the unit square $(\{\alpha + i\beta:\ 0 \leq \alpha, \beta \leq 1\})$ with $\partial\Omega_n = \partial\Omega_1$ *for all* $n = 1,2,\ldots$, *such that* $\mathrm{meas}_2(\Omega_n) < \varepsilon/2^n$, $n = 1,2,\ldots$. In this case,

$$\mathrm{meas}_2(\Omega_n^-) > \mathrm{meas}_2(\partial\Omega_n) > 1-\varepsilon \text{ for all } n = 1,2,\ldots\ !$$

The following theorem provides a representation for T modulo compact operators, "symmetric" to the representation (4.6). Its proof follows by the same arguments as in Theorem 4.11 and will be omitted.

THEOREM 4.13. *Given* $T \in L(H)$, *there exist* $C \in K(H)$ *and a decomposition* $H = H_- \oplus H_0 \oplus H_+$ *such that*

$$T + C \simeq \begin{pmatrix} N_-(T) & * & * \\ 0 & B & * \\ 0 & 0 & N_+(T) \end{pmatrix} \tag{4.7}$$

with respect to that decomposition, where B *is a smooth operator and* $\sigma(B) = \sigma_e(B) = \sigma_e(T)$.

REMARK 4.14. If T is essentially normal, then it directly follows from the BDF theorem that $T - F \simeq N_+(T) \oplus B \oplus N_-(T)$ (for a suitable compact operator F), where B is a normal operator such that $\sigma(B) = \sigma_e(B) = \sigma_e(T)$.

4.3 Spectra of compact perturbations of operators

Now we are in a position to improve the qualitative parts of Proposition 3.45 and Theorem 3.48.

THEOREM 4.15. *Let* $T \in L(H)$. *If* $\{\Omega_n\}_{n\in\mathbf{M}}$ *is an enumeration of the bounded components of* $\rho_{s-F}(T)$ *(where either* $\mathbf{M} = \{1,2,\dots,m\}$ *or* $\mathbf{M}$ *is the set of all natural numbers),* $\{d_n\}_{n\in M}$ *is a family of non-negative integers,* $\{\lambda_k\}_{k\in K}$ *(where either* $K = \{1,2,\dots,r\}$ *or* K *is the set of all natural numbers) is an at most denumerable subset of* $\rho_{s-F}(T)$ *with no accumulation point in* $\rho_{s-F}(T)$ *and* $\{J_k\}_{k\in K}$ *is a family of operators acting on spaces of positive finite dimension such that* $\sigma(J_k) = \{\lambda_k\}$ *(*$k \in K$*), then there exists a compact operator* C *such that*

(i) $\sigma(T+C) = \sigma_W(T) \cup \{\lambda_k\} \cup [\cup\{\Omega_n: \ d_n > 0\}]$;

(ii) $\rho^s_{s-F}(T+C) = \{\lambda_k\}$ *and* $(T+C)_0|H_0(\lambda_k;T+C) \sim J_k$ *for all* $k \in K$ *(where* $(T+C)_0$ *is the central operator in the Apostol's triangular representation* (3.8) *of* $T+C$*); and*

(iii) $\min.\mathrm{ind}(T+C-\lambda) = d_n$ *for all* $\lambda \in \Omega_n \setminus \{\lambda_k\}$.

COROLLARY 4.16. *Let* $T \in L(H)$ *and let* σ *be the union of* $\sigma_W(T)$ *and some bounded components of* $\mathbb{C}\setminus\sigma_W(T)$*; then there exists* $K \in K(H)$ *such that* $\sigma(T+K) = \sigma$.

In particular, there exists $K_0 \in K(H)$ *such that* $\mathbb{C}\setminus\sigma(T+K_0)$ *is connected.*

PROOF OF THEOREM 4.15. By Theorem 3.40 there exists $C_1 \in K(H)$ such that

$$T+C_1 \simeq \begin{pmatrix} M & * & * \\ 0 & B_1 & * \\ 0 & 0 & M \end{pmatrix},$$

where $\mathbf{M}$ is a diagonal normal operator such that $\mathbf{M} \simeq \mathbf{M}^{(\infty)}$ and $\sigma(\mathbf{M}) = \sigma_e(\mathbf{M}) = \sigma_{\ell re}(T)$, and $\sigma_{\ell re}(B_1) = \sigma_{\ell re}(T)$.

By Theorem 3.48, we can find $C_2 \in K(H)$ (which only modifies the action of B_1) such that

$$T+C_1+C_2 \simeq \begin{pmatrix} M & * & * \\ 0 & B_2 & * \\ 0 & 0 & M \end{pmatrix}, \text{ and}$$

B_2 is a smooth compact perturbation of B_1. Clearly, $T+C_1+C_2$ is also smooth.

Let $i_n = \mathrm{ind}(T+C_1+C_2-\lambda) = \mathrm{ind}(B_2-\lambda)$, $\lambda \in \Omega_n$ $(-\infty \le i_n \le \infty)$. Since the λ_k's do not accumulate in $\rho_{s-F}(T)$, we can find points μ_k $(k \in K)$ in $\sigma_p(M)$ such that $|\lambda_k - \mu_k| \to 0$ $(k \to \infty)$. Let $J_+ = \oplus\{J'_k: \ \lambda_k \in \rho_r(B_2)\}$ and $J_- =$

$\oplus\{J_k': \lambda_k \in \rho_\ell(B_2)\}$, where $J_k' \sim J_k$ is any operator such that $\|J_k' - \lambda_k\| < 1/k$ for all $k \in K$. It is easily seen that there exist compact perturbations M^+ and M^- of M such that $M^+ \simeq M\oplus J_+$ and $M^- \simeq M\oplus J_-$, respectively.

Now we can proceed as in the proof of Theorem 4.11 in order to obtain a compact operator C_3 such that

$$T+C_1+C_2+C_3 \simeq \begin{pmatrix} J_- & 0 & 0 & * & * & * & * \\ 0 & N_- & 0 & * & * & * & * \\ 0 & 0 & N_-' & * & * & * & * \\ 0 & 0 & 0 & B_2 & * & * & * \\ 0 & 0 & 0 & 0 & N_+' & 0 & 0 \\ 0 & 0 & 0 & 0 & 0 & N_+ & 0 \\ 0 & 0 & 0 & 0 & 0 & 0 & J_+ \end{pmatrix} = \begin{pmatrix} J_-\oplus N_- & * & * \\ 0 & B_3 & * \\ 0 & 0 & N_+\oplus J_+ \end{pmatrix},$$

where $N_+ \simeq \oplus\{N_+(\Omega_n)^{(d_n)}: i_n \geq 0, d_n > 0\}$, $N_- \simeq \oplus\{N_+(\Omega_n^*)^{*(d_n)}: i_n < 0, d_n > 0\}$, $J_-\oplus N_-\oplus N_-'$ and $N_+'\oplus N_+\oplus J_+$ are compact perturbations of M and

$$B_3 = \begin{pmatrix} N_-' & * & * \\ 0 & B_2 & * \\ 0 & 0 & N_+' \end{pmatrix}.$$

Finally, we can use Theorem 3.48 (as in the proof of Theorem 4.11) in order to find a compact operator C_4 (which only modifies the action of B_3) such that

$$T+C_1+C_2+C_3+C_4 \simeq \begin{pmatrix} J_-\oplus N_- & * & * \\ 0 & B & * \\ 0 & 0 & N_+\oplus J_+ \end{pmatrix} \text{ (B is smooth)}.$$

Let $C = C_1+C_2+C_3+C_4$ (clearly, $C \in K(H)$). It can be easily checked that $\sigma(J_-) = \sigma_p(J_-)^- \subset \{\lambda_k\}\cup\sigma_{\ell re}(T)$, $\sigma(J_+) = \sigma_p(J_+)^- \subset \{\lambda_k\}\cup\sigma_{\ell re}(T)$, $\rho_{s-F}^s(T+C) \subset \{\lambda_k\}$ and (since B is smooth) $\min.\mathrm{ind}(T+C-\lambda) = d_n$ for all $\lambda \in \Omega_n\setminus\{\lambda_k\}$, so that

$$\sigma_W(T)\cup[\cup\{\Omega_n: d_n > 0\}] \subset \sigma(T+C) \subset \sigma_W(T)\cup\{\lambda_k\}\cup[\cup\{\Omega_n: d_n > 0\}].$$

On the other hand, if $\lambda_k \in \sigma_p(J_-)$ then $\lambda_k \in \rho_\ell(B)\cap\rho(N_+\oplus J_+) = \rho_\ell\left(\begin{pmatrix} B & * \\ 0 & N_+\oplus J_+ \end{pmatrix}\right)$ and Corollary 3.22 implies that $T \sim J_k\oplus T_k$, where $\lambda_k \in \rho_{s-F}^r(T_k)$. A similar argument implies that the same result holds for $\lambda_k \in \sigma_p(J_+)$, whence we obtain (ii).

Since (i) and (iii) follow immediately from (ii) and our previous observations, we are done. □

4.4 Voiculescu's theorem

An abstract C*-algebra is a Banach algebra C with an operation $*: C \to C$ satisfying the conditions:

(i) $(a^*)^* = a$, $a \in C$;

(ii) $(a+b)^* = a^*+b^*$, $a,b \in C$;

(iii) $(\lambda a)^* = \bar{\lambda} a^*$, $\lambda \in \mathbb{C}$, $a \in C$;

(iv) $(ab)^* = b^*a^*$, $a,b \in C$; and

(v) $||a||^2 = ||a^*a||$, $a \in C$.

(An operation $*: C \to C$ satisfying (i) - (iv) is called an *involution*.)

The most obvious example of a C*-algebra is $L(H)$, with * equal to the usual 'adjoint operation'. Given a (finite or denumerable) family $\{T_n\}$ of operators in $L(H)$, we shall denote by $C^*(\{T_n\})$ the C*-algebra generated by the T_n's and the identity operator, i.e., the minimal norm-closed algebra containing $1, T_1, T_2, \ldots, T_n, \ldots$ and $T_1^*, T_2^*, \ldots, T_n^*, \ldots$. Every separable C*-subalgebra of $L(H)$ (containing the identity) has this form. If $\{T_n\} = \{T\}$ consists of a single operator, then we shall simply write $C^*(T)$ instead of $C^*(\{T\})$.

A *-representation ρ of a C*-algebra C on a Hilbert space H_ρ is an operator ρ from C into $L(H_\rho)$ which preserves products (i.e., ρ is a homomorphism of rings: $\rho(ab) = \rho(a)\rho(b)$) and involutions (i.e., $\rho(a^*) = \rho(a)^*$ in $L(H_\rho)$). If C has an identity e and $\rho(e) = 1$, then ρ is called *unital*; ρ is *faithful* if it has a trivial kernel. It is well-known that a *-representation ρ is faithful if and only if it is an isometric mapping from C onto $\rho(C) \subset L(H_\rho)$ (see, e.g., [37], [72], [176]). The following classical result can be found in any of these references.

THEOREM 4.17 (Gelfand-Naimark-Segal theorem). *Every abstract C*-algebra C with identity admits a faithful unital *-representation ρ in $L(H_\rho)$ for a suitable Hilbert space H_ρ, i.e., C is isometrically *-isomorphic to a C*-algebra of operators. Furthermore, if C is separable, then H_ρ can be chosen separable.*

Since we are only interested in *separable* C*-algebras, every Hilbert space (H, H_ρ, etc) considered here will be separable.

Two *-representations $\rho_1: C \to L(H_1)$ and $\rho_2: C \to L(H_2)$ are unitarily equivalent (or, simply, *equivalent*) if there is a unitary operator U from H_1 onto H_2 such that $\rho_2(a) = U\rho_1(a)U^*$ for all $a \in C$; ρ_1 and ρ_2 are *approximately equivalent* ($\rho_1 \simeq_a \rho_2$) if there is a sequence of unitary operators $U_n: H_2 \to H_1$ such that

$$\begin{cases} \rho_1(a) - U_n\rho_2(a)U_n^* \in K(H_1) \quad (n=1,2,\ldots) \text{ and} \\ \lim(n \to \infty) \, ||\rho_1(a) - U_n\rho_2(a)U_n^*|| = 0 \end{cases} \tag{4.8}$$

for each $a \in C$. It is easily seen that $\simeq_a$ is an equivalence relation.

THEOREM 4.18. *(i) Let* C *be a separable* C^**-algebra with identity and let* $\rho: C \to L(H_\rho)$ *be a unital *-representation of* C. *Let* γ *be a unital *-representation of* $\pi\circ\rho(C)$ $(\subset A(H_\rho))$ *on a Hilbert space* H_γ. *Then*

$$\rho \simeq_a \rho \oplus \gamma\circ\pi\circ\rho.$$

(ii) Let ρ_1 *and* ρ_2 *be two unital *-representations of* C *on Hilbert spaces* H_1 *and* H_2, *respectively, and let* $H_j' = \{\rho_j(a)x: a \in \ker \pi\circ\rho_j, x \in H_j\}^-$, $j = 1,2$. *The the following are equivalent:*

(1) $\rho_1 \simeq_a \rho_2$.

(2) There is a sequence of unitary operators $U_n: H_2 \to H_1$ $(n = 1,2, \ldots)$ *such that*

$$\lim(n \to \infty) \|\rho_1(a) - U_n\rho_2(a)U_n^*\| = 0, \text{ for each } a \in C.$$

(3) There exist sequences of unitary operators $U_n: H_2 \to H_1$ *and* $V_n: H_1 \to H_2$ $(n = 1,2,\ldots)$ *such that*

$$U_n\rho_2(a)U_n^* \to \rho_1(a) \quad \text{and} \quad V_n\rho_1(a)V_n^* \to \rho_2(a),$$

in the weak operator topology, for each $a \in C$.

(4) $\ker \rho_1 = \ker \rho_2$, $\ker \pi\circ\rho_1 = \ker \pi\circ\rho_2$ *and the *-representations of* C *on* H_1' *and* H_2' *induced by* ρ_1 *and* ρ_2, *respectively, are equivalent.*

(5) $\ker \rho_1 = \ker \rho_2$, $\ker \pi\circ\rho_1 = \ker \pi\circ\rho_2$ *and the *-representations of* $\ker \pi\circ\rho_1$ *on* H_1' *and* H_2' *induced by* ρ_1 *and* ρ_2, *resp., are equivalent.*

Recall that a *-representation ρ of a C^*-algebra C on H_ρ is *irreducible* if there is no non-trivial subspace M of H_ρ reducing all operators in $\rho(C)$.

COROLLARY 4.19. *(i) Every unital *-representation of a separable* C^**-algebra with identity is approximately equivalent to a direct sum of irreducible *-representations.*

(ii) Let ρ *be a unital *-representation of a separable* C^**-algebra* C *on a Hilbert space* H_ρ. *In order that all unital *-representations of* C *approximately equivalent to* ρ *be actually equivalent to* ρ *it is necessary and sufficient that* $\rho(C)$ *be finite-dimensional.*

COROLLARY 4.20. *Let* ρ_1, ρ_2 *be infinite dimensional unital *-representations of a separable* C^**-algebra* C *with identity; then*

(i) If $\ker \rho_1 = \ker \rho_2$ *and neither* $\rho_1(C)$ *nor* $\rho_2(C)$ *contains any non-zero compact operator, then* $\rho_1 \simeq_a \rho_2$.

(ii) If $\ker \rho_2 \supset \ker \pi\circ\rho_1$, *then* $\rho_1 \oplus \rho_2 \simeq_a \rho_1$.

4.5 Closures of unitary orbits

Let $T \in L(H)$ and let $C^*(T)$ be the C^*-algebra generated by T and the identity. Denote by ρ_0 the identity *-representation of $C^*(T)$ on H ($\rho_0(A) = A$ for each A in $C^*(T)$) and let S be any operator in $U(T)^-$. It is easily seen that

$$C'(T) = \{L \in L(H): \ \{U_n L U_n^*\}_{n=1}^{\infty} \text{ is a Cauchy sequence in } L(H) \text{ for each sequence } \{U_n\} \text{ in } U(H) \text{ such that } \{U_n T U_n^*\}_{n=1}^{\infty} \text{ is Cauchy}\}$$

is a C^*-algebra containing $C^*(T)$ and therefore there is a (necessarily unique) unital *-representation ρ_1 of $C^*(T)$ on H such that $\rho_1(T) = S$. (This representation is defined as follows: If $\|S - U_n T U_n^*\| \to 0$ $(n \to \infty)$ for a suitable sequence $\{U_n\}_{n=1}^{\infty}$ in $U(H)$, then

$$\rho_1(A) = (\text{norm}) \lim(n \to \infty)\ U_n A U_n^*,\ A \in C^*(T).)$$

Since $T \in U(S)^-$, it is not difficult to check that ρ_1 is also faithful; indeed, $\rho_1 \simeq_a \rho_0$. Conversely, if ρ_1 is a unital *-representation of $C^*(T)$ on H and $\rho_1 \simeq_a \rho_0$ (clearly, this ρ_1 is faithful), then $\rho_1(T)$ belongs to the closure of the unitary orbit of $T = \rho_0(T)$.

4.5.1 Operator-valued spectrum and unitary orbits

The above remarks and the results of Section 4.4 have the following direct consequences.

PROPOSITION 4.21. *Let* $T \in L(H)$; *then*

(i) $S \in U(T)^-$ *if and only if there exist sequences* $\{U_n\}_{n=1}^{\infty}$ *and* $\{V_n\}_{n=1}^{\infty}$ *of unitary operators such that*

$$U_n T U_n^* \to S \text{ (weakly)} \quad \text{and} \quad V_n S V_n^* \to T \text{ (weakly)}.$$

(ii) If ρ *is a unital *-representation of* $C^*(\tilde{T})$ *on* H_ρ *and* $\rho(\tilde{T}) = A$, *then there exist* $S_1, S_2 \in U(T)^-$ *such that* $S_1 \simeq T \oplus A$, $S_2 \simeq T \oplus A^{(\infty)}$. *(If* ρ_0 *is the identity *-representation of* $C^*(T)$, *then* $\rho_0 \simeq_a \rho_0 \oplus \rho \simeq_a \rho_0 \oplus \rho^{(\infty)}$.)

(iii) $B = \rho(T)$ *for some unital *-representation* ρ *of* $C^*(T)$ *if and only if either* T *has a reducing subspace* M *such that* $T|M \simeq B$ *or there exists a sequence* $\{T_n\}_{n=1}^{\infty}$ *in* $L(H)$ *such that* $T_n \simeq B^{(\infty)} \oplus T_n'$ *(for suitable operators* T_n', $n = 1, 2, \ldots$*) and* $\|T - T_n\| \to 0$ $(n \to \infty)$.

(iv) If $S \in U(T)^-$ *and* $\varepsilon > 0$, *there is a unitary operator* U *such that* $S - UTU^* \in K(H)$ *and* $\|S - UTU^*\| < \varepsilon$.

(v) There exists $S \in U(T)^-$ *such that* S *is an infinite direct sum of irreducible operators.*

PROPOSITION 4.22. *The following are equivalent for* T *in* $L(H)$:

(i) $U(T)$ *is closed.*

(ii) $C^*(T)$ *is finite dimensional.*

(iii) $T \simeq A \oplus B^{(\infty)}$, *where* $A \in L(\mathbb{C}^m)$, $B \in L(\mathbb{C}^n)$ *for some pair* (m,n), $0 \le m < \infty$, $1 \le n < \infty$.

COROLLARY 4.23. *The set of all reducible operators is dense in* $L(H)$. *Furthermore, given* T *in* $L(H)$ *and* $\varepsilon > 0$, *there exists* $K \in K(H)$, $\|K\| < \varepsilon$, *such that* T+K *is reducible.*

REMARK 4.24. Given an n-tuple $\bar{T} = (T_1, T_2, \ldots, T_n) \in L(H)^{(n)}$, consider its unitary orbit

$$\{(UT_1U^*, UT_2U^*, \ldots, UT_nU^*) \in L(H)^{(n)} : U \in U(H)\}.$$

We shall say that $\bar{T}$ is reducible if $T_1, T_2, \ldots, T_n$ have a joint reducing subspace. Clearly, in this framework, the results of Section 4.4 imply that the analogues of Propositions 4.21 and 4.22 and Corolary 4.23 hold for n-tuples.

In order to determine $U(T)^-$ in concrete terms we shall need the following definition: The *operator-valued spectrum* $\Sigma(T)$ of $T \in L(H)$ is the set of all those operators A acting on some subspace of dimension n, $1 \le n < \infty$, or $A \in L(H)$, and T is the (norm) limit of a sequence $\{T_n\}_{n=1}^{\infty}$ in $L(H)$ with $T_n \simeq A \oplus T_n'$ (for suitable operators T_n').

If $S \in U(T)^-$, we shall say that T and S are *approximately unitarily equivalent operators* ($T \simeq_a S$). It is easily seen that $\simeq_a$ is an equivalence relation and that $T \simeq_a S$ implies that $T \underset{\text{sim}}{\to} S$ and $S \underset{\text{sim}}{\to} T$. With the above definition in mind, we have the following characterization of $U(T)^-$:

THEOREM 4.25. *Suppose* T, S $\in L(H)$. *Then* $T \simeq_a S$ *if and only if* $\Sigma(T) = \Sigma(S)$.

4.5.2 Concrete examples of closures of unitary orbits

From Proposition 4.21(iv), we obtain

PROPOSITION 4.26. *If* $N \in L(H)$ *is normal,* L *is a diagonal normal operator in* $U(N)^-$ *and* $\varepsilon > 0$, *then there exists* $U \in U(H)$ *such that*

$$N - ULU^* \in K(H) \quad \textit{and} \quad \|N - ULU^*\| < \varepsilon.$$

Hence, every normal operator is equal to a diagonal operator up to a compact perturbation of arbitrarily small norm

It is not difficult to check that if $T \in L(H)$ then H can be written as $H_{Nor(T)} \oplus H_{Abnor(T)}$, where $H_{Nor(T)}$ and $H_{Abnor(T)}$ reduce T, $T_{Nor} = T|H_{Nor(T)}$ is normal (= the normal part of T) and $T_{Abnor} = T|H_{Abnor(T)}$ (= the abnormal part of T) has no nontrivial reducing subspace M such that $T|M$ is normal; more precisely, $H_{Nor(T)} = \cap_{n=1}^{\infty} \ker[T^{*n}, T^n]$. (Of course, $H_{Nor(T)}$ and $H_{Abnor(T)}$ can be infinite dimensional, finite dimensional or even $\{0\}$.)

PROPOSITION 4.27. *If* $T, S \in L(H)$ *and* T *is essentially normal, then* $T \simeq_a S$ *if and only if* $S_{Abnor} \simeq T_{Abnor}$, $\sigma_e(S) = \sigma_e(T)$ *and* $\text{nul}(\lambda - S) = \text{nul}(\lambda - T)$ *for all* $\lambda \in [\sigma(S) \cup \sigma(T)] \setminus \sigma_e(T)$. *In particular, if* $K \in K(H)$, *then* $U(K)^- = \{C \in K(H): C \simeq K_{Abnor} \oplus 0_d$, *where* 0_d *is the* 0 *operator acting on a subspace of dimension* d, $0 \le d \le \infty$, *and* $d + \dim H_{Abnor(K)} = \infty\}$.

PROPOSITION 4.28. *If* T *is hyponormal and* $T = T_{Abnor}$, *then*

$U(T)^- \supset \{S \in L(H): S \simeq T \oplus N$, N *is normal and* $\sigma(N) \subset \sigma_{\ell e}(T)\}$.

4.5.3 On normal and quasinilpotent restrictions

Let $T \in L(H)$ and let $\{\lambda_n\}$ be a (finite or denumerable) sequence of distinct points of $\sigma_{\ell e}(T)$. According to Theorem 3.40, there exists a compact operator K_1 such that

$$T + K_1 = \begin{pmatrix} N_\ell & * \\ 0 & A_1 \end{pmatrix} \begin{matrix} H_1 \\ H_1^{\perp}, \end{matrix}$$

where $N_\ell \in L(H_1)$ is a diagonal normal operator such that $N_\ell \simeq (\text{diag}\{\lambda_1, \lambda_2, \ldots, \lambda_n, \ldots\})^{(\infty)}$ and $\sigma_e(A_1) = \sigma_e(T)$, $\sigma_{re}(A_1) = \sigma_{re}(T)$.

Let $\{\mu_m\}$ be a (finite or denumerable) sequence of distinct points of $\sigma_{re}(T) = \sigma_{re}(A_1)$. Applying the same theorem to A_1^*, we can find a compact operator K such that

$$T + K = \begin{pmatrix} N_\ell & * & * \\ 0 & A & * \\ 0 & 0 & N_r \end{pmatrix} \begin{matrix} H_1 \\ H_2 \\ H_3 \end{matrix},$$

where $N_r \simeq (\text{diag}\{\mu_1, \mu_2, \ldots, \mu_m, \ldots\})^{(\infty)}$.

Let $\Gamma_\ell = \{\lambda_n\}^- = \sigma(N_\ell) = \sigma_e(N_\ell)$ and $\Gamma_r = \{\mu_m\}^- = \sigma(N_r) = \sigma_e(N_r)$ and consider the (separable!) C*-algebra $C = C^*(\{\tilde{T}; \tilde{P}_1, \tilde{P}_2, \ldots, \tilde{P}_n, \ldots; \tilde{R}_1, \tilde{R}_2, \ldots, \tilde{R}_m, \ldots\})$ generated by $\tilde{I}$, $\tilde{T}$ and the images in the Calkin algebra of the orthogonal projections P_n (onto $\ker(\lambda_n - N_\ell)$; $n = 1, 2, \ldots$) and R_m (onto

$\ker(\mu_m - N_r)$; $m = 1,2,\ldots$).

Let $\rho: C \to L(H)$ be a faithful unital *-representation. Since

$$(1-\sum_{j=1}^{n} P_n)(T+K)\sum_{j=1}^{n} P_n = 0,\ \sum_{k=1}^{m} R_k A(1-\sum_{k=1}^{m} R_k) = 0,$$

it is not difficult to conclude that

$$\rho(\tilde{T})^{(\infty)} \simeq \begin{pmatrix} N_\ell & * & * \\ 0 & B & * \\ 0 & 0 & N_r \end{pmatrix}.$$

By Proposition 4.21, we have

PROPOSITION 4.29. *Let* $T \in L(H)$ *and let* Γ_ℓ *and* Γ_r *be closed subsets of* $\sigma_{\ell e}(T)$ *and* $\sigma_{re}(T)$*, respectively. Then there exists* $S \simeq_a T$ *such that*

$$S \simeq T \oplus \begin{pmatrix} N_\ell & * & * \\ 0 & B & * \\ 0 & 0 & N_r \end{pmatrix},$$

where N_ℓ *and* N_r *are diagonal normal operators of uniform infinite multiplicity such that* $\sigma(N_\ell) = \sigma_e(N_\ell) = \Gamma_\ell$ *and* $\sigma(N_r) = \sigma_e(N_r) = \Gamma_r$*, respectively.*

COROLLARY 4.30. *Let* $T \in L(H)$. *If* Γ *is a closed subset of* $\sigma_e(T)$ *and* N *is a normal operator such that* $\sigma(N) = \sigma_e(N) = \Gamma$*, then there exists an operator* A *such that*

$$\begin{cases} \rho_{s-F}(A) = \rho_{s-F}(T) \setminus \Gamma, \\ \text{min.ind}(\lambda - A)^k = \text{min.ind}(\lambda - T)^k \text{ for all } \lambda \in \rho_{s-F}(A) \\ k = 1,2,\ldots, \end{cases} \tag{4.9}$$

and

$$T \underset{sim}{\to} N \oplus A.$$

PROOF. Clearly Γ can be written as $\Gamma = \Gamma_\ell \cup \Gamma_r$, where $\Gamma_\ell \subset \sigma_{\ell e}(T)$ and $\Gamma_r \subset \sigma_{re}(T)$ are closed sets. Let S be the operator constructed in Proposition 4.28; then, by Lemma 3.36, we have

$$T \simeq_a S \underset{sim}{\to} T \oplus (N_\ell \oplus B \oplus N_r) \simeq (N_\ell \oplus N_r) \oplus (T \oplus B).$$

Clearly, $\sigma(N_\ell \oplus N_r) = \sigma_e(N_\ell \oplus N_r) = \Gamma$, so that $N_\ell \oplus N_r \simeq_a N$ (Lemma 4.8). Let A be an operator unitarily equivalent to $N \oplus T \oplus B$. It is completely apparent that $\rho_{s-F}(A) \subset \rho_{s-F}(T) \setminus \Gamma$. On the other hand, since ρ is faithful, it readily follows that $\rho_{s-F}(B) = \rho_{s-F}^{\infty}(B) \cup \rho_{s-F}^{-\infty}(B)$, $\rho_{s-F}^{\infty}(B) \supset \rho_{s-F}^{\infty}(T) \setminus \Gamma$, $\rho_{s-F}^{-\infty}(B) \supset \rho_{s-F}^{-\infty}(T) \setminus \Gamma$ and $\rho_{s-F}^{s}(B) \cap (\rho_{s-F}(T) \setminus \Gamma) = \emptyset$, whence we immediately conclude that A satisfies (4.9).

Since $(N\ \oplus N_r) \oplus (T \oplus B) \simeq_a N \oplus A$, we are done. □

Applying Corollary 3.50 to the direct summand of S unitarily equivalent to $\rho(\tilde{T})^{(\infty)}$ (in Proposition 4.28), we obtain

COROLLARY 4.31. *Let* $T \in L(H)$ *and let* Γ_ℓ *and* Γ_r *be closed subsets of* $\sigma_{\ell e}(T)$ *and* $\sigma_{re}(T)$*, respectively. Given* $\varepsilon > 0$ *and normal operators* N_ℓ *and* N_r *such that* $\Gamma_\ell \subset \sigma(N_\ell) = \sigma_e(N_\ell) \subset (\Gamma_\ell)_\varepsilon$ *and* $\Gamma_r \subset \sigma(N_r) = \sigma_e(N_r) \subset (\Gamma_r)_\varepsilon$*, there exists* $L \in L(H)$ *such that*

$$L \simeq \begin{pmatrix} N_\ell & * & * \\ 0 & L_1 & * \\ 0 & 0 & N_r \end{pmatrix},$$

$\rho_{s-F}(L) = \rho_{s-F}(T)\setminus[\sigma(N_r)\cup\sigma(N_\ell)]$, $\min.\mathrm{ind}(\lambda - L)^k = \min.\mathrm{ind}(\lambda - T)^k$ *for all* $\lambda \in \rho_{s-F}(L)$, $k = 1,2,\ldots$, *and* $\|T - L\| < 2\varepsilon$.

(Apply Lemma 4.8 to the result of the first modification.)

Let $T \in L(H)$ and assume that μ is an isolated point of $\sigma_e(T)$. If $\varepsilon > 0$ is small enough to guarantee that $\{\lambda: |\lambda - \mu| \le \varepsilon\}\cap\sigma_e(T) = \{\mu\}$, then

$$\tilde{R} = \frac{1}{2\pi i}\int_{|\lambda-\mu|=\varepsilon} (\lambda - \tilde{T})^{-1}\, d\lambda$$

defines an idempotent commuting with $\tilde{T}$ (in the Calkin algebra). This idempotent can be lifted to an idempotent $R \in L(H)$ such that $\sigma_e(RTR|\mathrm{ran}\, R) = \{\mu\}$ ($RTR = RT = TR$ modulo $K(H)$) and $\sigma_e((1-R)T(1-R)|\ker R) = \sigma_e(T)\setminus\{\mu\}$. Let $W \in G(H)$ be such that $WRW^{-1} = P$ is an orthogonal projection and let $L = WTW^{-1}$; then

$$L = B\oplus(\mu+Q) + (\text{compact})$$

for some B such that $\mu \notin \sigma_e(B)$ and Q is a quasinilpotent operator acting on the infinite dimensional reducing subspace H_μ of L.

Let $\rho: C^*(\mu+\tilde{Q}) \to L(H_\rho)$ be a faithful unital *-representation of $C^*(\mu+\tilde{Q})$. Clearly, $0\oplus\rho$ is a (not faithful, unless B acts on a finite dimensional space!) unital *-representation of $C^*(L)$. By Proposition 4.21(ii), $L \simeq_a L\oplus(\mu+G)^{(\infty)}$, where $G = \rho(\tilde{Q})$.

It readily follows that $T \underset{\text{sim}}{\to} T\oplus(\mu+G)^{(\infty)}$; furthermore, since ρ is faithful, we can use Corollary 1.8 to show

PROPOSITION 4.32. *Let* $T \in L(H)$ *and assume that* $\sigma_e(T)$ *has an isola-ted point* μ. *Then there exists a quasinilpotent* G *such that*

$$T \underset{\text{sim}}{\to} T\oplus(\mu+G)^{(\infty)}.$$

Moreover, if

$$f_k(\lambda) = \begin{cases} (\lambda-\mu)^k & \textit{in some open neighborhood of } \mu, \\ 0, & \textit{in some open neighborhood of } \sigma_e(T)\setminus\{\mu\}, \end{cases}$$

($k = 1,2,\ldots$), then $G^k = 0$ *if and only if* $f_k(\tilde{T}) = 0$.

4.6 Irreducible operators

According to Corollary 4.23, every operator has an arbitrarily small compact perturbation which is reducible. If reducible is replaced by *irreducible*, the result is still true. In fact, in this case the answer is much stronger and the proof is definitely simpler. More precisely, we have

LEMMA 4.33. *The set of all irreducible operators is dense in* $L(H)$. *Furthermore, given* T *in* $L(H)$, $p > 1$ *and* $0 < \varepsilon < 1$, *there exists* K_p *in* $C^p(H)$, $|K_p|_p < \varepsilon$, *such that* $T+K_p$ *is irreducible.*

PROOF. Let $T = A+iB$ (Cartesian decomposition). According to the Weyl-von Neumann-Kuroda theorem (see, e.g., [153, p. 525, Theorem 2.3]), there exists $K_1 \in C^p(H)$, $|K_1|_p < \varepsilon/3$ such that $A+K_1$ is a diagonal hermitian operator with respect to some ONB $\{e_n\}_{n=1}^{\infty}$ of H. Let $\{\lambda_n\}_{n=1}^{\infty}$ be the sequence of corresponding eigenvalues (i.e., $(A+K_1)e_n = \lambda_n e_n$, $n = 1, 2,\ldots$). It is easily seen that now we can find $K_2 \in C^1(H)$, $|K_2|_1 < \varepsilon/3$. K_2 an hermitian diagonal operator (with respect to the same ONB) such that $(A+K_1+K_2)e_n = \mu_n e_n$, $n = 1,2,\ldots$, and the μ_n's are all distinct.

It is not difficult to check that $A'(A+K_1+K_2) = A(A+K_1+K_2) = \{B \in L(H)\colon B = \sum_{n=1}^{\infty}\gamma_n e_n\otimes e_n$ (*strong* limit of partial sums)$\}$ is the set of all diagonal normal operators with respect to that basis. In particular, a subspace M reduces $A+K_1+K_2$ if and only if $P_M = \sum\{e_j\otimes e_j\colon j \in N(M)\}$, where $N(M)$ is a subset of the natural numbers ($N(M) = \{j\colon e_j \in M\}$). Assume that $j \in N(M)$, but $h \notin N(M)$ and let $R \in L(H)$ be such that either $\langle Re_j,e_h\rangle \neq 0$ or $\langle Re_h,e_j\rangle \neq 0$. It is easily seen that R cannot commute with P_M, i.e., M does not reduce R.

Let $B = (b_{jk})_{j,k=1}^{\infty}$ be the matrix of B with respect to the ONB $\{e_n\}_{n=1}^{\infty}$ and let $0 < \eta < \varepsilon/6$. Define K_η by

$$K_\eta = \begin{pmatrix} \eta & \eta^2/2 & \eta^3/3 & \eta^4/4 & . & . & . \\ \eta^2/2 & \eta^3/3 & \eta^4/4 & . & . & . & \\ \eta^3/3 & \eta^4/4 & . & . & . & & \\ \eta^4/4 & . & . & . & & & \\ . & . & . & & & & \\ . & . & & & & & \\ . & & & & & & \end{pmatrix} \begin{matrix} e_1 \\ e_2 \\ e_3 \\ e_4 \\ . \\ . \\ . \end{matrix} .$$

Then $|K_\eta|_1 \le \sum_{n=1}^{\infty} n(\eta^n/n) = \eta/(1-\eta) < \varepsilon/3$. Since the set $\{b_{jk}\}_{j,k=1}^{\infty}$ is at most denumerable, there exists some η such that

$$b_{jk} + \eta^{j+k}/(j+k) \neq 0, \text{ for all } j,k = 1,2,\ldots . \qquad (4.10)$$

Let $K_3 = K_\eta$ for some η, $0 < \eta < \varepsilon/6$, satisfying (4.10). Then $B+K_3$ cannot be reduced by any nontrivial reducing subspace of $A+K_1+K_2$. Thus,

$$T+(K_1+K_2+iK_3) \text{ is irreducible}$$

and

$$|K_1+K_2+iK_3|_p \le |K_1|_p+|K_2|_p+|iK_3|_p \le |K_1|_p+|K_2|_1+|K_3|_1 \quad . \square$$

THEOREM 4.34. *(i) The set of all irreducible operators is a* G_δ *dense subset of* $L(H)$.

(ii) The set of all reducible operators is an F_σ *dense subset of* $L(H)$.

PROOF. According to Corollary 4.23 and Lemma 4.33, it only remains to show that the set $R(H)$ of all irreducible operators is an F_σ.

Let P be the set of all those hermitian operators H for which $0 \le H \le 1$. Recall that P is exactly the weak closure of the set of all projections. Let $P_o = P\setminus\{\lambda: \lambda \in \mathbb{C}\}$. Since P is a weakly closed subset of the closed unit ball of $L(H)$, it is weakly compact, and hence the weak topology for P is metrizable. Since the set of all multiples of the identity is weakly closed, it follows that P_o is weakly locally compact. Since the weak topology has a countable base, the same is true for P_o, and therefore P_o is weakly σ-compact. Let $P_1, P_2, \ldots,$ be weakly compact subsets of P_o such that $\cup_{n=1}^{\infty} P_n = P_o$.

Let $\hat{P}_n = \{A \in L(H): AH = HA \text{ for some } H \in P_n\}$ $(n = 1,2,\ldots)$; the spectral theorem implies that $\cup_{n=1}^{\infty} \hat{P}_n = R(H)$.

Suppose that $A_k \in \hat{P}_n$ $(k = 1,2,\ldots)$ and $||A_k - A|| \to 0$ $(k \to \infty)$. For each k, find an $H_k \in P_n$ such $A_kH_k = H_kA_k$. Since P_n is weakly compact and metrizable, there is no loss of generality in assuming that the sequence $\{H_k\}_{k=1}^{\infty}$ is weakly convergent to $H \in P_n$.

CLAIM: $AH = HA$.

Indeed, if $x,y \in H$, then

$$|\langle A_kH_kx,y\rangle - \langle AHx,y\rangle| \le |\langle A_kH_kx,y\rangle - \langle AH_kx,y\rangle|+|\langle AH_kx,y\rangle - \langle AHx,y\rangle|$$

$$\le ||A_k - A||\cdot||x||\cdot||y||+|\langle (H_k - H)x,A^*y\rangle| \to 0 \ (k \to \infty).$$

Hence, $A_kH_k \to AH$ (weakly, as $k \to \infty$). By taking adjoints, we can similarly check that $H_kA_k \to HA$ and therefore $A \in \hat{P}_n$.

Hence, $\hat{P}_n$ is closed $(n = 1,2,\ldots)$ and therefore $R(H)$ is an F_σ (in the norm topology). $\square$

4.7 Notes and remarks

The BDF theorem and the Berger-Shaw trace inequality have been published in the "Proccedings of a Conference on Operator Theory, Halifax, Nova Scotia, 1973", Lecture Notes in Mathematics Volume 345, Springer Verlag, Berlin-Heidelberg-New York, 1973 (see [60], [50]). The interested reader can also consult [61], [62], [49] and W. B. Arveson's survey article [38].

Lemma 4.3 is a portion of Proposition 2 of D. A. Herrero's article [146]. The present proof of the Berger-Shaw trace inequality (Theorem 4.4) combines four ingredients: (I) is an argument due to D. Voiculescu [196,Proposition 1]. This argument has been independently discovered by C. Berger (unpublished result; see also comments in [196]); (II) is a classical result (see, e.g., [50,Theorem 1]); (III) is an "ad hoc" modification of the second computational lemma of C. A. Berger and B. I. Shaw [49]; and (IV) is an observation of D. A. Herrero. An extended version of the Berger-Shaw trace inequality can be found in the above mentioned article of D. Voiculescu. The reader is referred to [66], [167] and [168] for a detailed account of the theory of hyponormal operators.

Proposition 4.9 is [133,Lemma 1]. The matrix models for operators are due to C. Apostol [6] and Theorem 4.15 is a consequence of those constructions and Apostol's triangular representation. Corollary 4.16 (in fact, a stronger form of it) was proved by C. Apostol, C. M. Pearcy and N. Salinas in [28].

The notion (4.8) of approximately equivalent *-representations and all the results of Sections 4.4 and 4.5.1 are contained in the article [195] by D. Voiculescu, except for the equivalence between (1) and (3) of Theorem 4.18, which has been independently obtained by W. Arveson in the above mentioned survey article (see [38,Theorem 5]) and by D. W. Hadwin [112,Corollary 4.2] (indeed, Hadwin established it in a slightly weaker form), and Theorem 4.25 is also due to Hadwin [112,Corollary 3.6].

Hadwin's reducing operator-valued spectrum $\Sigma(T)$ has been modelled on J. Ernest's operator-valued spectrum [80] and the reducing essential matricial spectra of operators of C. M. Pearcy and N. Salinas. (See [160], [164], [165], [166], [179]. The article [63] of J. Bunce and N. Salinas is also relevant in this context.) The different pieces of $\Sigma(T)$ in [112] correspond to the possibilities offered by Proposition 4.21 (iii): namely: 1) A is in the *essential reducing operator spectrum* of T, denoted by $\Sigma_e(T)$, if $A^{(\infty)} \in \Sigma(T)$; 2) If n is a positive integer, then the n-*dimensional reducing operator spectrum* of T, denoted by $\Sigma^n(T)$, is $\Sigma(T)\cap L(\mathbb{C}^n)$ and the n-*dimensional essential reducing operator spectrum* of T, denoted by $\Sigma_e^n(T)$, is $\Sigma_e(T)\cap L(\mathbb{C}^n)$; 3) The *infinite-dimensional reducing operator spectrum* of T, denoted by $\Sigma^\infty(T)$, is $\Sigma(T)\cap L(H)$, and the *infinite-dimensional essential reducing operator spectrum* of T, denoted by $\Sigma_e^\infty(T)$, is $\Sigma_e(T)\cap L(H)$.

In [112] Hadwin also proved (Theorem 3.9) that $U(T)^-$ is always arcwise connected, thus extending the corresponding result for normal operators, previously obtained by R. Gellar and L. Page in [104] (see also [85]).

Proposition 4.26 is essentially the I. D. Berg's extension to normal operators of the classical Weyl-von Neumann theorem on diagonalization of hermitian operators via small compact perturbations [47].(This result has also been independently obtained by W. Sikonia in his doctoral dissertation; see [186].) Propositions 4.27 and 4.28 are essentially due to D. W. Hadwin [111] and to C. M. Pearcy and N. Salinas [162] (see also [160]). Proposition 4.29 was proved in [97] in connection with a problem of quasisimilarity of operators; see also [58], [160].

The problem of characterizing closures of unitary orbits has been developed by D. W. Hadwin in his doctoral dissertation [111] and in [112]. His ideas and techniques go beyond this particular problem. The interested reader can also consult Hadwin's papers [113], [114] and [54] (joint work with A. Brown and C.-K. Fong), and the closely related articles [101] and [191].

Lemma 4.33 is a mild improvement of a result of P. R. Halmos. That result and Theorem 4.34 are contained in [120]. The proof of Lemma 4.33 given here is due to H. Radjavi and P. Rosenthal [170]. It is convenient to remark that Voiculescu's Corollary 4.23 affirmatively answers in the most complete form Halmos' Problem 8:

Is every operator the norm-limit of reducible ones? [123,p.919].

In [116], D. W. Hadwin extended Voiculescu's theorem (and many other related results) to cover the case of non-separable C*-algebras and operators acting on non-separable spaces.

5 Limits of nilpotent and algebraic operators

As a first application of the tools introduced in the previous two chapters, it will be shown that the closures of the set of all nilpotent operators and the set of all algebraic operators can be characterized very simply in terms of the different parts of the spectrum.

Essentially the same kind of arguments can be used to obtain two important results about closures of similarity orbits.

5.1 Limits of nilpotent operators

THEOREM 5.1. *The closure of the set* $N(H)$ *of all nilpotent operators acting on* H *coincides with the set of all those* A *in* $L(H)$ *satisfying the conditions:*

(i) $\sigma(A)$ *is connected and contains the origin;*

(ii) $\sigma_e(A)$ *is connected and contains the origin; and*

(iii) $\mathrm{ind}(\lambda-A) = 0$ *for all* $\lambda \in \rho_{s-F}(A)$.

Since $\sigma(Q) = \sigma_e(Q) = \{0\}$ for each $Q \in N(H)$, the necessity of (i) and (ii) follows at once from Corollary 1.6(i) and (iii), respectively. Since $\lambda-Q$ is invertible (hence $\mathrm{ind}(\lambda-Q) = 0$) for all $\lambda \neq 0$, the necessity of (iii) follows from Theorem 1.13(iii). The sufficiency of conditions (i), (ii) and (iii) will be proven in two steps.

REMARK 5.2. Theorem 5.1 implies, in particular, that every quasinilpotent operator is a limit of nilpotent ones.

5.1.1 Normal limits of nilpotents

We have already shown (Proposition 2.28(ii)) that $N(H)^-$ contains a normal operator M such that $\sigma(M) = D(0,1)^-$, where $D(\mu,r) = \{\lambda \in \mathbb{C}: |\lambda-\mu| < r\}$ is the disk of center μ and radius $r > 0$. By Lemma 4.8,

$$U(M)^- = \{N \in L(H): N \text{ is normal and } \sigma(N) = D(0,1)^-\} \subset N(H)^-.$$

Since $rQ \in N(H)$ for each Q in $N(H)$ and each $r > 0$, it readily follows that $N(H)^-$ contains every normal operator whose spectrum is a disk centered at the origin.

LEMMA 5.3. *Let* N *be a normal operator such that* $\sigma_e(N)$ *has an accumulation point* λ_o *and let* $Q \in N(H)$; *then* $N \underset{\text{sim}}{\to} N\oplus(\lambda_o+Q)$.

PROOF. Let $\{\lambda_n\}_{n=1}^{\infty}$ be a sequence of distinct points of $\sigma_e(N)$ such that $\lambda_n \to \lambda_o$ $(n \to \infty)$. By Lemma 4.8, $N \simeq_a N_1 = N\oplus(\text{diag}\{\lambda_o,\lambda_1,\lambda_2,\ldots\})^{(\infty)}$.

Assume that $Q \in N_k(H)$; then Q admits a canonical $k \times k$ operator matrix representation of the form

$$Q = \begin{pmatrix} 0 & Q_{12} & Q_{13} & \cdots & Q_{1,k-1} & Q_{1k} \\ & 0 & Q_{23} & \cdots & Q_{2,k-1} & Q_{2k} \\ & & 0 & \cdots & Q_{3,k-1} & Q_{3k} \\ & & & \ddots & \vdots & \vdots \\ & 0 & & & 0 & Q_{k-1,k} \\ & & & & & 0 \end{pmatrix} \begin{matrix} \ker Q^1 \ominus \ker Q^0 \\ \ker Q^2 \ominus \ker Q^1 \\ \ker Q^3 \ominus \ker Q^2 \\ \vdots \\ \ker Q^{k-1} \ominus \ker Q^{k-2} \\ \ker Q^k \ominus \ker Q^{k-1} \end{matrix} \qquad (5.1)$$

(Clearly, $\ker Q^0 = \{0\}$ and $\ker Q^k = H$.)

Replacing, if necessary, Q by $Q\oplus\varepsilon q_k^{(\infty)}$ (given by (2.1); $0 < \varepsilon << 1$) we can directly assume that $\ker Q^j \ominus \ker Q^{j-1}$ is infinite dimensional for all $j = 1,2,\ldots,k$.

Given $n \geq 1$, let R_n be the operator obtained by replacing the diagonal entries λ_o in λ_o+Q by $\lambda_{n+1},\lambda_{n+2},\ldots,\lambda_{n+k}$. It is easily seen that

$$\|(\lambda_o+Q)-R_n\| = \max\{|\lambda_o-\lambda_{n+j}|: j = 1,2,\ldots,k\} \to 0 \ (n \to \infty).$$

By Corollary 3.22 and the above observation,

$$N_1 \simeq N_1\oplus(\oplus_{j=1}^{k} \lambda_{n+j})^{(\infty)} \sim N_1\oplus[(\oplus_{j=1}^{k} \lambda_{n+j})^{(\infty)}+Q] \simeq N_1\oplus R_n \to N_1\oplus(\lambda_o+Q)$$

$(n \to \infty)$, i.e., $N \simeq_a N_1 \underset{\text{sim}}{\to} N\oplus(\lambda_o+Q)$. □

Recall that the *Hausdorff distance* in the family $BP(X)$ of all *bounded nonempty* subsets of a metric space (X,d) is the pseudometric defined by

$$d_H(A,B) = \inf\{\varepsilon > 0: A \subset B_\varepsilon,\ B \subset A_\varepsilon\}. \qquad (5.2)$$

Moreover, the restriction of d_H to the family $B_c(X)$ of all *closed bounded nonempty subsets* of X is a metric space: $(B_c(X),d_H)$.

Lemma 4.8 and the "spreading the spectral measure argument" (that we have already applied, e.g., in Corollary 3.50) applied to normal op erators with spectra equal to their essential spectra, yields the following simple (but very useful!) result:

LEMMA 5.4. *If* M *and* N *are normal operators such that* $\sigma(M) = \sigma_e(M)$, $\sigma(N) = \sigma_e(N)$ *and* $d_H[\sigma(M),\sigma(N)] < \varepsilon$, *then*

$$\operatorname{dist}[M,\mathcal{U}(N)] < \varepsilon, \qquad \operatorname{dist}[N,\mathcal{U}(M)] < \varepsilon.$$

COROLLARY 5.5. *If* M *and* N *are normal operators such that* $\sigma(N)$ *is a perfect set,* $\sigma(M) \supset \sigma(N)$ *and each component of* $\sigma(M)$ *intersects* $\sigma(N)$, *then* $N \underset{\text{sim}}{\to} M$.

PROOF. It is completely apparent that $\sigma_e(N) = \sigma(N)$ and $\sigma_e(M) = \sigma(M)$. Let $\varepsilon > 0$ and let $\sigma(M) \subset \cup_{j=1}^{m} D(\lambda_j,\varepsilon/4)$ be a covering of $\sigma(M)$ by open disks of radii $\varepsilon/4$ centered at points of $\sigma(M)$. It is easily seen that if the λ_j's are properly ordered, we can find points μ_j in $D(\lambda_j,\varepsilon/4)$, $j = 1,2,\ldots,m$ such that, if $\Lambda_1 = \{\mu_1,\mu_2,\ldots,\mu_{r_1}\}$, $\Lambda_2 = \{\mu_{r_1+1},\ldots,\mu_{r_2}\}$, ... , $\Lambda_p = \{\mu_{r_{p-1}+1},\ldots,\mu_{r_p}\}$ $(r_p = m)$, then $\Lambda_1 \subset \sigma(N)$, and

$$\max\{\operatorname{dist}[\mu_j,\Lambda_{h-1}]:\ \mu_j \in \Lambda_h\} < \varepsilon/2,\ h = 1,2,\ldots,p-1.$$

Let $\{M_j\}_{j=1}^{m}$ be a finite set of normal operators such that $\sigma(M_j) = \sigma_e(M_j) = D(\lambda_j,\varepsilon/2)^-$ and let $\{Q_j\}_{j=1}^{m}$ be nilpotent operators. By Lemma 5.3,

$$N \underset{\text{sim}}{\to} N\oplus\{\oplus_{j=1}^{r_1} (\mu_j+Q_j)\}$$

and, by using Propositions 2.28(ii) and our previous observations, we see that the Q_j's can be chosen so that $\|M_j-(\mu_j+Q_j)\|$ is arbitrarily small for all $j = 1,2,\ldots,p_1$. It readily follows that

$$N \underset{\text{sim}}{\to} N\oplus(\oplus_{j=1}^{p_1} M_j)$$

and, by an obvious inductive argument, that $N \underset{\text{sim}}{\to} M_\varepsilon = N\oplus(\oplus_{j=1}^{m} M_j)$, where M_ε is a normal operator such that

$$\sigma(M) \subset \sigma(M_\varepsilon) = \sigma_e(M_\varepsilon) = [\cup_{j=1}^{m} D(\mu_j,\varepsilon/2)]^- \subset \sigma(M)_{(3\varepsilon/4)},$$

so that $d_H[\sigma(M),\sigma(M_\varepsilon)] < \varepsilon$. By Lemma 5.4, we can find $M'_\varepsilon \simeq M_\varepsilon$ such that $\|M-M'_\varepsilon\| < \varepsilon$. Since ε can be chosen arbitrarily small, we conclude that $N \underset{\text{sim}}{\to} M$. □

PROPOSITION 5.6. *If* M *is normal, then* $M \in \mathcal{N}(\mathcal{H})^-$ *if and only if* $\sigma(M)$ *is a connected set containing the origin.*

PROOF. It was already observed that $\sigma(M)$ must necessarily satisfy the given conditions for each M in $\mathcal{N}(\mathcal{H})^-$.

Assume that $\sigma(M)$ is connected and contains 0. Given $\varepsilon > 0$, it fol lows from Lemma 5.4 that there is a normal operator M_ε such that $\sigma(M_\varepsilon)$ is the closure of a connected open neighborhood of the origin and $\|M-M_\varepsilon\| < \varepsilon$. Clearly, it suffices to show that $M_\varepsilon \in N(H)^-$. Let $\eta > 0$ be small enough to guarantee that $D(0,\eta) \subset \sigma(M_\varepsilon)$ and let N_η be a normal operator such that $\sigma(N_\eta) = D(0,\eta)^-$. By Proposition 2.28(ii), Lemma 4.8 and Corollary 5.5, $N_\eta \in N(H)^-$ and $N_\eta \underset{sim}{\to} M_\varepsilon$.

Since $N(H)$ is invariant under similarities, it follows from Proposition 1.15 that $M_\varepsilon \in N(H)^-$. □

5.1.2 Spectral characterization of $N(H)^-$

Recall that an operator R is called *algebraic* if it satisfies an algebraic equation. If R is algebraic with minimal monic polynomial p, $p(\lambda) = \Pi_{j=1}^{m} (\lambda-\lambda_j)^{k_j}$, then it is easily seen that $\sigma(R) = \{\lambda_1,\lambda_2,\ldots,\lambda_m\}$ and, by Corollary 3.22,

$$R = \begin{pmatrix} \lambda_1+Q_1 & & & & * \\ & \lambda_2+Q_2 & & & \\ & & \cdot & & \\ & & & \cdot & \\ 0 & & & & \cdot \\ & & & & \lambda_m+Q_m \end{pmatrix} \begin{matrix} H_1 \\ H_2 \\ \cdot \\ \cdot \\ \cdot \\ H_m \end{matrix} \sim \oplus_{j=1}^{m} (\lambda_j+Q_j), \qquad (5.3)$$

where $H_1 = \ker(R-\lambda_1)^{k_1}$, and $H_r = \{\ker \Pi_{j=1}^{r}(R-\lambda_j)^{k_j}\}\ominus\{\ker \Pi_{j=1}^{r-1}(R-\lambda_j)^{k_j}\}$ for $r = 2,3,\ldots,m$, and $Q_r \in N_{k_r}(H_r)\setminus N_{k_r-1}(H_r)$, $r = 1,2,\ldots,m$.

Let $Alg(H)$ denote the set of all algebraic operators acting on H. It is easy to see, by using the expressions (5.1), (5.3) and Corollary 3.22 that the algebraic operator R is the limit of a sequence $\{R_n\}_{n=1}^{\infty}$ of operators such that R_n is similar to a normal operator and $\sigma(R_n)$ consists of exactly $\sum_{j=1}^{m} k_j$ (= the degree of the minimal polynomial of R) points.

LEMMA 5.7. *Let* $N,T \in L(H)$, *where* N *is a normal operator such that* $\sigma(N)$ *is a perfect set and* $\sigma(T) \subset \sigma(N)$. *Then* $N \underset{sim}{\to} N\oplus T$.

PROOF. Let $\varepsilon > 0$ and let Ω_1, Ω be analytic Cauchy domains such that $\sigma(N) \subset \Omega_1 \subset \Omega_1^- \subset \Omega \subset \sigma(N)_{\varepsilon/2}$ and each component of Ω intersects $\sigma(N)$. If M_1, M are normal operators such that $\sigma(M_1) = \Omega_1^-$ and $\sigma(M) = \Omega^-$, and R is an algebraic operator such that $\sigma(R) \subset \Omega\setminus\Omega_1^-$, then it follows from Lemmas 5.3 and 5.4, and Corollary 4.10 that

$$N \underset{sim}{\to} M\oplus M_1 \underset{sim}{\to} (M\oplus R)\oplus(M_1\oplus M_+(\partial\Omega_1)^{(\infty)}\oplus T) \simeq_a M\oplus T\oplus(R\oplus M_+(\partial\Omega_1)^{(\infty)}).$$

By Propositions 5.6 and 1.15, and Corollary 4.10, R can be chosen so that

$$R = \begin{pmatrix} M+R_{11} & R_{12} & R_{13} \\ R_{21} & M_1+R_{22} & R_{23} \\ R_{31} & R_{32} & M_-(\partial\Omega_1)^{(\infty)}+R_{33} \end{pmatrix} = M\oplus M_1\oplus M_-(\partial\Omega_1)^{(\infty)}+S,$$

where $\|S\| < \varepsilon/2$.

Since $\sigma(R)\cap\sigma(M_+(\partial\Omega_1)^{(\infty)}) \subset (\Omega\backslash\Omega_1^-)\cap\Omega_1^- = \emptyset$, it follows from Corollary 3.22 that

$$R\oplus M_+(\partial\Omega_1)^{(\infty)} \sim \begin{pmatrix} R & 0 \\ X & M_+(\partial\Omega_1)^{(\infty)} \end{pmatrix}$$

for any X. Thus, if we choose $X = 0\oplus 0\oplus Z^{(\infty)}$ (where Z is defined by (3.2)), then

$$M\oplus T\oplus R\oplus M_+(\partial\Omega_1)^{(\infty)} \sim M\oplus T\oplus[M\oplus M_1\oplus \begin{pmatrix} M_-(\partial\Omega_1)^{(\infty)} & 0 \\ Z^{(\infty)} & M_+(\partial\Omega_1)^{(\infty)} \end{pmatrix} +S\oplus 0]$$

$$= M\oplus T[M\oplus M_1\oplus M(\partial\Omega_1)^{(\infty)}+S\oplus 0].$$

Since $M\oplus M\oplus M_1\oplus M(\partial\Omega_1)^{(\infty)} \simeq_a \mathbf{M}$ and $d_H[\sigma(N),\sigma(\mathbf{M})] < \varepsilon/2$, it follows from Lemma 5.4 that $S(N)$ contains an operator unitarily equivalent to $N\oplus T+S'$, where $\|S'\| < \varepsilon/2+\|S\| < \varepsilon$. Since ε can be chosen arbitrarily small, we conclude that $N \underset{sim}{\to} N\oplus T$. □

Now we are in a position to complete the proof of Theorem 5.1. Suppose that $A \in L(H)$ satisfies (i), (ii) and (iii) and let $\varepsilon > 0$.

Since $\sigma_o(A) = \emptyset$, it follows from Theorem 3.48 and (iii) that there exists $K_1 \in K(H)$, $\|K_1\| < \varepsilon/4$, such that $A_1 = A-K_1$ satisfies $\sigma(A_1) = \sigma_{\ell re}(A_1)$. Clearly, A_1 also satisfies (i).

Let ρ be a faithful unital *-representation of $C^*(\tilde{A}_1)$ on H_ρ and let $A_2' = \rho(\tilde{A}_1)$. It readily follows that if $A_2 = (A_2')^{(\infty)}$, then $\sigma(A_2) = \sigma_{\ell re}(A_2) = \sigma(A_1)$. By Proposition 4.21(ii) and (iv), there exists $K_2 \in K(H)$, $\|K_2\| < \varepsilon/4$, such that $A_3 = A_1-K_2 \simeq A_1\oplus A_2$. Finally, by Propositions 4.29 and 4.21(ii) and (iv), there exists $K_3 \in K(H)$, $\|K_3\| < \varepsilon/4$, $K_3 = K_{3,1}\oplus K_{3,2}$, such that

$$A_1-K_{3,1} \simeq \begin{pmatrix} N & C_1 \\ 0 & B_1 \end{pmatrix} \quad \text{and} \quad A_2-K_{3,2} \simeq \begin{pmatrix} B_2 & C_2 \\ 0 & N \end{pmatrix},$$

where N is normal and $\sigma(N) = \sigma_{\ell re}(A_1) = \sigma(A_1)$.

Hence, if $K = K_1+K_2+K_3$, then

$$A-K \simeq \begin{pmatrix} N & C_1 \\ 0 & B_1 \end{pmatrix} \oplus \begin{pmatrix} B_2 & C_2 \\ 0 & N \end{pmatrix} \simeq \left(\begin{array}{cc|cc} N & 0 & C_1 & 0 \\ 0 & B_2 & 0 & C_2 \\ \hline 0 & 0 & B_1 & 0 \\ 0 & 0 & 0 & N \end{array}\right). \tag{5.4}$$

Since $\sigma(B_1)\cup\sigma(B_2) \subset \sigma(N)$, it follows from Proposition 5.6, Lemma 5.7 and Proposition 1.15 that there exist nilpotent operators Q_j such that $\|B_1\oplus N - Q_1\| < \varepsilon/4$ and $\|N\oplus B_2 - Q_2\| < \varepsilon/4$. Hence, there exists R in $L(H)$, $\|R\| < \varepsilon/4$ such that

$$A-(K+R) \simeq Q = \begin{pmatrix} Q_2 & C_1\oplus C_2 \\ 0 & Q_1 \end{pmatrix}.$$

A straighforward computation shows that if $Q_j^{k_j} = 0$, then $Q^{k_1+k_2} = 0$, i.e., $Q \in N(H)$ and therefore $\operatorname{dist}[A,N(H)] < \|K\|+\|R\| < \varepsilon$. Since ε can be chosen arbitrarily small, it readily follows that $A \in N(H)^-$.

The proof of Theorem 5.1 is complete now. □

5.2 Closures of similarity orbits of normal operators with perfect spectra

THEOREM 5.8. *Let* $N \in L(H)$ *be a normal operator such that* $\sigma(N)$ *is a perfect set; then* $S(N)^-$ *is the set of all those operators* A *in* $L(H)$ *satisfying the following conditions*

(i) $\sigma(A) \supset \sigma(N)$ *and each component of* $\sigma(A)$ *intersects* $\sigma(N)$;

(ii) $\sigma_e(A) \supset \sigma_e(N)$ *(*$=\sigma(N)$*) and each component of* $\sigma_e(A)$ *intersects* $\sigma_e(N)$; *and*

(iii) $\operatorname{ind}(\lambda-A) = 0$ *for all* $\lambda \in \rho_{s-F}(A)$.

PROOF. Since $\sigma(N) = \sigma_e(N) = \sigma_{\ell re}(N)$, the necessity of the conditions (i), (ii) and (iii) follows from Corollary 1.6(i) and (iii) and Theorem 1.13(iii), exactly as in Theorem 5.1.

Assume that $A \in L(H)$ satisfies (i), (ii) and (iii) and let $\varepsilon > 0$. Minor modifications of the proof of Theorem 5.1 (formula (5.4)) indicate that we can find a compact operator K, $\|K\| < \varepsilon/3$ such that

$$A-K \simeq \left(\begin{array}{cc|cc} N_2 & 0 & C_1 & 0 \\ 0 & B_2 & 0 & C_2 \\ \hline 0 & 0 & B_1 & 0 \\ 0 & 0 & 0 & N_1 \end{array}\right) \oplus \left(\begin{array}{cc|cc} N_4 & 0 & C_3 & 0 \\ 0 & B_4 & 0 & C_4 \\ \hline 0 & 0 & B_3 & 0 \\ 0 & 0 & 0 & N_3 \end{array}\right), \tag{5.5}$$

where $N_j \simeq N$ and $\sigma(B_j) \subset \sigma(N_j)$, $j = 1,2,3,4$.

By Lemma 5.4 we can find an operator M such that $N \underset{sim}{\to} M$, $\sigma(N) \subset \sigma(M) = \sigma_e(M) \subset \sigma(N)_\varepsilon$, $\sigma(M)$ is the closure of a Cauchy domain and $\|N-M\|$

$< \varepsilon/3$. A fortiori, we can find $R \in L(H)$, $\|R\| < \varepsilon/3$, such that $B = A - (K+R)$ is the operator obtained from (5.5) by replacing the N_j's by normal operators $M_j \simeq M$ such that $\|N_j - M_j\| < \varepsilon/3$, $j = 1,2,3,4$.

Since $\sigma(M)$ is the closure of a Cauchy domain, it is very easy to prove that $M = A_1 \oplus A_2 \oplus A_3 \oplus A_4$, where the A_j's are normal operators such that $\sigma(A_1) \cap \sigma(A_2) = \sigma(A_3) \cap \sigma(A_4) = \emptyset$, $\sigma(A_j)$ is the closure of a Cauchy domain and each component of $\sigma(M)$ contains a component of $\sigma(A_j)$ for all $j = 1,2,3,4$. By Corollary 5.5 and Lemma 5.7, $N \underset{\text{sim}}{\to} M = (A_1 \oplus A_2) \oplus (A_3 \oplus A_4)$ and $A_j \underset{\text{sim}}{\to} M_j \oplus B_j$, $j = 1,2,3,4$. Thus, we can find operators $A_j' \sim A_j$ such that $\|M_j \oplus B_j - A_j'\| < \varepsilon/3$, $j = 1,2,3,4$. Since $\sigma(A_1') \cap \sigma(A_2') = \sigma(A_3') \cap \sigma(A_4') = \emptyset$, it follows from Corollary 3.22 that

$$M \sim C = \begin{pmatrix} A_2' & C_1 \oplus C_2 \\ 0 & A_1' \end{pmatrix} \oplus \begin{pmatrix} A_4' & C_3 \oplus C_4 \\ 0 & A_3' \end{pmatrix}.$$

Hence, $N \underset{\text{sim}}{\to} C$ and

$$\|A-C\| < \|A-B\| + \max\{\|M_j \oplus B_j - A_j'\| : j = 1,2,3,4\} < \|K\| + \|R\| + \varepsilon/3 < \varepsilon.$$

Since ε can be chosen arbitrarily small, we conclude that $N \underset{\text{sim}}{\to} A$. □

5.3 Limits of algebraic operators

THEOREM 5.9. *The closure of the set* $\mathcal{A}\ell g(H)$ *of all algebraic operators acting on* H *coincides with the set of all those operators* A *in* $L(H)$ *satisfying the following condition*

$$\operatorname{ind}(\lambda - A) = 0 \quad \textit{for all } \lambda \in \rho_{s-F}(A). \tag{5.6}$$

PROOF. The necessity of (5.6) follows from Theorem 1.13(iii). On the other hand, if A satisfies (5.6) and $\varepsilon > 0$, then we can proceed as in the proof of Theorem 5.1 in order to show that $A-K$ has the form (5.4) for a suitable compact operator K, $\|K\| < \varepsilon/3$, where N is a normal operator such that $\sigma(N) = \sigma_e(N) = \sigma_e(A) = \sigma(A-K) \supset \sigma(B_j)$, $j = 1,2$.

By Lemma 5.4 we can find a normal operator M such that $\sigma(M)$ is the closure of a Cauchy domain, $\sigma(N) \subset \sigma(M) \subset \sigma(N)_{\varepsilon/3}$ and $\|N-M\| < \varepsilon/3$. Thus, if B is a suitable operator in $L(H)$ unitarily equivalent to the result of replacing the N's by M's in the 4×4 operator matrix of (5.4), then $\|A-B\| < \|K\| + \|N-M\| < 2\varepsilon/3$. Since M is normal and $\sigma(M)$ has only finitely many components, it readily follows from Theorem 5.1 that $M \in \mathcal{A}\ell g(H)^-$. Since $\sigma(M)$ is a perfect set and $\sigma(B_j) \subset \sigma(M)$, $j = 1,2$, it follows from Lemma 5.7 that $M \underset{\text{sim}}{\to} M \oplus B_j$, $j = 1,2$. Since $\mathcal{A}\ell g(H)$ is invariant under similarities, it follows from Proposition 1.15 that $M \oplus B_j \in \mathcal{A}\ell g(H)^-$, $j = 1,2$.

Thus, we can find algebraic operators R_1, R_2 such that $\|M \oplus B_2 - R_2\|$

$< \varepsilon/3$ and $\|B_1 \oplus M - R_1\| < \varepsilon/3$.

It is completely apparent that if R_1, R_2 are algebraic operators, then so is

$$\begin{pmatrix} R_2 & C_1 \oplus C_2 \\ 0 & R_2 \end{pmatrix},$$

whence we deduce that

$$\text{dist}[A, \mathcal{Alg}(H)] < 2\varepsilon/3 + \max\{\|M \oplus B_2 - R_2\|, \|B_1 \oplus M - R_1\|\} < \varepsilon.$$

Since ε can be chosen arbitrarily small, we conclude that A belongs to $\mathcal{Alg}(H)^-$. □

Observe that the condition (5.6) is invariant under compact perturbations. By using this observation, our remarks at the beginning of Section 5.1.2 and Theorem 5.8, we obtain the following

COROLLARY 5.10. $\mathcal{Alg}(H)^- = \mathcal{Alg}(H)^- + K(H) = [\mathcal{Alg}(H) + K(H)]^- = \{R \in L(H)$: *R is similar to a normal operator with finite spectrum*$\}^- = \{A \in L(H)$: $\text{ind}(\lambda - A) = 0$ *for all* $\lambda \in \rho_{s-F}(A)\}$.

From Theorem 5.1 and Remark 5.2, we obtain

COROLLARY 5.11. *Suppose that* $A \in L(H)$, $\text{ind}(\lambda - A) = 0$ *for all* $\lambda \in \rho_{s-F}(A)$ *and* $\sigma_0(A)$ *is a finite set and* $\sigma_e(A)$ *has finitely many components* $\sigma_1, \sigma_2, \ldots, \sigma_m$. *Let* λ_j *be an arbitrary point of* σ_j, $j = 1,2,\ldots,m$. *Then there exists a sequence* $\{R_n\}_{n=1}^{\infty}$ *of algebraic operators such that* $\sigma(R_n) = \{\lambda_j\}_{j=1}^{m} \cup \sigma_0(A)$, $R_n|H(\sigma_0(A);R_n) = A|H(\sigma_0(A);A)$ *for all* $n = 1,2,\ldots$ *and* $\|A - R_n\| \to 0$ $(n \to \infty)$.

COROLLARY 5.12. *Let* Λ *be a finite nonempty subset of* $\mathbb{C}$ *and let* $S(\Lambda) = \{A \in L(H): \sigma(A) \subset \Lambda\}$; *then* $S(\Lambda)^-$ *is the set of all operators* T *satisfying the conditions*

(i) *Each component of* $\sigma(T)$ *or* $\sigma_e(T)$ *intersects* Λ; *and*

(ii) $\text{ind}(\lambda - T) = 0$ *for all* $\lambda \in \rho_{s-F}(T)$.

Furthermore, $S(\Lambda)^- = \{A \in S(\Lambda)$: A *is algebraic*$\}^-$.

5.4 Normal operators in closures of similarity orbits

PROPOSITION 5.13. *Let* $T \in L(H)$ *and let* N *be a normal operator such that* $\sigma(N) = \sigma(T)$ *and* $\dim H(\lambda;N) = \dim H(\lambda;T)$ *for each isolated point* λ *of* $\sigma(T)$. *Then* $T \underset{\text{sim}}{\to} N$.

It is completely apparent that if T and N are related as in the above proposition, then $\sigma_o(N) = \sigma_o(T)$ and $\sigma_e(N) = \sigma_B(T) = \sigma(T)\backslash\sigma_o(T)$. It will be convenient to provide a separate proof for the case when $\sigma(T)$ is totally disconnected.

LEMMA 5.14. *If* $T \in L(H)$, $\sigma(T)$ *is totally disconnected and* N *is a normal operator such that* $\sigma(N) = \sigma(T)$ *and* $\dim H(\lambda;N) = \dim H(\lambda;T)$ *for each isolated point* λ *of* $\sigma(T)$, *then* $T \underset{\text{sim}}{\rightarrow} N$.

PROOF. Let T and N be as indicated and let $\varepsilon > 0$. Then we can find a finite covering $\{\Omega_j\}_{j=1}^m$ of $\sigma(T)$ by pairwise disjoint open sets such that $\Omega_j \cap \sigma(T) \neq \emptyset$ and diameter $\Omega_j < \varepsilon/4$, for all $j = 1,2,\ldots,m$. Thus, if $\lambda_j \in \Omega_j$, then $\Omega_j \subset D(\lambda_j,\varepsilon/2)$ $(j = 1,2,\ldots,m)$.

By Corollary 3.22, $T \sim \oplus_{j=1}^m T_j$, where $T_j = T|H(\sigma(T)\cap\Omega_j;T)$. Let $N_j = N|H(\sigma(T_j);N)$. It is easily seen that $\dim H(\sigma(T)\cap\Omega_j;T) = \dim H(\sigma(T)\cap\Omega_j;N)$ for all $j = 1,2,\ldots,m$. By Corollary 3.35, there exists $T_j' \in L[H(\sigma(T)\cap\Omega_j;N)]$, $T_j' \sim T_j$ such that $\|T_j'-\lambda_j\| < \varepsilon/2$. On the other hand, it is completely apparent that $\|N_j-\lambda_j\| < \varepsilon/2$.

Hence, $T \sim \oplus_{j=1}^m T_j'$ and

$$\begin{aligned}\|N-\oplus_{j=1}^m T_j'\| &= \max\{\|N_j-T_j'\|: \ j = 1,2,\ldots,m\}\\ &\leq \max\{\|N_j-\lambda_j\|+\|\lambda_j-T_j'\|: \ j = 1,2,\ldots,m\} < \varepsilon,\end{aligned}$$

whence we readily conclude that $T \underset{\text{sim}}{\rightarrow} N$. □

PROOF OF PROPOSITION 5.13. Let σ denote the set of all non-isolated points of $\sigma_e(T)$. Given $\varepsilon > 0$, we can use Corollary 3.22 to obtain that $T \sim A\oplus B$, where $A = T|H(\sigma_{\varepsilon/2};T)$ and $B = T|H(\sigma(T)\backslash\sigma_{\varepsilon/2};T)$.

It is easily seen that $\sigma(B)$ is at most denumerable and therefore totally disconnected. By Lemma 5.14, $B \underset{\text{sim}}{\rightarrow} N_B = N|H(\sigma(B);N)$. Let $N_A = N|H(\sigma(A);N)$ and let M be a normal operator such that $\sigma(M) = \sigma(N_A)_{\varepsilon/2}$ and $\|N_A-M\| < \varepsilon$. Since $\|N-M\oplus N_B\| = \|N_A-M\| < \varepsilon$ and ε can be chosen arbitrarily small, it will be enough to show that $A \underset{\text{sim}}{\rightarrow} M$.

By Corollary 4.30, Lemma 5.3 and Corollary 5.5 and its proof, if M_A is a normal operator such that $\sigma(M_A) = \sigma_e(A)$, then we have

$$A \underset{\text{sim}}{\rightarrow} M_A\oplus C \underset{\text{sim}}{\rightarrow} M\oplus C,$$

where C is an operator such that $\sigma(C) \subset \sigma(A)$. A fortiori, $\sigma(C) \subset \sigma(M)$, which is a perfect set.

Given $\eta > 0$, let Ω_1, Ω be analytic Cauchy domains such that $\sigma(M) \subset \Omega_1 \subset \Omega_1^- \subset \Omega \subset \sigma(M)_\eta$ and each component of Ω intersects $\sigma(M)$. Let M_1 be a normal operator such that $\sigma(M_1) = \Omega_1^-$. By Lemma 5.4 we can find M_η normal such that $\sigma(M_\eta) = \sigma_e(M_\eta) = \Omega^-$ and $\|M-M_\eta\| < 2\eta$ and, by Corollary 5.12 there exist algebraic operators R_1, R_2 such that $\sigma(R_1)\cup\sigma(R_2)$

$\subset \Omega\backslash\Omega_1^-$, $\sigma(R_1)\cap\sigma(R_2) = \emptyset$, $\|R_1 - M_\eta \oplus M_1 \oplus M_+(\partial\Omega_1)^{(\infty)}\| < \eta$ and $\|R_2 - M_\eta \oplus M_1 \oplus M_-(\partial\Omega_1)^{(\infty)}\| < \eta$. By Lemma 5.3 and Corollary 5.5,

$M \underset{\text{sim}}{\to} M \underset{\text{sim}}{\to} M_\eta \oplus R_1 \oplus R_2$, so that $A \underset{\text{sim}}{\to} M_\eta \oplus R_1 \oplus C \oplus R_2$.

Since $\sigma(R_1)$, $\sigma(R_2)$ and $\sigma(C)$ are pairwise disjoint, it follows from Corollary 3.22 that $M_\eta \oplus (R_1 \oplus C \oplus R_2) \sim M_\eta \oplus L(C';X_1,X_2,X_3)$, where

$$L(C';X_1,X_2,X_3) = \begin{pmatrix} R_1 & X_1 & X_2 \\ 0 & C' & X_3 \\ 0 & 0 & R_2 \end{pmatrix},$$

$C' \sim C$ and X_1, X_2, X_3 are arbitrarily chosen operators.

It follows from (3.2), Theorem 3.23 and our choice of R_1 and R_2 that X_1, X_2, X_3 and C' can be chosen so that

$$\|L(C';X_1,X_2,X_3) - M_1^{(2)} \oplus M(\partial\Omega_1)^{(\infty)}\| < \varepsilon.$$

Hence, $S(A)^-$ contains an operator A' such that $\|A' - M_\eta \oplus M_1^{(2)} \oplus M(\partial\Omega_1)^{(\infty)}\| < \eta$. Since $M_\eta \oplus M_1^{(2)} \oplus M(\partial\Omega_1)^{(\infty)} \simeq_a M_\eta$ (Lemma 4.8) and η can be chosen arbitrarily small, we conclude that $A \underset{\text{sim}}{\to} M$. □

5.5 Sums of two nilpotents

THEOREM 5.15. *$N(H)^- + [N(H)^- \cap Nor(H)] = L(H)$; however, $1 \notin N(H)+N(H)+K(H)$.*

PROOF. Let $T \in L(H)$ and $\mu \in \partial\sigma_e(T)$. By Proposition 4.29 there exist $K \in K(H)$ and a decomposition $H = H_0 \oplus H_1^{(\infty)}$ (H_0, H_1 are infinite dimensional spaces) such that

$$T = A \oplus R^{(\infty)} + K,\ A \in L(H_0),\ R \in L(H_1),\ A \simeq T,\ R \simeq R^{(\infty)}, \text{ and}$$

$$R = \begin{pmatrix} \mu & * & * \\ 0 & B & * \\ 0 & 0 & \mu \end{pmatrix} \begin{matrix} M_1 \\ M_2, \\ M_3 \end{matrix}$$

where $\sigma(B) \subset \sigma_e(T)$.

Let $\varepsilon > 0$ and let M_k be a normal operator such that $\sigma(M_k) = \{\lambda: |\lambda| \le sp(T) + (1-2^{-k})\varepsilon\}$ $(k = 1,2,\ldots)$. Define

$$M_\varepsilon = 0 \oplus (0 \oplus \ldots \oplus 0 \oplus M_1' \oplus 0 \oplus \ldots \oplus 0 \oplus M_2' \oplus 0 \oplus \ldots \oplus 0 \oplus M_k' \oplus 0 \oplus \ldots)$$

(with respect to the decomposition $H = H_0 \oplus H_1^{(\infty)}$), where the direct summand M_k' occupies the n_k-th coördinate,

$$M_k' = \begin{pmatrix} M_k & 0 & 0 \\ 0 & 0 & 0 \\ 0 & 0 & M_k \end{pmatrix} \begin{matrix} M_1 \\ M_2, \\ M_3 \end{matrix} \quad k = 1,2,\ldots,$$

and $1 \le n_1 < n_2 < \dots < n_k < \dots$.

Let

$$L_k = \begin{pmatrix} M_k & * & * \\ 0 & B & * \\ 0 & 0 & M_k \end{pmatrix} \begin{matrix} M_1 \\ M_2 \\ M_3 \end{matrix}$$

be the operator obtained from R by replacing the μ's in the (1,1) and (3,3) entries by two copies of M_k, $k = 1,2,\dots$, let L be the operator obtained from $A \oplus R^{(\infty)}$ by replacing R by L_k in the n_k-th coördinate ($k = 1,2,\dots$) and define $S_\varepsilon = L+K$.

It is completely apparent that $T = S_\varepsilon + (\mu - M_\varepsilon)$. Since $\mu - M_\varepsilon$ is normal and $\sigma(\mu - M_\varepsilon) = \{\lambda \in \mathbb{C}: |\lambda - \mu| \le a+\varepsilon\}$ is a closed disk containing the origin, it follows from Proposition 5.6 that $M_\varepsilon \in N(H)^-$. On the other hand, it is clear that $\sigma_e(S_\varepsilon) = \sigma_{\ell re}(S_\varepsilon) = \{\lambda \in \mathbb{C}: |\lambda| \le a+\varepsilon\}$. If $\sigma(S_\varepsilon) = \sigma_e(S_\varepsilon)$, then $S_\varepsilon \in N(H)^-$ (by Theorem 5.1) and we are done. Thus, in order to complete the proof of the first statement, we only have to show that if $n_k \to \infty$ $(k \to \infty)$ fast enough, then $\sigma(S_\varepsilon) = \sigma_e(S_\varepsilon)$.

Let P_n denote the orthogonal projection of H onto the n-th copy of H_1 and let P_0 be the orthogonal projection onto H_0. Clearly, if

$$K_n = K - (1 - \textstyle\sum_{j=0}^{n-1} P_j) K (1 - \sum_{j=0}^{n-1} P_j),$$

then $\|K_n\| \to 0$ $(n \to \infty)$.

By Corollary 1.2(ii) there exist constants δ_k, $\delta_1 > \delta_2 > \dots > \delta_k > \dots > 0$ and $\delta_k \to 0$ $(k \to \infty)$, such that $\sigma(T \oplus [\oplus_{j=1}^{k} L_j] \oplus C_k) \subset \sigma(M_{k+1})$, provided $\|C_k\| < \delta_k$ $(k = 1,2,\dots)$. Let $n_1 \ge 1$ be the first index such that $\|K_{n_1}\| < \delta_1/2$ and define n_k inductively so that $1 \le n_1 < n_2 < \dots < n_k < \dots$ and $\|K_{n_k}\| < \delta_k/2$ for all $k = 1,2,\dots$. Define S_ε as above with this particular sequence. It is obvious that $\sigma(S_\varepsilon) = \sigma_e(S_\varepsilon) \cup \sigma_0(S_\varepsilon) = \sigma(M_\varepsilon) \cup \sigma_0(S_\varepsilon)$. Assume that $\sigma_0(S_\varepsilon) \ne \emptyset$ and let $\gamma \in \sigma_0(S_\varepsilon)$; then $|\gamma| = a+\varepsilon+\eta$ for some $\eta > 0$. By Corollary 1.2(i) there exists $\delta > 0$ such that $\sigma(S_\varepsilon + F) \cap D(\gamma, \eta/2) \ne \emptyset$ for all F in $L(H)$ with $\|F\| < \delta$. Let h be large enough to guarantee that $\delta_h < 2\delta$ and observe that (if n_0 is defined to be equal to 0)

$$S_\varepsilon - K_{n_h} = L + (K - K_{n_h}) = \{(A \oplus [\oplus_{j=1}^{h} \{R^{(n_j - n_{j-1} - j)} \oplus L_j\}]) + (K - K_{n_h}) | \operatorname{ran} \textstyle\sum_{j=0}^{n_h} P_j\}$$

$$\oplus [\oplus_{j=h+1}^{\infty} \{R^{(n_j - n_{j-1} - j)} \oplus L_j\}].$$

Since $\|K_{n_h}\| < \delta$ and $\sigma(\oplus_{j=h+1}^{\infty} \{R^{(n_j - n_{j-1} - j)} \oplus L_j\}) = \sigma(M_\varepsilon)$, we conclude that

$$\sigma(A \oplus [\oplus_{j=1}^{h} \{R^{(n_j - n_{j-1} - j)} \oplus L_j\}] + (K - K_{n_h}) | \operatorname{ran} \textstyle\sum_{j=0}^{n_h} P_j)$$

is not contained in $\sigma(M_\varepsilon)$.

Since $\sigma(R) \subset \sigma(M_\varepsilon)$, it easily follows that if

$$V = (A\oplus \oplus_{j=1}^{h}\{R_j^{(n_j-n_{j-1}-j)} \oplus L_j\} + (K-K_{n_h})|\operatorname{ran} \textstyle\sum_{j=0}^{n_h} P_j)\oplus R^{(\infty)},$$

then $\sigma(V)$ is not contained in $\sigma(M_\varepsilon)$. But, on the other hand, it is not difficult to check that $V \simeq T\oplus[\oplus_{j=1}^{h} L_j]+C_h$, where C_h is compact and $\|C_h\| \le 2\|K_{n_h}\| < \delta_h$. Hence, by our definition of δ_h, $\sigma(V) \subset \sigma(M_{k+1})$ $\sigma(M_\varepsilon)$, a contradiction.

Hence, $\sigma(S_\varepsilon) = \sigma_e(S_\varepsilon) = \sigma(M_\varepsilon)$, whence it follows that $T = S_\varepsilon+(\mu-M_\varepsilon) \in N(H)^- +[N(H)^- \cap Nor(H)]$.

Finally, observe that if $1 = Q_1+Q_2+C$, where $\sigma_e(Q_1)\cup\sigma_e(Q_2) \subset D(0,\frac{1}{2})$ and C is compact, then $1-Q_2 = Q_1+C$ and therefore $\tilde{1}-\tilde{Q}_2 = \tilde{Q}_1$, so that $\sigma_e(Q_1)=\sigma_e(1-Q_2) = \sigma_e(Q_1)\cap\sigma_e(1-Q_2)\subset D(0,\frac{1}{2})\cap D(1,\frac{1}{2}) = \emptyset$, a contradiction. Hence, $1 \notin N(H)+N(H)+K(H)$. □

A simpler construction yields the following

PROPOSITION 5.16. *For each* $T \in L(H)$, $\operatorname{dist}[T,N(H)] \le \operatorname{sp}(T)$.

PROOF. Let $\varepsilon > 0$ and let $\{\mu_n\}_{1\le n\le m}$ $(1 \le m \le \infty)$ be a dense subset of $\sigma_e(T)$ $(\subset \sigma_{\ell re}(T))$. By Proposition 4.29, there exists $K \in K(H)$, $\|K\| < \varepsilon/2$, and a decomposition $H = H_0\oplus H_1$ (H_0,H_1 are infinite dimensional spaces) such that

$T = A\oplus R+K$, $A \in L(H_0)$, $R \in L(H_1)$, $A \simeq T$, $\sigma(R) = \sigma_e(T)$ and

$$R = \begin{pmatrix} N & * & * \\ 0 & B & * \\ 0 & 0 & N \end{pmatrix},$$

where $N = (\operatorname{diag}\{\mu_1,\mu_2,\ldots,\mu_n,\ldots\})^{(\infty)} = \oplus_{1\le n\le m} \mu_n 1_n$ (1_n denotes the identity on a subspace $R_n \simeq H_1$).

By Lemma 5.4 we can find normal operators $M_n \in L(R_n)$ such that $\sigma(M_n) = \sigma_e(M_n) = \{t\mu_n: \ 0 \le t \le 1\}_{\varepsilon/4}$ and $\|\mu_n 1_n-M_n\| < |\mu_n|+\varepsilon/2$.

Let $M = \oplus_{1\le n\le m} M_n$; clearly,

$$\sigma(M) = \sigma_e(M) = [\cup_{1\le n\le m} \sigma(M_n)]^- = \cup_{\mu\in\sigma(T)} \{t\mu: \ 0 \le t \le 1\}_{\varepsilon/4}$$

is connected and contains the origin. Thus, if S is the operator obtained from R by replacing the two N's by M's, then $\sigma(S) = \sigma_{\ell re}(S) = \sigma(M)$ and therefore (by Theorem 5.1) $S \in N(H_1)^-$.

It is easily seen that $\Sigma = \sigma(A)\backslash\sigma(S) = \sigma(T)\backslash\sigma(M)$ is a finite (eventually empty) subset $\{\lambda_1,\lambda_2,\ldots,\lambda_p\}$ of $\sigma_0(A)$. Hence,

$$A = \begin{pmatrix} A|H(\Sigma;A) & * \\ 0 & A_0 \end{pmatrix} \begin{matrix} H(\Sigma;A) \\ H_0\ominus H(\Sigma;A) \end{matrix},$$

where $H(\Sigma;A)$ is finite dimensional and $\sigma(A_0) \subset \sigma(S)$. It readily follows that $A|H(\Sigma;A) = L+Q$, where L is a normal diagonal operator such that $\|L\| = \max\{|\lambda_j|: 1 \le j \le p\} \le sp(T)$ and Q is a nilpotent.

Since $\sigma(A_o) \subset \sigma(S)$, it follows that

$$\sigma((A-L\oplus 0)\oplus S) = \sigma(Q)\cup\sigma(A_o)\cup\sigma(S) \subset \sigma(S).$$

Since $S \in N(H)^-$, the above inclusion implies that $T_1 = (A-L\oplus 0)\oplus S \in N(H)^-$ (Use Theorem 5.1). On the other hand, by construction

$$\|T-T_1\| = \max\{\|L\|, \|R-S\| + \|K\|\} < sp(T)+\varepsilon.$$

Since ε can be chosen arbitrarily small, we conclude that

$$\operatorname{dist}[T,N(H)] \le sp(T).$$

It readily follows from Theorem 5.1 that the upper estimate given by Proposition 5.16 is very poor, in general; namely, if H is hermitian, $\sigma(H) = [0,1]$ and $\varepsilon > 0$, then there exist nilpotent operators Q_1 and Q_2 such that $\|H-Q_1\| < \varepsilon$ and $\|H-(1-Q_2)\| < \varepsilon$. Hence, $\sigma(1-Q_2) = \{1\}$ and $sp(1-Q_2) = 1$, but $\operatorname{dist}[1-Q_2, N(H)] < 2\varepsilon$!

Better estimates will be given in Chapter XI (second monograph).

5.6 The Apostol-Salinas approach: An estimate for the distance to $N_k(H)$

LEMMA 5.17. *Let* $T \in N(H)$. *Then for every* $\alpha > 0$, $\beta > sp(T)$ *and every positive integer* k *there exists* $Q \in N_k(H\oplus H)$ *such that*

$$\|(T\oplus 0) - Q\| < \alpha\|T\| + \beta + \frac{\|T^k\|}{\alpha\beta^{k-1}}.$$

PROOF. Let U and V be two isometries in $L(H)$ such that $VV^*+UU^* = 1$. We define for $1 \le j \le k$ the subspace M_j of $H\oplus H$ given by $M_j = \{T^{j-1}x\oplus \alpha\beta^{j-1}V^{j-1}Ux: x \in H\}$. (Note that

$$M_j = \begin{pmatrix} 1 & 0 \\ 0 & V^{j-1}U \end{pmatrix}\{(1/\alpha\beta^{j-1})T^{j-1}x\oplus x: x \in H\},$$

i.e., M_j is the image under an isometry of the graph of the transpose of $(1/\alpha\beta^{j-1})T^{j-1}$. Hence M_j is closed, i.e., it is a subspace.)

It is easy to see that $M_j\cap M_h = \{0\}$ if $1 \le j,h \le k$, $j \ne h$. This is due to the fact that the second components of the elements in M_j and M_h are orthogonal. Let $M = \sum_{j=1}^k M_j$ and let $\{y_m\}_{m=1}^\infty$ be a sequence in M such that $\lim(m\to\infty)\|y_m\| = 0$. Since $M_j\cap M_h = \{0\}$, $j \ne h$, we can write uniquely $y_m = \sum_{j=1}^k y_{m,j}$, where $y_{m,j} = T^{j-1}x_{m,j}\oplus\alpha\beta^{j-1}V^{j-1}Ux_{m,j}$, for some $x_{m,j} \in H$, $1 \le j \le k$, $m = 1,2,\ldots$. Since $\lim(m\to\infty)\ \|y_m\| = 0$, $\lim(m\to\infty) \|\sum_{j=1}^k \alpha\beta^{j-1}V^{j-1}Ux_{m,j}\| = 0$. Then $\lim(m\to\infty)\|V^{j-1}Ux_{m,j}\| = 0$, $1 \le j \le k$, and hence $\lim(m\to\infty)\|x_{m,j}\| = 0$, $1 \le j \le k$. Therefore, $\lim(m\to\infty)\|y_{m,j}\|$

$= 0$, $1 \leq j \leq k$, and hence the algebraic direct sum M is closed, i.e., it is a subspace. Now define $Q \in L(H \oplus H)$ by

$$Q | M_k \oplus M^{\perp} = 0, \quad Q | \sum_{j=1}^{k-1} M_j = (T \oplus \beta V) | \sum_{j=1}^{k-1} M_j.$$

Thus, the representing matrix of $Q|M$ on $\sum_{j=1}^{k} M_j$ is of the form

$$\begin{pmatrix} 0 & 0 & 0 & . & . & . & 0 & 0 \\ * & 0 & 0 & . & . & . & 0 & 0 \\ 0 & * & 0 & . & . & . & 0 & 0 \\ . & . & . & . & & & . & . \\ . & . & . & . & & & . & . \\ . & . & . & . & & & . & . \\ 0 & 0 & 0 & . & . & . & 0 & 0 \\ 0 & 0 & 0 & . & . & . & * & 0 \end{pmatrix}.$$

Therefore it is clear that $Q^k = 0$. Let P_k be the projection onto $\ker V^{*k}$. Then

$$\begin{aligned} &[(T \oplus \beta P_k V) - Q] \sum_{j=1}^{k} (T^{j-1} x_j \oplus \alpha\beta^{j-1} V^{j-1} U x_j) \\ &\quad = [T \sum_{j=1}^{k} T^{j-1} x_j - T \sum_{j=1}^{k-1} T^{j-1} x_j] \oplus [\beta P_k V \sum_{j=1}^{k} \alpha\beta^{j-1} V^{j-1} U x_j \\ &\quad\quad - \beta V \sum_{j=1}^{k} \alpha\beta^{j-1} V^{j-1} U x_j] = T^k x_k \oplus [P_k \alpha\beta^k V^k U x_k + \sum_{j=1}^{k-1} \alpha\beta^{j-1} V^{j-1} U x_j] \\ &\quad = T^k x_k \oplus 0. \end{aligned}$$

Since

$$\begin{aligned} \|x_k\| &= \frac{1}{\alpha\beta^{k-1}} \|\alpha\beta^{k-1} V^{k-1} U x_k\| \leq \frac{1}{\alpha\beta^{k-1}} \|\sum_{j=1}^{k} \alpha\beta^{j-1} V^{j-1} U x_j\| \\ &\leq \frac{1}{\alpha\beta^{k-1}} \|\sum_{j=1}^{k} T^j x_j \oplus \alpha\beta^{j-1} V^{j-1} U x_j\|, \end{aligned}$$

we conclude that $\|[(T \oplus \beta P_k V) - Q] | M\| \leq \|T^k\| / \alpha\beta^{k-1}$. Hence

$$\begin{aligned} \|(T \oplus 0) - Q\| &\leq \|(T \oplus 0) P_{M^{\perp}}\| + \|[(T \oplus 0) - Q] P_M\| \leq \|T\| \cdot \|P_{H \oplus \{0\}} P_{M^{\perp}}\| \\ &\quad + \|[(T \oplus \beta P_k V) - Q] P_M\| + \|0 \oplus \beta P_k V\| \leq \|T\| \cdot \|P_{H \oplus \{0\}} P_{M_1^{\perp}}\| + \beta + \|T^k\| / \alpha\beta^{k-1}. \end{aligned}$$

Thus in order to complete the proof it suffices to show that $\|P_{H \oplus \{0\}} P_{M_1^{\perp}}\| \leq \alpha$. Notice that $M_1^{\perp} = \{y \oplus z : \langle y, x \rangle + \alpha \langle z, Ux \rangle = 0 \text{ for all } x \in H\}$. Hence $M_1^{\perp} = \{(-\alpha U^*) z \oplus z : z \in H\}$ and therefore $\|P_{H \oplus \{0\}} P_{M_1^{\perp}}\| = \|P_{H \oplus \{0\}} | M_1^{\perp}\| \leq \alpha$. □

THEOREM 5.18. *Let* $T \in L(H)$ *and suppose that* $0 \in \sigma_{\ell re}(T)$. *Then for every* $\alpha > 0$, $\beta > sp(T)$, $\gamma > 1$ *and every positive integer* k *there exists* $Q \in N_{2k}(H)$ *such that*

$$\|T - Q\| \leq \gamma (\alpha \|T\| + \beta + \frac{\|T^k\|}{\alpha\beta^{k-1}}). \tag{5.7}$$

PROOF. Let $0 < \varepsilon < \min\{\alpha\beta, (\gamma - 1)\beta/(1+\alpha)\}$. Since $0 \in \sigma_{\ell re}(T)$, we can proceed as in the proof of Theorem 5.1 (formula (5.4)) in order to find $K \in K(H)$, $\|K\| < \varepsilon$ such that

$$T - K = \begin{pmatrix} 0 & 0 & C_1 & 0 \\ 0 & R_2 & 0 & C_2 \\ 0 & 0 & R_1 & 0 \\ 0 & 0 & 0 & 0 \end{pmatrix} \begin{matrix} H_2 \\ \\ H_1 \\ \end{matrix},$$

where $\sigma(R_1) \cup \sigma(R_2) \subset \sigma(T)$. Furthermore, if ε is small enough, then $\max\{\|R_1^k\|, \|R_2^k\|\} \le \|(T-K)^k\| < \gamma\|T^k\|$.

By Lemma 5.17 we can find $Q_j \in N_k(H_j)$, $j = 1,2$, such that

$$\|(R_j \oplus 0) - Q_j\| < \alpha\|R_j\| + \beta + \|R_j^k\|/\alpha\beta^{k-1} < \alpha(\|T\| + \|K\|) + \beta + \gamma\|T^k\|/\alpha\beta^{k-1}$$
$$< \gamma(\alpha\|T\| + \beta + \|T^k\|/\alpha\beta^{k-1}) - \alpha\beta, \quad j = 1,2.$$

Thus, if

$$Q = \begin{pmatrix} Q_2 & C_1 \oplus C_2 \\ 0 & Q_1 \end{pmatrix} \begin{matrix} H_2 \\ H_1 \end{matrix},$$

then $Q^{2k} = 0$ and

$$\|T - Q\| \le \max\{\|(R_j \oplus 0) - Q_j\| : \ j = 1,2\} + \|K\| < \gamma(\alpha\|T\| + \beta + \|T^k\|/\alpha\beta^{k-1}).$$

□

The above theorem is especially useful to estimate the distance from T to $N_k(H)$ for the case when T is a quasinilpotent. Namely, if $\sigma(T) = \{0\}$ and $\|T^k\|^{1/k}$ does not decrease very fast, then there exists a sequence $\{Q_n\}_{n=1}^{\infty}$ in $N(H)$ where Q_n has order k_n ($k_n \to \infty$, as $n \to \infty$) such that the rate of decrease of the sequence $\|T^{k_n}\|^{1/k_n}$ is the same as the rate of decrease of the sequence $\|T - Q_n\|$.

COROLLARY 5.19. *Let* $T \in L(H)$ *be a quasinilpotent operator and let* $\delta > 1$. *Then there exists a sequence* $\{Q_k\}_{k=1}^{\infty}$ *in* $N(H)$ *such that* $Q_k^{2k} = 0$ *and for some constant* $c = c(T,\delta) > 0$,

$$\|T - Q_k\| \le c(\|T\|\delta^{-k} + \|T^k\|^{1/k}).$$

PROOF. It is a direct consequence of Theorem 5.18: Take $\alpha = \delta^{-k}$ and $\beta = \delta\|T^k\|^{1/k}$ in (5.7). □

If $k \ge 2$ and we choose $\beta = [(k-1)^2\|T\|.\|T^k\|]^{1/(k+1)}$ and $\alpha = \beta[(k-1)\|T\|]^{-1}$ in (5.7), then the left side of this formula takes the form

$$\gamma C(k)(\|T\|.\|T^k\|)^{1/(k+1)},$$

where $C(k) = (k+1)(k-1)^{-(k-1)/(k+1)}$. Observe that $1 < C(k) \le 3$ for all $k \ge 2$ and $C(k) \to 1$ ($k \to \infty$). Hence,

COROLLARY 5.20. *Let* $T \in L(H)$ *be a quasinilpotent operator; then*

$$\operatorname{dist}[T, N_{2k}(H)] \le C(k)(\|T\|.\|T^k\|)^{1/(k+1)} \quad \textit{for all } k \ge 2.$$

In the converse direction, we have the following

PROPOSITION 5.21. *Let* $T \in L(H)$ *and suppose that there exists a sequence* $\{Q_n\}_{n=1}^{\infty}$ *in* $N(H)$ *such that* $Q_n^{k_n} = 0$, $n = 1,2,\ldots$ *and*

$$\liminf(n \to \infty)\|T - Q_n\|^{1/k_n} = 0.$$

Then T *is quasinilpotent.*

PROOF. Since $Q_n^{k_n} = 0$, it follows as in the proof of Proposition 1.10(i) that

$$\|T\|^{1/k_n} \le \|T-Q_n\|^{1/k_n}\{\sum_{j=0}^{k_n-1} \|T\|^{k_n-j-1} (\|T\| + \|T - Q_n\|)^j\}^{1/k_n}$$

$$\le (\|T\|^{1+1/k_n})\|T - Q_n\|^{1/k_n} \to 0 \ (n \to \infty).$$ □

REMARKS 5.22(i) Let T, α, β and k be as in Lemma 5.17 and let $R \in N_m(H_o)$, where H_o is an infinite dimensional space. Identify $H_1 \simeq H$ with $\ker R \subset H_o$ and let $H_2 = H_o \ominus \ker R$. If

$$R = \begin{pmatrix} 0 & R_{12} & R_{13} & \cdot & \cdot & \cdot & R_{1,m-1} & R_{1m} \\ & 0 & R_{23} & \cdot & \cdot & \cdot & R_{2,m-1} & R_{2m} \\ & & 0 & \cdot & \cdot & \cdot & R_{3,m-1} & R_{3m} \\ & & & \cdot & & & \cdot & \cdot \\ & & & & \cdot & & \cdot & \cdot \\ & & 0 & & & \cdot & \cdot & \cdot \\ & & & & & & 0 & R_{m-1,m} \\ & & & & & & & 0 \end{pmatrix} \begin{matrix} \ker R^1 \ominus \ker R^0 = H_1 \\ \ker R^2 \ominus \ker R^1 \\ \ker R^3 \ominus \ker R^2 \\ \cdot \\ \cdot \\ \cdot \\ \ker R^{m-1} \ominus \ker R^{m-2} \\ \ker R^m \ominus \ker R^{m-1} \end{matrix}$$

and $Q \in N_k(H \oplus H_1)$ is constructed as in that lemma, then

$$L = \begin{pmatrix} Q & R'_{12} & R'_{13} & \cdot & \cdot & \cdot & R'_{1,m-1} & R'_{1m} \\ & 0 & R_{23} & \cdot & \cdot & \cdot & R_{2,m-1} & R_{2m} \\ & & 0 & \cdot & \cdot & \cdot & R_{3,m-1} & R_{3m} \\ & & & \cdot & & & \cdot & \cdot \\ & & & & \cdot & & \cdot & \cdot \\ & & 0 & & & \cdot & \cdot & \cdot \\ & & & & & & 0 & R_{m-1,m} \\ & & & & & & & 0 \end{pmatrix} \begin{matrix} H \oplus H_1 \\ \ker R^2 \ominus \ker R^1 \\ \ker R^3 \ominus \ker R^2 \\ \cdot \\ \cdot \\ \cdot \\ \ker R^{m-1} \ominus \ker R^{m-1} \\ \ker R^m \ominus \ker R^{m-1} \end{matrix},$$

where $R'_{1j} \simeq \begin{pmatrix} 0 \\ R_{1j} \end{pmatrix}$ maps $\ker R^j \ominus \ker R^{j-1}$ into $H \oplus H_1$, is a nilpotent of order at most $k+m-1$ acting on $H \oplus H_o$ and

$$\|(T \oplus R) - L\| < \alpha\|T\| + \beta + \frac{\|T^k\|}{\alpha\beta^{k-1}}.$$

(ii) Assume that $T \in L(\mathbb{C}^d)$, $0 < d < \infty$, $R \in N_m(\mathbb{C}^p)$, $\alpha > 0$ and $\beta > sp(T)$, and let k be a positive integer. Then we can apply the above construction to $T \oplus 0$ and $R^{(s)} \oplus 0$ (where $1 \le s < \infty$ and 0 acts on an infinite dimensional space). If $s = (k-1)d$, then the operator L admits a reducing subspace N of dimension $[(k-1)p+1]d$ such that $L|N^\perp = 0$ and

$$\max\{\|x - P_N x\| : x \in \mathbb{C}^d, \|x\| = 1\}$$

is small enough to guarantee the existence of a unitary operator U mapping $P_N(\mathbb{C}^d)$ onto $\mathbb{C}^d$ such that $\|U - 1\| < \alpha\sqrt{2}$. If $M = UN$, then $M \supset \mathbb{C}^d$, $\dim M = [(k-1)p+1]d$ and there exists $L' \in N_{k+m-1}(M)$, $L' \simeq L|N$ such that

$$\|Q \oplus R^{(k-1)} - L'\| < 2\alpha\sqrt{2}\,\|T\| + (1+2\alpha\sqrt{2})\{\alpha\|T\| + \beta + \frac{\|T^k\|}{\alpha\beta^{k-1}}\} .$$

Furthermore, if nul $R = n \geq 2$, then M can be chosen so that

$$\dim M = d + \left[\frac{(k-1)d+1}{n}\right]p,$$

where $[r]$ denotes the integral part of the real number r.

The details of the construction can be found in [149, Lemma 5.10] and [150] (Lemma 2.3 and remark following it).

5.7 Salinas' pseudonilpotents

We shall say that $T \in L(H)$ is a *pseudonilpotent operator* if for every $\varepsilon > 0$ there exists a decomposition $H = \oplus_{j=1}^{n} H_j$ of H into the direct sum of a finite orthogonal family of subspaces $H_1, H_2, \ldots, H_n$ such that if

$$T = \begin{pmatrix} T_{11} & T_{12} & \cdot & \cdot & \cdot & T_{1,n-1} & T_{1n} \\ T_{21} & T_{22} & \cdot & \cdot & \cdot & T_{2,n-1} & T_{2n} \\ \cdot & \cdot & & & & \cdot & \cdot \\ \cdot & \cdot & & & & \cdot & \cdot \\ \cdot & \cdot & & & & \cdot & \cdot \\ T_{n-1,1} & T_{n-1,2} & \cdot & \cdot & \cdot & T_{n-1,n-1} & T_{n-1,n} \\ T_{n1} & T_{n2} & \cdot & \cdot & \cdot & T_{n,n-1} & T_{nn} \end{pmatrix} \begin{matrix} H_1 \\ H_2 \\ \cdot \\ \cdot \\ \cdot \\ H_{n-1} \\ H_n \end{matrix},$$

then the norm of

$$T_{lower} = \begin{pmatrix} T_{11} & & & & & & \\ T_{21} & T_{22} & & & & 0 & \\ \cdot & \cdot & \cdot & & & & \\ \cdot & \cdot & & \cdot & & & \\ \cdot & \cdot & & & \cdot & & \\ T_{n-1,1} & T_{n-1,2} & & \cdot & \cdot & T_{n-1,n-1} & \\ T_{n1} & T_{n2} & \cdot & \cdot & \cdot & T_{n,n-1} & T_{nn} \end{pmatrix} \begin{matrix} H_1 \\ H_2 \\ \cdot \\ \cdot \\ \cdot \\ H_{n-1} \\ H_n \end{matrix}$$

is less than ε.

It is clear from the definition that a pseunilpotent operator is an operator that can be "obviously" approximated by nilpotents. On the other hand, if Q is a nilpotent of order n, then it is completely apparent (see (5.1)) that Q admits a decomposition of the above type with $Q_{lower} = 0$. Thus, if we denote by $\Psi(H)$ the set of all pseudonilpotents, then

$$N(H) \subset \Psi(H) \subset N(H)^- \tag{5.8}$$

We shall see immediately that both inclusions are proper.

By interchanging the subspaces $H_1, H_2, \ldots, H_n$ in the definition of pseudonilpotency it is easy to see that an operator T is in $\Psi(H)$ if and only if for every $\varepsilon > 0$ there exists a decomposition $H = \oplus_{j=1}^{n} H_j$ such that the norm of T_{upper} is smaller than ε. (T_{upper} is the triangular operator matrix with (i,j)-entries T_{ij} for $i > j$ and 0 for $i \le j$.) It readily follows that $\Psi(H) = \Psi(H)^* = (\text{def}) \{T \in L(H): T^* \in \Psi(H)\}$.

THEOREM 5.23. *Let* $A = H+K$, *where* H *is hermitian and* K *is compact; then* $A \in \Psi(H)$ *if and only if* A *is a compact quasinilpotent.*

PROOF. Suppose that $A = H+K \in \Psi(H)$ is not compact and let $0 < \varepsilon < \|\tilde{A}\|/3$. Let $H = \oplus_{j=1}^{n} H_j$ be a decomposition of the space with respect to which $\|A_{lower}\| < \varepsilon$; then $\|\tilde{H}_{lower}\| = \|\tilde{A}_{lower}\| \le \|A_{lower}\| < \varepsilon$.

Since H is hermitian, $\|\tilde{H}_{upper}\| = \|\tilde{H}_{lower}\| < \varepsilon$. Let $H_{diag} = H_{11} \oplus H_{22} \oplus \ldots \oplus H_{nn}$ be the diagonal of H with respect to the above decomposition. It is easily seen that $\|\tilde{H}_{diag}\| \le \|\tilde{H}_{lower}\| < \varepsilon$ and, a fortiori,

$$\|\tilde{A}\| = \|\tilde{H}\| = \|\tilde{H}_{lower} + \tilde{H}_{upper} - \tilde{H}_{diag}\| < 3\varepsilon.$$

Since ε can be chosen arbitrarily small, we conclude that $\|\tilde{A}\| = 0$, i.e., A is compact, a contradiction.

On the other side, if $A \in K(H) \cap \Psi(H)$, then $A \in N(H)^-$ and therefore it must be a compact quasinilpotent (Theorem 5.1).

Conversely, if A is a compact quasinilpotent and $\{P_n\}_{n=1}^{\infty}$ is an increasing sequence of finite rank orthogonal projections such that $P_n \to 1$ (strongly, $n \to \infty$), then $\|A - P_n A\| \to 0$ $(n \to \infty)$. From the upper semicontinuity of the spectrum (Corollary 1.2(ii)), given $\varepsilon > 0$ there exists a positive integer n_0 such that if $n > n_0$, then $\sigma(P_n A P_n) \subset D(0, \varepsilon/3)$. Let $m > n_0$ so that $\|A - P_m A\| < \varepsilon/3$. Since ran P_m is finite dimensional there exists a basis $e_1, e_2, \ldots, e_k$ of $P_m H$ on which the representing matrix of the operator $P_m A|\text{ran } P_m$ is in upper triangular form. Observe that the diagonal entries of this matrix are in absolute value less than $\varepsilon/3$. Letting $H_j = \bigvee\{e_j\}$, $j = 1, 2, \ldots, k$, and defining $H_{k+1} = \text{ran}(1 - P_m)$, we deduce that with respect to the decomposition $H = \oplus_{j=1}^{k+1} H_j$, $\|A_{lower}\| < \varepsilon$. Since ε is arbitrary we conclude that $A \in \Psi(H)$. □

It follows from Theorems 5.1 and 5.23 that if H is hermitian and $\sigma(H) = [0,1]$, then $H \in N(H)^- \setminus \Psi(H)$. On the other hand, if $B \in L(H)$ is the unilateral weighted shift, with respect to some ONB $\{e_n\}_{n=1}^{\infty}$, defined by $Be_n = \alpha_n e_{n+1}$, $n = 1,2,\ldots$, $\alpha_n \to 0$ $(n \to \infty)$ then B is a compact quasinilpotent, but $B^k \neq 0$ for all $k = 1,2,\ldots$, so that $B \in \Psi(N) \setminus N(H)$. Hence, we have

COROLLARY 5.24. $N(H) \subsetneq \Psi(H) \subsetneq N(H)^-$.

The following result provides a simple way to construct operators in $\Psi(H)$ with positive spectral radius.

LEMMA 5.25. *Let* $T \in L(H)$ *and suppose that there exists a sequence* $\{Q_n\}_{n=1}^{\infty}$ *in* $N(H)$ *such that* $Q_n^{k_n} = 0$ *and*

$$\liminf(n \to \infty) \sqrt{k_n}\|T - Q_n\| = 0.$$

Then $T \in \Psi(H)$.

PROOF. Given $\varepsilon > 0$, let n be sufficiently large so that $\sqrt{k_n}\|T-Q_n\| < \varepsilon$. For each j, $1 \le j \le k_n$, let P_j be the orthogonal projection of H onto $\ker Q_n^j \ominus \ker Q_n^{j-1}$. Also let $x \in H$; then

$$\|\textstyle\sum_{j=1}^{k_n}\sum_{i=j}^{k_n} P_i T P_j x\| = \|\sum_{j=1}^{k_n}\sum_{i=j}^{k_n} P_i (T-Q_n) P_j x\|$$

$$\le \textstyle\sum_{j=1}^{k_n} \|(\sum_{i=j}^{k_n} P_i)(T-Q_n)P_j\| \cdot \|P_j x\|$$

$$\le \|T-Q_n\| \textstyle\sum_{j=1}^{k_n}\|P_j x\| \le \|T-Q_n\|\sqrt{k_n}\|x\|.$$

Therefore $\|\sum_{j=1}^{k_n}\sum_{i=j}^{k_n} P_i T P_j\| \le \sqrt{k_n}\|T-Q_n\| < \varepsilon$, as desired. □

COROLLARY 5.26. *Let* $q_k \in L(\mathbb{C}^k)$ *(*$k = 1,2,\ldots$*) be defined as in* (2.1) *and let* S *be the unilateral shift of multiplicity one. If either* $\alpha = \beta = \infty$ *or* $\alpha_n \neq 0$ *(*$0 \le \alpha_n \le \infty$*) for infinitely many* n's, *then the operator*

$$T = \{\oplus_{n=1}^{\infty} q_n^{(\alpha_n)}\} \oplus S^{(\alpha)} \oplus S^{*(\beta)}$$

is a pseudonilpotent with spectral radius $sp(T) = \|T\| = 1$.

PROOF. It is clear that $sp(T) = \|T\| = 1$. On the other hand, by Corollary 2.24 (or minor modifications of its proof), there exist operators $Q_k \simeq q_k^{(\infty)} \oplus \{\oplus_{n=1}^{k-1} q_n^{(\alpha_n)}\}$ such that $\|T-Q_k\| < 4/k$ $(k = 1,2,\ldots)$. Thus $\sqrt{k}\|T - Q_k\| \to 0$ $(k \to \infty)$ and since also $Q_k^k = 0$, Lemma 5.25 implies $T \in \Psi(H)$. □

PROPOSITION 5.27. *(i) Let* $T \in L(H)$ *be such that*

$$\liminf(n \to \infty)\ k\|T^k\|^{1/k} = 0.$$

Then T *is pseudonilpotent and quasinilpotent.*

(ii) However, given a sequence $\Sigma = \{c_k\}_{k=1}^{\infty}$ *of positive reals increasing to* ∞ *arbitrarily slowly, there exists a compact quasinilpotent unilateral weighted shift* $B = B_\Sigma$ *in* $\Psi(H)$ *such that*

$$\lim(k \to \infty)\ c_k \|B^k\|^{1/k} = \infty.$$

PROOF. The first statement is an immediate consequence of Corollary 5.19 and Lemma 5.25.

Define B as in the proof of Corollary 5.24 with weights $\alpha_1 = \alpha_2 = \dots = \alpha_{n_1} = 1$, $\alpha_{n_1+1} = \dots = \alpha_{n_2} = 1/2, \dots, \alpha_{n_{j-1}+1} = \dots = \alpha_{n_j} = 1/j, \dots$, where $\{n_j\}_{j=1}^{\infty}$ is a strictly increasing sequence of natural numbers. Given the sequence $\{c_k\}_{k=1}^{\infty}$, it is not difficult to define the sequence $\{n_j\}_{j=1}^{\infty}$ inductively so that $\lim(j \to \infty) c_{n_{j-1}}/j = \infty$. It follows that for $j > 1$, if $n_{j-1} < k \le n_j$, then

$$c_k \|B^k\|^{1/k} > c_{n_{j-1}} \|B^{n_j}\|^{1/n_j} > c_{n_{j-1}}/j.$$

Since B is a compact quasinilpotent, $B \in \Psi(H)$ (Theorem 5.23) but

$$\lim(k \to \infty)\ c_k \|B^k\|^{1/k} = \infty.$$

REMARK 5.28. Let $B = B_\Sigma$ be as in Proposition 5.27 and let $A = B^{(\infty)}$, then $A \in \Psi(H)$ and $\lim(k \to \infty)\ c_k \|A^k\|^{1/k} = \infty$, but A^k is compact for no value of k.

Very little is known about the structure of the pseudonilpotent operators. In [149,Lemma 5.1], D. A. Herrero proved that if H is hermitian and $\sigma(H) = [-1,1]$, then

$$\operatorname{dist}[H, N_k(H)] \le 5/\log k$$

for all k large enough. On the other hand, Theorem 5.23 and Lemma 5.25 imply that

$$\operatorname{dist}[H, N_k(H)] \ne o(1/\sqrt{k}).$$

But the gap between $O(1/\log k)$ and $o(1/\sqrt{k})$ is still very large. (See also the results of Section 2.3.3.)

The following proposition follows immediately from the definition of pseudonilpotency.

PROPOSITION 5.29. *(i) If* $A \in L(H)$ *has the form of Lemma 3.36 and* $A_{jj} \in \Psi(H_j)$ *for all* $j = 1,2,\dots,k$, *then* $A \in \Psi(H)$.

(ii) If $A \in \Psi(H)$, *then* $U(A)^{-} \subset \Psi(H)$.

COROLLARY 5.30. *If* N *is normal and* $\sigma(N)$ *is equal to the closed unit disk, then* N *is a pseudonilpotent.*

PROOF. Let M be the normal operator defined in the proof of Proposition 2.28(ii). Since the matrix of $(L_k - Q_k)_{upper}$ has no more than four non-zero entries in each column or row, a straightforward computation shows that if $M = (M_{ij})_{i,j=1}^{k}$ with respect to the decomposition $H = \oplus_{j=1}^{k} H_j$, where H_j is the subspace defined by

$$H_j = \ker[(\oplus_{m=1}^{\infty} \lambda_m Q_k)^{(\infty)}]^j \ominus \ker[(\oplus_{m=1}^{\infty} \lambda_m Q_k)^{(\infty)}]^{j-1}$$

$(j = 1,2,\ldots,k)$, then $\|M_{upper}\| < 20(\pi/k)^{1/2}$, whence we conclude that M is a pseudonilpotent.

Now the result follows from Lemma 4.8 and Proposition 5.29(ii) □

Theorem 5.23 and Corollary 5.30 suggest the following

PROBLEM 5.31. Which normal operators are pseudonilpotents?

PROBLEM 5.32. Let $A \in \Psi(H)$.

(i) If $W \in G(H)$, is $WAW^{-1} \in \Psi(H)$?

(ii) Is $Ap(A) \in \Psi(H)$ for all polynomials p?

PROBLEM 5.33. Is every quasinilpotent a pseudonilpotent?

5.8 Limits of nilpotent and algebraic elements in the Calkin algebra

Most of the results of Sections 5.1 through 5.7 induce analogous results in the Calkin algebra; namely, we have

THEOREM 5.34. *The closure of the set* $N(A(H))$ *of all nilpotent elements of the Calkin algebra coincides with the set of all those* $\tilde{A} \in A(H)$ *satisfying the conditions:*

(ii') $\sigma_e(A)$ *is connected and contains the origin; and*

(iii') $\operatorname{ind}(\lambda - A) = 0$ *for all* $\lambda \in \rho_{s-F}(A)$.

Moreover, $N(A(H))^- = \pi[N(H)]^- = \pi[N(H)^-]$ *and* $N(H)^- + K(H) = [N(H) + K(H)]^- = \pi^{-1}[N(A(H))^-]$ *is closed in* $L(H)$.

PROOF. The necessity of the conditions (ii') and (iii') follows exactly as in Theorem 5.1. On the other hand, if A satisfies (ii') and (iii'), then it follows from Theorem 3.48 that A+K satisfies the conditions (i), (ii) and (iii) of Theorem 5.1 for some K in $K(H)$. Therefore $A+K \in N(H)^-$ and, a fortiori, $\tilde{A} = \tilde{A}+\tilde{K} \in \pi[N(H)^-] \subset \pi[N(H)]^- \subset N(A(H))^-$, whence the results follow. □

Similarly, from Sections 5.2 through 5.6, we obtain the following corollaries.(All the proofs follow exactly the above scheme and will be omitted.) Of course, similarity orbits in $A(H)$ are defined with respect to the invertible group $G(A(H)) = \{\tilde{A}: A \in L(H)$ is a Fredholm operator$\}$.

THEOREM 5.35. *Let* $N \in L(H)$ *be a normal operator such that* $\sigma_e(N)$ *is a perfect set; then* $S(\tilde{N})^-$ *is the set of all those* $\tilde{A} \in A(H)$ *satisfying the following conditions:*

(ii') $\sigma_e(A) \supset \sigma_e(N)$ *and each component of* $\sigma_e(A)$ *intersects* $\sigma_e(N)$; *and*

(iii') $\mathrm{ind}(\lambda - A) = 0$ *for all* $\lambda \in \rho_{s-F}(A)$.

THEOREM 5.36. *(i) The closure of the set* $A\ell g(A(H))$ *of all algebraic elements of* $A(H)$ *coincides with the set of all those* $\tilde{A}$ *in* $A(H)$ *satisfying the condition* (5.6).

(ii) $A\ell g(A(H))^- = \{\tilde{R} \in A(H)$: R *is similar to a normal operator with finite essential spectrum*$\} = \pi[A\ell g(H)]^- = \pi[A\ell g(H)^-]$.

(iii) If Λ *is a finite nonempty subset of* $\mathbb{C}$ *and* $A(H;\Lambda) = \{\tilde{A} \in A(H): \sigma_e(A) \subset \Lambda\}$, *then* $A(H;\Lambda)^-$ *is the set of all those* $\tilde{T}$ *such that each component of* $\sigma_e(T)$ *intersects* Λ *and* T *satisfies* (5.6). *Furthermore,* $A(H;\Lambda)^- = \{\tilde{A} \in A(H;\Lambda): A$ *is algebraic*$\}^-$.

PROPOSITION 5.37. *If* $\tilde{T} \in A(H)$ *and* N *is a normal operator such that* $\sigma_e(N) = \sigma_W(T)$, *then* $\tilde{T} \underset{sim}{\rightarrow} \tilde{N}$.

THEOREM 5.38. *(i)* $N(A(H))^- + \{N(A(H)) \cap \pi[Nor(H)]\}^- = A(H)$; *however,* $\tilde{I} \notin N(A(H)) + N(A(H))$.

(ii) For each $\tilde{T} \in A(H)$, $\mathrm{dist}[\tilde{T}, N(A(H))] \le \mathrm{sp}(\tilde{T})$.

It will be shown later (Theorem 7.2) that if $\tilde{Q} \in N_k(A(H))$, then $\tilde{Q}$ can be lifted to an operator $Q \in N_k(H)$. By using this result, we have

THEOREM 5.39. *If* $\tilde{T} \in A(H)$ *and* $0 \in \sigma_{\ell re}(T)$, *then for every* $\alpha > 0$, $\beta > \mathrm{sp}(T)$, $\gamma > 1$ *and every positive integer* k *there exists* $\tilde{Q} \in N_{2k}(A(H))$ *such that*

$$\|\tilde{T} - \tilde{Q}\| \le \gamma\left(\alpha\|\tilde{T}\| + \beta + \frac{\|\tilde{T}^k\|}{\alpha\beta^{k-1}}\right).$$

Finally, observe that pseudonilpotency admits the following expression in $A(H)$: $\tilde{T} \in A(H)$ is a *pseudonilpotent* if for every $\varepsilon > 0$ there exists a decomposition $\tilde{I} = \sum_{j=1}^n \tilde{P}_j$, where $\{\tilde{P}_j\}_{j=1}^n$ is a finite family of pairwise orthogonal hermitian idempotents, such that

$$\|\sum_{j=1}^{n}\sum_{i=j}^{n} \tilde{P}_i\tilde{T}\tilde{P}_j\| < \varepsilon.$$

With this definition in mind and using the well-known fact that every hermitian element $\tilde{H}$ of $A(H)$ can be lifted to an hermitian operator H_o in $L(H)$ such that $\sigma_e(H_o) = \sigma(\tilde{H})$, we have the following

THEOREM 5.40. *(i)* $N(A(H)) \subsetneq \Psi(A(H)) \subsetneq N(A(H))^-$. *Indeed,* $\Psi(A(H))\setminus N(A(H))$ *contains quasinilpotents as well as elements* $\tilde{T}$ *such that* $\mathrm{sp}(\tilde{T}) = \|\tilde{T}\| = 1$*; if* $\tilde{H}$ *is a non-zero hermitian, then* $\tilde{H} \in N(A(H))^-\setminus\Psi(A(H))$.

(ii) If $\lim(n \to \infty)\sqrt{k_n}\|\tilde{T} - \tilde{Q}_n\| = 0$ *for a sequence* $\{\tilde{Q}_n\}_{n=1}^{\infty} \subset N(A(H))$ *such that* $\tilde{Q}_n^{k_n} = 0$, *then* $\tilde{T} \in \Psi(A(H))$.

(iii) If $\lim(k \to \infty)\sqrt{k}\|\tilde{T}^k\|^{1/k} = 0$, *then* $\tilde{T}$ *is pseudonilpotent and quasinilpotent.*

(iv) However, given a sequence $\Sigma = \{c_k\}_{k=1}^{\infty}$ *of positive reals increasing to* ∞ *arbitrarily slowly, there exists a quasinilpotent* $\tilde{B}$ *in* $\Psi(A(H))$ *such that* $\lim(k \to \infty)\sqrt{c_k}\|\tilde{B}^k\|^{1/k} = \infty$.

PROPOSITION 5.41. *If* $\tilde{A} \in \Psi(A(H))$*; then* $U(\tilde{A})^- \subset \Psi(A(H))$.

5.9 On the spectra of infinite direct sums of operators

THEOREM 5.42. *Let* $\{T_n\}_{n=1}^{\infty}$ $(T_n \in L(H_n),\ n = 1,2,\dots)$ *be a uniformly bounded family of operators and let* $T = \oplus_{n=1}^{\infty} T_n \in L(H)$, *where* $H = \oplus_{n=1}^{\infty} H_n$*; then*

(i) $\sigma_p(T) = \cup_{n=1}^{\infty} \sigma_p(T_n)$,

(ii) $\sigma(T) = [\cup_{n=1}^{\infty} \sigma(T_n)]\cup\sigma$ *(disjoint union), where* $\sigma = \{\lambda \notin \cup_{n=1}^{\infty} \sigma(T_n):\ \{\|(\lambda-T_n)^{-1}\|\}_{n=1}^{\infty}$ *is not uniformly bounded*$\}$,

(iii) moreover, $\sigma \subset \sigma_{\ell re}(T)$ *and each component of* σ^- *intersects* $[\cup_{n=1}^{\infty} \sigma(T_n)]^-$.

(iv) If there exists a function $\Phi(r)$ *defined on* $(0,\delta)$ *such that*

$$\|(\lambda-T_n)^{-1}\| \le \Phi(\mathrm{dist}[\lambda,\sigma(T_n)]) \tag{5.9}$$

for all $\lambda \in \sigma(T_n)$ *and all* $n = 1,2,\dots,$ *then* $\sigma(T) = [\cup_{n=1}^{\infty} \sigma(T_n)]^-$.

(v) The condition (5.9) *cannot be replaced by the weaker condition*

$$\|(\lambda-T_n)^{-1}\| = O(\Phi(\mathrm{dist}[\lambda,\sigma(T_n)])) \tag{5.10}$$

for all $\lambda \in \sigma(T_n)$ *and all* $n = 1,2,\dots,$ *even if* $\Phi(r)$ *has the minimal possible growth:* $\Phi(r) = 1/r$.

PROOF. (i) Let $x = \oplus_{n=1}^{\infty} x_n \in H$, $\lambda \in \mathbb{C}$ and assume that $(\lambda-T)x = 0$.

Then $0 = (\lambda-T)x = \oplus_{n=1}^{\infty} (\lambda-T_n)x_n$ so that either $x_n = 0$ or $\lambda \in \sigma_p(T_n)$ for each $n = 1,2,\ldots$, whence the result follows.

(ii) - (iii) The inclusion $\cup_{n=1}^{\infty} \sigma(T_n) \subset \sigma(T)$ is obvious. If $\lambda \in \sigma$, then there exists a sequence $\{x_{n_k}\}_{k=1}^{\infty}$, $x_{n_k} \in H_{n_k}$, $\|x_{n_k}\| = 1$, such that $\|(\lambda-T_{n_k})^{-1}x_{n_k}\| \to 0$ $(k \to \infty)$. Thus, if $y_k = (\oplus_{n=1}^{n_k-1} 0)\oplus x_{n_k} \oplus (\oplus_{n=n_k+1}^{\infty} 0)$, then $\|y_k\| = 1$ and $\|(\lambda-T)y_k\| \to 0$ $(k \to \infty)$. Hence, $\sigma \subset \sigma_\ell(T)$.

The same argument, applied to T^* indicates that $\sigma \subset \sigma_\ell(T^*)^* = \sigma_r(T)$. Since (by (i)) $\sigma \cap [\sigma_p(T) \cup \sigma_p(T^*)^*] = \emptyset$, we conclude that σ is included in $\sigma_{\ell re}(T)$.

Conversely, if $\|(\lambda-T_n)^{-1}\| \leq C$ (for some constant C depending only on λ), then it is easily seen that $\lambda \in \rho(T)$. Indeed, $\|(\lambda-T)^{-1}\| \leq C$. Hence, $\sigma(T) = [\cup_{n=1}^{\infty} \sigma(T_n)] \cup \sigma$.

Let Γ be a clopen subset of σ^- and assume that $\Gamma \cap [\cup_{n=1}^{\infty} \sigma(T_n)]^- = \emptyset$. Let Ω be a Cauchy domain containing Γ such that $\Omega^- \cap [\cup_{n=1}^{\infty} \sigma(T_n)]^- = \emptyset$; then there exists a constant C such that $\|(\lambda-T)^{-1}\| \leq C$ for all $\lambda \in \partial\Omega$. A fortiori, $\|(\lambda-T_n)^{-1}\| \leq C$ for all $\lambda \in \Omega^-$ and all $n = 1,2,\ldots$. Since $(\lambda-T_n)^{-1}$ is an analytic function defined (at least) in some neighborhood of Ω^-, it follows from the maximum modulus theorem that $\|(\lambda-T_n)^{-1}\| \leq C$ for all $\lambda \in \Omega^-$ and all $n = 1,2,\ldots$. It readily follows (as above) that $\lambda \in \rho(T)$ and $\|(\lambda-T)^{-1}\| \leq C$ for all $\lambda \in \Omega^-$, i.e., $\Gamma = \emptyset$.

Since σ^- is compact, it follows that each component of σ^- must intersect $[\cup_{n=1}^{\infty} \sigma(T_n)]^-$.

(iv) This follows immediately from (ii).

(v) Let H be an hermitian operator such that $\sigma(H) = [-1,1]$. By Theorem 5.8, $H \underset{\text{sim}}{\to} N$, where N is a normal operator such that $\sigma(N) = D(0,1)^-$. Let $\{T_n\}_{n=1}^{\infty}$, $T_n \sim H$ be such that $\|N-T_n\| \to 0$ $(n \to \infty)$ and let $T = \oplus_{n=1}^{\infty} T_n$. Since $\|(\lambda-H)^{-1}\| = 1/\text{dist}[\lambda,\sigma(H)]$ for all $\lambda \in \rho(H)$, it follows that the sequence $\{T_n\}_{n=1}^{\infty}$ satisfies the condition (5.10) with $\Phi(r) = 1/r$ and, moreover, $\sigma(T_n) = \sigma(H) = [-1,1]$ for all $n = 1,2,\ldots$. However, if $\lambda \in D(0,1)^-$, then

$$\lim(n \to \infty)\{\inf\|(\lambda-T_n)x_n\|: \ x_n \in H_n, \|x_n\| = 1\}$$
$$= \inf\{\|(\lambda-N)x\|: \ \|x\| = 1\} = 0,$$

so that

$$\sup\{\|(\lambda-T_n)^{-1}\|: \ n \geq 1\} = \infty$$

for all $\lambda \in D(0,1)^- \setminus [-1,1]$. Hence, $\sigma(T) = \sigma(N) = D(0,1)^-$. □

Now we are in a position to show that the result of Corollary 3.40 is the best possible. Indeed, we have

EXAMPLE 5.43. Given a nonempty finite or denumerable bounded set $d = \{\lambda_n\}_{n=1}^m$ and a compact set Γ such that every component of Γ intersects d^-, there exists a triangular operator A such that $d(A) = \sigma_p(A) = d$ and $\sigma(A) = \Gamma$. Furthermore, if d^- is a perfect set, then A can be chosen so that $\mathrm{nul}(\lambda_n - A) = 1$ for all $n = 1,2,\ldots$.

PROOF. Let N be a normal operator such that $\sigma(N) = \sigma_e(N) = \Gamma$. By Corollary 5.12 there exists a sequence $\{A_j\}_{j=1}^\infty$ of algebraic operators such that $\sigma(A_j) = \{\lambda_1,\lambda_2,\ldots,\lambda_j\}$ $(j = 1,2,\ldots)$ and $\|N-A_j\| \to 0$ $(j \to \infty)$.

Define $A = \oplus_{j=1}^\infty A_j$. By Theorem 5.42 (see, in particular, the proof of (v)) $\sigma_p(A) = \cup_{j=1}^\infty \sigma_p(A_j) = d$ and $\sigma(A) = d \cup \sigma$, where

$$\sigma = \{\lambda \notin d: \ \{\|(\lambda - A_j)^{-1}\|\}_{j=1}^\infty \text{ is not bounded below}\} = \sigma(N)\setminus d = \Gamma\setminus d,$$

so that $\sigma(A) = \Gamma$.

It is easily seen that A is triangular and that the only possible choice for $d(A)$ is $d(A) = d$.

This solves the first part of the problem for the case when d is denumerable. If d is finite, then the first statement follows from a trivial modification of the above argument. If d^- is perfect, then it follows from the Bernstein-Schröder theorem (see, e.g., [122]) and Proposition 3.35 (and its proof) that A admits a compact perturbation $A-K$, where $K \in K(H)$ is normal, such that $A-K$ is triangular, $\sigma_p(A-K) = d(A-K) = d$, $\mathrm{nul}(A-K-\lambda_n) = 1$ for all $n = 1,2,\ldots$, and $\sigma(A-K) = \Gamma$.□

5.10 Notes and remarks

A classical example due to S. Kakutani (see [119,Problem 87],[172, p. 282]) with weighted shift operators indicates that a limit of nilpotent operators can be an operator with positive spectral radius, so that neither the spectrum nor the spectral radius are continuous, in general. In his survey article "Ten problems in Hilbert space", P. R. Halmos raised the following

PROBLEM 7. Is every quasinilpotent operator the norm limit of nilpotent ones?

In view of Kakutani's example, Halmos remarked that the above question is "wrong" in the sense that the condition is already known to be *not* sufficient. The right question should be: "What is the closure of the set of nilpotent operators?" Can it be characterized in simple terms?

In [129], J. H. Hedlund found several interesting examples of operators with positive spectral radii contained in $N(H)^-$, including the operator $J = \oplus_{k=1}^{\infty} q_k^{(\infty)}$. This example was the key result for the first proof, due to D. A. Herrero [132], of Proposition 5.6. An alternative proof of this result was later given in [26]. (The proof included here is different from both of them.) N. Salinas [180] extended Herrero's result to certain classes of subnormal operators and C. Apostol and D. Voiculescu affirmatively answered Halmos' problem in [32]. (See also [5], [11].) A different proof of the same result was later given by C. Apostol and N. Salinas in [29]. The complete spectral characterization of $N(H)^-$ (Theorem 5.1) was obtained by C. Apostol, C. Foiaș and D. Voiculescu [26]. Partial results in that direction were independently obtained by D. A. Herrero [133], [134].

The spectral characterization of $A\ell g(H)^-$ is due to D. Voiculescu [194] and the results on closures of similarity orbits were obtained by D. A. Herrero in [140] and [141], respectively. Theorem 5.8 has been extended in [42] to cover the case when $\sigma_e(N) = \sigma(N)$ (N normal), but $\sigma(N)$ need not be perfect.

The original proofs of Theorems 5.1, 5.8 and 5.9 and Proposition 5.13 do not depend on Voiculescu's theorem, so they are rather different than the ones given here; e. g., the proof of Theorem 5.8 given in [140] strongly depends on D. A. Herrero's result on the existence of universal quasinilpotent operators [137].

In [135], [136], D. A. Herrero characterized $N(A)^-$ and $A\ell g(H)^{--}$ (in terms of parts of the weighted spectra; see, e.g., [78]) for the case when H is a non-separable Hilbert space.

Theorem 5.15 improves two previous results [133,Theorem 7], [149, Proposition 6.2] of D. A. Herrero. Proposition 5.16 is due to C. Apostol and N. Salinas [29,Theorem 3.5] and the results of Section 5.6 are contained in the same article, except for some mild improvements in Corollaries 5.19 and 5.20, based on Voiculescu's theorem. The observations of Remarks 5.22 are due to D. A. Herrero; they are related with approximation problems in the class of quasidiagonal operators ([149, Theorem 5.11], [150,Section 2]; see also Section 6.2.1 below).

The notion of pseudonilpotent was introduced by N. Salinas in his article [180], which also contains Theorem 5.23 (in fact, a slightly weaker form of the theorem; see comments at the beginning of [150,Section 6]), its corollary and Problem 5.32. The results going from Lemma 5.25 through Remark 5.28 are due to C. Apostol and N. Salinas [29]. Proposition 5.29 contains two unpublished observations of D. A. Herrero. Corollary 5.30 is an unpublished result of I. D. Berg, who also raised Problem 5.31.

The results of Section 5.8 are tacitly contained in the corresponding articles on approximation of operators. In particular, L. A. Fialkow proved in [83] that if N is normal, then

$$\{WNV: \ \tilde{W},\tilde{V} \text{ are Fredholm operators and } \tilde{W}^{-1} = \tilde{V}\}+K(H) = S(N)+K(H).$$

Finally, Theorem 5.42 is a particular case of [144] (Theorem 1 or Theorem 2; see also [84,Section 2]) and [147,Lemma 1], and Example 5.43 is essentially due to N. Salinas (see [84,Theorem 3.2]).

6 Quasitriangularity

Our next step will be the analysis of the all important notion of quasitriangularity and its characterization in spectral terms. Several consequences of this characterization will be developed as well.

An operator is *quasitriangular* if it can be approximated by triangular ones. The most obvious examples of triangular operators are the algebraic operators and the operators $M_-(\Gamma)$ (defined by (3.2)). The main step of the characterization of quasitriangularity says that all quasitriangular operators can be constructed out of algebraic operators, the operators $M_-(\Gamma)$ and approximation.

The introductory section of this chapter deals with a general argument of approximation of operators by operators with a very simple structure.

Most of the results of this chapter are "invariant under compact perturbations" and therefore they can be immediately "translated" to results about closures of subsets of the Calkin algebra.

6.1 Apostol-Morrel simple models

We shall say that an operator S is a *simple model* if it has the form

$$S = \begin{pmatrix} S_+ & * & * \\ 0 & A & * \\ 0 & 0 & S_- \end{pmatrix}, \tag{6.1}$$

where

(i) $\sigma(S_+)$, $\sigma(A)$ and $\sigma(S_-)$ are pairwise disjoint;

(ii) A is similar to a normal operator with finite spectrum;

(iii) S_+ is (either absent or) unitarily equivalent to

$$\oplus_{i=1}^{m} M_+(\partial\Omega_i)^{(k_i)}, \quad 1 \le k_i \le \infty,$$

where $\{\Omega_i\}_{i=1}^{m}$ is a finite family of analytic Cauchy domains with pairwise disjoint closures;

(iv) S_- is (either absent or) unitarily equivalent to

$$\oplus_{j=1}^{n} M_-(\partial\Phi_j)^{(h_j)}, \quad 1 \le h_j \le \infty,$$

where $\{\Phi_j\}_{j=1}^{n}$ is a finite family of analytic Cauchy domains with pairwise disjoint closures.

THEOREM 6.1. *The simple models are dense in* $L(H)$. *More precisely: Given* $T \in L(H)$ *and* $\varepsilon > 0$ *there exists a simple model* S *such that* $\sigma(S_+) \subset \rho^-_{s-F}(T) \subset \sigma(S_+)_\varepsilon$, $\sigma(S_-) \subset \rho^+_{s-F}(T) \subset \sigma(S_-)_\varepsilon$, $\sigma(A) \subset \sigma(T)_\varepsilon$, $\operatorname{ind}(\lambda-S) = \operatorname{ind}(\lambda-T)$ *for all* $\lambda \in \rho^-_{s-F}(S_+)\cup\rho^+_{s-F}(S_-)$ *and* $\|T - S\| < \varepsilon$.

Let S be the simple model given by (6.1) and let N be a normal operator such that $A \sim N$. It is easily seen that (use (i) and Corollary 3.22) $S \sim S_+\oplus N\oplus S_-$.

It follows from (3.4) that $\Gamma^+ = \sigma_{\ell e}(S_+) = \partial\sigma(S_+) = \cup_{i=1}^{m} \partial\Omega_i$ and $\sigma_{re}(S_-) = \partial\sigma(S_-) = \cup_{j=1}^{n} \partial\Phi_j = \Gamma^-$. By using the results of Section 4.1.3 and Proposition 4.27, we see that if N^+, N^- are normal operators such that $\sigma(N^+) \subset \Gamma^+$ and $\sigma(N^-) \subset \Gamma^-$, respectively, then

$$S_+\oplus N^+ \simeq_a S_+ \quad \text{and} \quad S_-\oplus N^- \simeq_a S_-.$$

Let Σ^+, Σ^- be compact subsets of $(\Gamma^+)_\varepsilon$ and $(\Gamma^-)_\varepsilon$, respectively, and let M^+, M^- be normal operators such that $\sigma(M^+) = \Sigma^+$ and $\sigma(M^-) = \Sigma^-$, resp.. If S has the form (6.1) and $\|T - S\| < \varepsilon$, then we can use our previous observations and Lemma 5.4 in order to show that there exists an operator S' such that $\|T - S'\| < 2\varepsilon$ and S' admits a 3×3 matrix decomposition of the form (6.1) with S_+ and S_- replaced by $S_+\oplus M^+$ and $S_-\oplus M^-$, respectively. Thus, we have the following.

COROLLARY 6.2. *Given* $T \in L(H)$ *and* $\varepsilon > 0$, *there exists*

$$S' \simeq \begin{pmatrix} S_+\oplus M^+ & * & * \\ 0 & A & * \\ 0 & 0 & S_-\oplus M^- \end{pmatrix}, \tag{6.2}$$

where

(i) S_+, A *and* S_- *have the form of* (6.1);

(ii) M^+ *and* M^- *are normal operators such that* $\sigma(M^+)\cap\sigma(S_+) = \partial\sigma(S_+)$ *and* $\sigma(M_-)\cap\sigma(S_-) = \partial\sigma(S_-)$, *resp.*;

(iii) $\operatorname{ind}(\lambda - S') = \operatorname{ind}(\lambda - T)$ *for all* $\lambda \in \rho^-_{s-F}(S_+)\cup\rho^+_{s-F}(S_-)$;

(iv) $\sigma(S_+\oplus M^+)$ *(* $\sigma(S_-\oplus M^-)$ *) is the closure of an analytic Cauchy domain* Ω *(*Φ*, resp.) such that each component of* Ω^- *(*Φ^-*, resp.) intersects* $\sigma(S_+)$ *(*$\sigma(S_-)$*, resp.) and* $\sigma(S_+) \subset \Omega^- \subset \sigma(S_+)_\varepsilon$ *(*$\sigma(S_-) \subset \Phi^- \subset \sigma(S_-)_\varepsilon$*)*;

(v) Ω^-, Φ^- *and* $\sigma(A)$ *are pairwise disjoint sets;*

(vi) $S' \sim (S_+ \oplus M^+) \oplus A \oplus (S_- \oplus M^-)$, *and*

(vii) $\|T - S'\| < 2\varepsilon$.

Similarly, we can replace A by a more general kind of algebraic operator, or by an operator similar to a normal operator with "large" spectrum, etc, i.e., Theorem 6.1 must be regarded as an argument to construct very general distinct families of "models" dense in $L(H)$. Each of the terms S_+, A or S_- can be replaced by an operator in a very large class; some other examples of this situation will be given later.

PROOF OF THEOREM 6.1. Let $T \in L(H)$ and $\varepsilon > 0$. If Ψ is an analytic Cauchy domain such that $\sigma_{\ell re}(T) \subset \Psi \subset \sigma_{\ell re}(T)_{\varepsilon/8}$, we can use Corollary 3.50 to obtain normal operators N_ℓ and N_r such that $\sigma(N_\ell) = \sigma_e(N_\ell) = \sigma(N_r) = \sigma_e(N_r) = \Psi^-$ and an operator

$$T_1 = \begin{pmatrix} N_\ell & * & * \\ 0 & L & * \\ 0 & 0 & N_r \end{pmatrix}$$

such that $\text{min.ind}(\lambda - T_1)^k = \text{min.ind}(\lambda - T)^k$ for all $\lambda \in \rho_{s-F}(T_1)$, $k = 1,2,\dots$, and $\|T - T_1\| < \varepsilon/4$.

Clearly, $\sigma_o(T_1)$ is a finite subset of $\sigma_o(T)$ and therefore

$$T_1 = \begin{pmatrix} C_1 & * \\ 0 & L_1 \end{pmatrix},$$

where $C_1 = T_1 | H(\sigma_o(T_1);T_1)$ and $\sigma_o(L_1) = \emptyset$. By Theorem 3.48, there exists $K_1 \in K(H)$, $K_1 = 0 \oplus K_1'$ (with respect to the same decomposition), $\|K_1\| < \varepsilon/4$ and

$$T_2 = T_1 + K_1 = \begin{pmatrix} C_1 & * \\ 0 & L_2 \end{pmatrix},$$

where $L_2 = L_1 + K_1'$ is a smooth operator. It is not difficult to check that $\rho^+_{s-F}(L_2) = \rho^+_{s-F}(T_2) = \rho^+_{s-F}(T) \setminus \Psi^-$, $\rho^-_{s-F}(L_2) = \rho^-_{s-F}(T_2) = \rho^-_{s-F}(T) \setminus \Psi^-$ and $\text{ind}(L_2 - \lambda) = \text{ind}(T_2 - \lambda) = \text{ind}(T - \lambda)$ for all $\lambda \in \rho_{s-F}(T_2)$.

Let $\Omega_1, \Omega_2, \dots, \Omega_m$ be the components of $\rho^-_{s-F}(T_2) \cap \sigma(T_2)$ and let $\Phi_1, \Phi_2, \dots, \Phi_n$ be the components of $\rho^+_{s-F}(T_2) \cap \sigma(T_2)$. It is easy to see that $\Omega_i^- \cap \Omega_h^- = \emptyset$ for all $i \neq h$, $\Phi_j^- \cap \Phi_h^- = \emptyset$ for all $j \neq h$, $(\cup_{i=1}^m \Omega_i)^- \cap (\cup_{j=1}^n \Phi_j)^- = \emptyset$, $\Omega_i^- \cap \Psi^- = \partial \Omega_i$ $(i = 1,2,\dots,m)$ and $\Phi_j^- \cap \Psi^- = \partial \Phi_j$ $(j = 1,2,\dots,n)$.

Define

$$M^+ = \oplus_{i=1}^m M(\partial \Omega_i)^{(k_i)}, \quad S_+ = \oplus_{i=1}^m M_+(\partial \Omega_i)^{(k_i)},$$

where $k_i = -\text{ind}(\lambda - T)$, $\lambda \in \Omega_i$, $i = 1,2,\dots,m$, and

$$M^- = \oplus_{j=1}^n M(\partial \Phi_j)^{(h_j)}, \quad S_- = \oplus_{j=1}^n M_-(\partial \Phi_j)^{(h_j)},$$

where $h_j = \mathrm{ind}(\lambda - T)$, $\lambda \in \Phi_j$, $j = 1,2,\ldots,n$, and let

$$T_2 = \begin{pmatrix} T_r & * & * \\ 0 & T_o & * \\ 0 & 0 & T_\ell \end{pmatrix}$$

be the Apostol's triangular representation (3.8) of T_2. Then $\sigma(M^+) \cup \sigma(M^-) \subset \partial\sigma_e(T) \subset \sigma_{\ell re}(T_2)$ and Propositions 4.29 and 4.26 imply that we can find $K_2 \in K(H)$, such that $K_2 = K_r \oplus 0 \oplus K_\ell$ (with respect to the same decomposition as for T_2), $\|K_2\| < \varepsilon/4$ and

$$T_3 = T_2+K_2 = \begin{pmatrix} T_r+K_r & * & * \\ 0 & T_o & * \\ 0 & 0 & T_\ell+K_\ell \end{pmatrix} = \begin{pmatrix} M^+ & * & * & * & * \\ 0 & B_r & * & * & * \\ 0 & 0 & T_o & * & * \\ 0 & 0 & 0 & B_\ell & * \\ 0 & 0 & 0 & 0 & M^- \end{pmatrix} = \begin{pmatrix} M^+ & * & * \\ 0 & B_o & * \\ 0 & 0 & M^- \end{pmatrix},$$

where

$$B_o = \begin{pmatrix} B_r & * & * \\ 0 & T_o & * \\ 0 & 0 & B_\ell \end{pmatrix},$$

and $\begin{pmatrix} M^+ & * \\ 0 & B_r \end{pmatrix}$ and $\begin{pmatrix} B_\ell & * \\ 0 & M^- \end{pmatrix}$ are smooth operators.

According to (3.2), we have

$$T_3 \simeq \begin{pmatrix} S_+ & * & * & * & * \\ 0 & \oplus_{i=1}^m M_-(\partial\Omega_i)^{(k_i)} & * & * & * \\ 0 & 0 & B_o & * & * \\ 0 & 0 & 0 & \oplus_{j=1}^n M_+(\partial\Phi_j)^{(h_j)} & * \\ 0 & 0 & 0 & 0 & S_- \end{pmatrix} = \begin{pmatrix} S_+ & T_{12} & T_{13} \\ 0 & R & T_{23} \\ 0 & 0 & S_- \end{pmatrix}, \tag{6.3}$$

where

$$R = \begin{pmatrix} \oplus_{i=1}^m M_-(\partial\Omega_i)^{(k_i)} & * & * \\ 0 & B_o & * \\ 0 & 0 & \oplus_{j=1}^n M_+(\partial\Phi_j)^{(h_j)} \end{pmatrix}.$$

It follows from our construction that $\sigma_o(T_3) = \sigma_o(T_1) = \sigma_o(B_o) = \sigma_o(R)$, $T_3|H(\sigma_o(T_1);T_3) = T|H(\sigma_o(T_1);T) = C_1$, $B_o|H(\sigma_o(T_1);B_o) \sim C_1$ and $R|H(\sigma_o(T_1);R) \sim C_1$. Moreover, T_3 is a compact pertubation of T_1, so that $\rho_{s-F}(T_3) = \rho_{s-F}(T_1)$ and $\mathrm{ind}(\lambda-T_3) = \mathrm{ind}(\lambda-T_1)$ for all $\lambda \in \rho_{s-F}(T_3)$.

Assume that $\lambda \in \Omega_i$ and $\mathrm{ind}(\lambda-T_3) = -k_i$ is finite; then $\mathrm{ind}(\lambda-S_+) = -k_i$ and $\lambda-S_-$ is invertible, so that $(\lambda-S_+)$ and $(\lambda-S_-)$ are invertible in the Calkin algebra. It follows that $(\lambda-\tilde{R})$ must be invertible too, i.e., $(\lambda-R)$ is a Fredholm operator; moreover,

$$\mathrm{ind}(\lambda-R) = \mathrm{ind}(\lambda-T_3) - \mathrm{ind}(\lambda-S_+) - \mathrm{ind}(\lambda-S_-) = -k_i+k_i+0 = 0.$$

If $\mathrm{ind}(\lambda-T_3) = -k_i = -\infty$, then $\mathrm{ind}(\lambda-S_+) = -\infty$ and $\lambda-S_-$ is invertible. In this case, $\mathrm{nul}(\lambda-R) = \mathrm{nul}(\lambda-M_-(\partial\Omega_i)^{(\infty)}) = \infty$ so that $\lambda \in \sigma_{\ell e}(R)$. On the other hand, $\lambda \in \sigma_{\ell re}(T_1)$ and therefore $\lambda \in \sigma_{\ell re}(B_o)$, by construction. Since $\lambda - S_-$ is invertible and $\lambda \in \rho_r(\oplus_{t=1}^m M_-(\partial\Omega_t)^{(k_t)})$, we conclude (Lemma 3.43) that $\lambda \in \sigma_{re}(R)$. Hence $\lambda \in \sigma_{\ell re}(R)$.

Similarly, if $\lambda \in \Phi_j$, then we conclude that either $\mathrm{ind}(\lambda - T_3) = h_j$ is finite and $\lambda - R$ is a Fredholm operator of index 0, or $\mathrm{ind}(\lambda - T_3) = h_j = \infty$ and $\lambda \in \sigma_{\ell re}(R)$.

Hence $\sigma(R)$ is the disjoint union of $\sigma_o(R) = \sigma_o(T_1)$, $[\sigma(R) \cap \rho_{s-F}(R)] \setminus \sigma_o(R)$, $\Omega = \{\lambda \in (\cup_{i=1}^m \Omega_i) \cup (\cup_{j=1}^n \Phi_j): \mathrm{ind}(\lambda - T_3)$ is finite$\}$ and $\sigma_{\ell re}(R) = \Psi^- \cup \{\lambda \in (\cup_{i=1}^m \Omega_i) \cup (\cup_{j=1}^n \Phi_j): \mathrm{ind}(\lambda - T_3) = \pm\infty\}$,

$$R \simeq \begin{pmatrix} C_1 & R_{12} \\ 0 & R_1 \end{pmatrix} \sim C_1 \oplus R_1,$$

where $\sigma(R_1) = \sigma(R) \setminus \sigma_o(R)$, $\sigma_e(R_1) = \sigma_{\ell re}(R_1) = \sigma_{\ell re}(R)$ and $\mathrm{ind}(\lambda - R_1) = 0$ for all $\lambda \in \rho_{s-F}(R_1)$.

By Corollary 5.11, we can find an algebraic operator H_1 such that $\sigma(H_1) = \sigma_e(H_1) \subset \Psi$ and $\|R_1 - H_1\| < \varepsilon/4$. A fortiori,

$$H = \begin{pmatrix} C_1 & R_{12} \\ 0 & H_1 \end{pmatrix} \text{ is algebraic,}$$

$\sigma(H) = \sigma(C_1) \cup \sigma(H_1)$ does not intersect $\sigma(S_+) \cup \sigma(S_-)$ and

$$\left\| \begin{pmatrix} C_1 & R_{12} \\ 0 & R_1 \end{pmatrix} - \begin{pmatrix} C_1 & R_{12} \\ 0 & H_1 \end{pmatrix} \right\| = \|R_1 - H_1\| < \varepsilon/4.$$

Furthermore, by using our remarks at the beginning of Section 5.1.2, we can find an operator A similar to a normal operator with finite spectrum, such that $\sigma(A) \subset \sigma(C_1)_\varepsilon \cup [\sigma(H_1)_\varepsilon \cap \Psi]$ (so that $\sigma(A) \cap [\sigma(S_+) \cup \sigma(S_-)] = \emptyset$) and $\|R - A\| < \varepsilon/4$.

Set

$$S = \begin{pmatrix} S_+ & T_{12} & T_{13} \\ 0 & A & T_{23} \\ 0 & 0 & S_- \end{pmatrix}.$$

Then $\|T - S\| \le \|T - T_1\| + \|K_1\| + \|K_2\| + \|R - A\| < \varepsilon$. It is completely apparent that S is a simple model which satisfies all our requirements. □

REMARK 6.3. It follows from (6.3) and the properties of the operator R that if Σ is an arbitrary perfect subset of Ψ^-, then R can be uniformly approximated by operators similar to $C_1 \oplus N$, where N is a normal operator such that $\sigma(N) = \Sigma$ (use Theorem 5.8). This indicates that the central piece A of the model S can be replaced by many other operators (for the purposes of approximation; see also Corollary 6.2 and remarks following it).

6.2 Quasitriangular operators

It is completely apparent that an operator A is triangular (see (3.6)) if and only if there exists an increasing sequence $\{P_n\}_{n=1}^{\infty}$ of finite rank projections such that $P_n \to 1$ (strongly, as $n \to \infty$) and $AP_n - P_nAP_n = (1-P_n)AP_n = 0$ for all $n = 1,2,\ldots$.

This formulation suggests an asymptotic generalization. An operator A is *quasitriangular* if there exists an increasing sequence $\{P_n\}_{n=1}^{\infty}$ of finite rank projections such that $P_n \to 1$ (strongly, as $n \to \infty$) and such that $\|(1-P_n)AP_n\| \to 0$ $(n \to \infty)$.

The set $PF(H)$ of all finite rank orthogonal projections, ordered by range inclusion, is a directed set. Thus, given T in $L(H)$, $P \to \|(1-P)TP\|$ is a net on that directed set and it makes sense to define

$$q(T) = \liminf(P \to 1)\,\|(1-P)TP\| \tag{6.4}$$

(where P runs over $PF(H)$; q(T) is the *modulus of quasitriangularity*).

Let (QT) (or (QT)(H)) denote the class of all quasitriangular operators. It will be shown that q(T) is equal to the distance from T to the set (QT), so that an operator A is quasitriangular if and only if $q(A) = 0$. In fact, (QT) admits many other different characterizations, the most important one being the characterization in terms of the different parts of the spectrum. This can be summarized as follows:

THEOREM 6.4. *The following are equivalent for* $A \in L(H)$.

(i) A *is quasitriangular.*

(ii) There exists a (not necessarily increasing) sequence $\{P_n\}_{n=1}^{\infty}$ *in* $PF(H)$ *such that* $P_n \to 1$ *(strongly) and* $\|(1-P_n)TP_n\| \to 0$, *as* $n \to \infty$.

(iii) $q(A) = 0$.

(iv) A *is the limit of a sequence of triangular operators.*

(v) $A = T+K$, *where* T *is triangular and* K *is compact.*

(vi) Given $\varepsilon > 0$, A *can be written as* $A = T_\varepsilon + K_\varepsilon$, *where* T_ε *is triangular,* K_ε *is compact and* $\|K_\varepsilon\| < \varepsilon$.

(vii) $\rho_{s-F}^{-}(A) = \emptyset$.

COROLLARY 6.5. (QT) *is a closed subset of* $L(H)$, *invariant under similarities and under compact perturbations. Furthermore, if* $A \in$ (QT) *and* p *is a polynomial, then* $p(A) \in$ (QT).

PROOF. Every triangular operator is obviously quasitriangular. Now the equivalence between (i) and (iv) indicates that $(QT) = \{T \in L(H): T \text{ is triangular}\}^-$ is a closed set.

The equivalence between (i) and (v) makes it completely apparent that $(QT)+K(H) = (QT)$. On the other hand, since every operator similar to a triangular operator is also triangular (to see this, we can use, e.g., the existence of a suitable increasing sequence of finite dimensional invariant subspaces), it readily follows that if A has the form of (v) and W is invertible, then $WAW^{-1} = WTW^{-1} + WKW^{-1}$, where WTW^{-1} is triangular and $WKW^{-1} \in K(H)$.

Finally, observe that $p(A) = p(T+K) = p(T)+C$, where $p(T)$ is triangular and $C \in K(H)$. □

REMARK 6.6. The proof of Corollary 6.5 only depends on the equivalence between (i), (iv) and (v) of Theorem 6.4.

6.2.1 Equivalence between the formal and the relaxed definitions of quasitriangularity

It is trivial that the definition of quasitriangularity implies the weakened form (ii) of Theorem 6.4. On the other hand, if $q(A) = 0$ and $\{e_n\}_{n=1}^{\infty}$ is an ONB of H, then there exists $P_1 \in PF(H)$ such that $e_1 \in$ ran P_1 and $\|(1-P_1)AP_1\| < 1$. Assume that we have used the condition $q(A) = 0$ inductively to obtain a sequence $P_1 \leq P_2 \leq \ldots \leq P_n$ in $PF(H)$ such that $e_j \in$ ran P_j and $\|(1-P_j)AP_j\| < 1/j$ for $j = 1,2,\ldots,n$. Then we can use the same condition in order to find $P_{n+1} \in PF(H)$ such that $P_n \leq P_{n+1}$, $e_{n+1} \in$ ran P_{n+1} and $\|(1-P_{n+1})AP_{n+1}\| < 1/(n+1)$. Hence, the condition $q(A) = 0$ implies that $A \in (QT)$.

(ii) => (iii) Assume that $\|(1-P_n)AP_n\| \to 0$ $(n \to \infty)$ for some sequence $\{P_n\}_{n=1}^{\infty}$ in $PF(H)$ such that $P_n \to 1$ (strongly) and let $P \in PF(H)$. Let ε, $0 < \varepsilon < \frac{1}{2}$, be given and let R_n be the orthogonal projection onto P_n(ran P). Since rank $P < \infty$ and since $\|P_n x - x\| \to 0$ $(n \to \infty)$ for all $x \in H$, there exists $n_0 = n_0(\varepsilon)$ such that $\|P_n x - x\| \leq (\varepsilon/4)\|x\|$ for all $x \in$ ran P and all $n \geq n_0$. Let $n \geq n_0$ and let $y \in P_n$(ran P), $\|y\| = 1$, $y = P_n x$, $x \in$ ran P. Then $\|x\| \leq \|x-P_n x\| + \|P_n x\| \leq (\varepsilon/4)\|x\| + 1$ so that $\|x\| \leq (1-\varepsilon)^{-1}$, and

$$\|R_n y - Py\| = \|y - Py\| = \|P_n x - PP_n x\| \leq \|P_n x - Px\| + \|Px - PP_n x\|$$

$$\leq \|P_n x - x\| + \|P\| . \|x - P_n x\| \leq (1+\|P\|)(\varepsilon/4)\|x\| \leq 2(\varepsilon/4)/(1-\varepsilon) < \varepsilon.$$

Furthermore, if $y \in$ ran P, $\|y\| = 1$, then $Py = y$ and $R_n y = P_n y$. Hence, for $n \geq n_0$, $\|R_n y - Py\| = \|P_n y - y\| < \varepsilon$. Combining this with the previous statement we have that $\|R_n y - Py\| < \varepsilon$ for all $y \in$ (ran P) $\vee P_n$(ran P), $\|y\| = 1$ and all $n \geq n_0$. If $y \in [(\text{ran } P) \vee P_n(\text{ran } P)]^{\perp} = (\text{ran } P)^{\perp} \cap [P_n(\text{ran } P)]^{\perp}$, then $R_n y = Py = 0$. Taking the supremum (for each fixed $n \geq n_0$) of $\|R_n y - Py\|$ over all $\|y\| = 1$, we obtain $\|R_n - P\| \leq \varepsilon$. Thus $\|R_n - P\| \to 0$.

Define $\mathcal{R}_n$ so that ran $P_n = P_n(\text{ran } P) \oplus \mathcal{R}_n$ and let L_n be the projection onto ran $P+R_n$. Then $L_n \in PF(H)$, $L_n \geq P$ and, since R_n is the projection onto $P_n(\text{ran } P)$, it follows that $\|P_n-L_n\| = \|R_n-P\| \to 0$ $(n \to \infty)$. Thus, since $\|(1-P_n)AP_n\| \to 0$ and $\|L_n-P_n\| \to 0$, we obtain $\|(1-L_n)AL_n\| \to 0$. Hence, $q(A) = 0$.

(iv) => (iii) Suppose that $\{A_n\}_{n=1}^{\infty}$ is a Cauchy sequence of quasitriangular operators and that $\|A - A_n\| \to 0$ $(n \to \infty)$. Since (i) =>(iii), we see that given $\varepsilon > 0$ and $P \in PF(H)$, we can find $n_0 = n_0(\varepsilon)$ so that $\|A-A_{n_0}\| < \varepsilon/2$ and $R \in PF(H)$ such that $R \geq P$ and $\|(1-R)A_{n_0}R\| < \varepsilon/2$. It follows that $\|(1-R)AR\| \leq \|(1-R)A_{n_0}R\| + \|(1-R)(A-A_{n_0})R\| \leq \|A-A_{n_0}\| + \|(1-R)A_{n_0}R\| < \varepsilon$.

(vi) => (iv) and (vi) => (v) => (i) The first two implications are trivial. On the other hand, it is completely apparent that if $K \in K(H)$, then

$$\limsup(P \to 1) \|(1-P)K\| = \limsup(P \to 1) \|K(1-P)\| = \limsup(P \to 1) \|(1-P)KP\| = 0. \tag{6.5}$$

Thus, if A has the form (v) and $\{P_n\}_{n=1}^{\infty}$ is the increasing sequence in $PF(H)$ naturally associated with the triangular matrix representation of T, so that $P_n \to 1$ (strongly) and $(1-P_n)TP_n = 0$ for all $n = 1,2,\ldots$, then

$$\liminf(n \to \infty) \|(1-P_n)AP_n\| = \liminf(n \to \infty) \|(1-P_n)KP_n\| = 0.$$

Hence, $A \in (QT)$. □

6.2.2 Two lower estimates for the distance to (QT)

LEMMA 6.7. *For each operator* $A \in L(H)$ *there is a sequence* $\{R_n\}_{n=1}^{\infty}$ *of pairwise orthogonal finite rank projections with* $\sum_{n=1}^{\infty} R_n = 1$ *(strongly) such that* $\|R_{n+1}AR_n\| \to q(A)$ $(n \to \infty)$ *and* $R_{n+k}AR_n = 0$ *for all* $k \geq 2$.

PROOF. Let $\{e_n\}_{n=1}^{\infty}$ be an ONB of H and let $\varepsilon > 0$. Choose $P_1 \in PF(H)$ (H) so that $e_1 \in \text{ran } P_1$ and $\|(1-P_1)AP_1\| < q(A)+\varepsilon$. Choose recursively $P_n \in PF(H)$ so that $\text{ran } P_n \supset \text{ran } P_{n-1} \vee \text{ran } AP_{n-1} \vee (\vee\{e_1,e_2,\ldots,e_n\})$ and $\|(1-P_n)AP_n\| < q(A)+\varepsilon/n$. It is completely apparent that $\{P_n\}_{n=1}^{\infty}$ is increasing and $P_n \to 1$. Since $q(A) \leq \liminf(n \to \infty) \|(1-P_n)AP_n\|$, it readily follows that $\|(1-P_n)AP_n\| \to q(A)$ $(n \to \infty)$. Furthermore, since $\text{ran } P_{n+1} \supset \text{ran } AP_n$, it follows that $(1-P_{n+1})AP_n = 0$.

Define $R_1 = P_1$, $R_n = P_n - P_{n-1}$ for $n \geq 2$. Then the R_n's are pairwise orthogonal finite rank projections, $\sum_{n=1}^{\infty} R_n = 1$ (strongly) and $R_{n+1}AR_n = (1-P_n)AP_n$ and therefore $\|R_{n+1}AR_n\| \to q(A)$ $(n \to \infty)$. Finally, observe that the equations $(1-P_{n+1})AP_n = 0$ $(n = 1,2,\ldots)$ imply that $R_{n+k}AR_n = 0$ when $k \geq 2$. □

PROPOSITION 6.8. *For each* $A \in L(H)$, $\text{dist}[A, (QT)] = q(A)$. *Furthermore, there exists* $T \in (QT)$ *such that* $\|A - T\| = \text{dist}[A, (QT)]$.

PROOF. If $q(A) = 0$, then choose $T = A$.

Assume that $q(A) > 0$ and choose $\{R_n\}_{n=1}^{\infty}$ as in the above lemma.

Define

$$s_n = \begin{cases} \frac{q(A)}{\|R_{n+1}AR_n\|}, & \text{if } R_{n+1}AR_n \neq 0 \\ 1 \end{cases}, \quad n = 1,2,\ldots,$$

and let $S_n = s_n R_{n+1} AR_n$. Observe that $s_n \to 1$ $(n \to \infty)$ and $\|S_n\| \le q(A)$ for all n. Now define

$$B = \sum_{n=1}^{\infty} S_n \text{ (strong sum)}, \quad T = A - B.$$

Since the R_n's are pairwise orthogonal, we can write $B^*B = \oplus_{n=1}^{\infty} S_n^*S_n | \text{ran } R_n$; therefore $\|B^*B\| = \sup\{\|S_n^*S_n\| : n \ge 1\} \le q(A)^2$, so that $\|B\| \le q(A)$.

Let $P_n = \sum_{k=1}^{n} R_k$. It is immediate that $(1-P_n)BP_n = S_n = s_n R_{n+1}AR_n$, so that

$$\|(1-P_n)TP_n\| = \|(1-P_n)(A-B)P_n\| = (1-s_n)\|R_{n+1}AR_n\| \to 0 \, (n \to \infty).$$

Hence $T \in (QT)$ and $\|A - T\| = \|B\| \le q(A)$. Thus, in order to complete the proof, it is enough to show that $\|A - X\| \ge q(A)$ for all $X \in (QT)$.

Assume that $\|A - X\| < q(A)$; then

$$\liminf(P \to 1)\|(1-P)AP\| \le \liminf(P \to 1)\{\|(1-P)(A-X)P\| + \|(1-P)XP\|\}$$
$$\le \|A-X\| + \liminf(P \to 1)\|(1-P)XP\| \le \|A-X\| + q(X),$$

so that $q(X) \ge q(A) - \|A - X\| > 0$, proving that X fails to be quasitriangular. □

PROPOSITION 6.9. *If* $A \in L(H)$, $\|(1-P_n)AP_n\| \to 0$ $(n \to \infty)$ *for some increasing sequence* $\{P_n\}_{n=1}^{\infty}$ *in* $PF(H)$ *such that* $P_n \to 1$ *(strongly,* $n \to \infty$*) and* $\varepsilon > 0$, *then there exists a compact operator* K_ε, $\|K_\varepsilon\| < \varepsilon$, *such that* $K_\varepsilon^2 = 0$ *and* $A - K_\varepsilon$ *is triangular.*

PROOF. Start with the sequence $\{R_n\}_{n=1}^{\infty}$ of Lemma 6.7. Since $q(A) = 0$, we shall have $\|R_{2n+1}AR_{2n}\| < \varepsilon/2n$ for all $n = 1,2,\ldots$. Since $R_{2n+1}AR_{2n}$ is a finite rank operator, we can define

$$K_\varepsilon = \sum_{n=1} R_{2n+1}AR_{2n} \in K(H).$$

(The sum converges in the norm.) The product of any two summands in the definition of K_ε is 0; therefore $K_\varepsilon^2 = 0$. The orthogonality of the R_n's implies that

$$K_\varepsilon^*K_\varepsilon = \oplus_{n=1}^{\infty}(R_{2n}A^*R_{2n+1})(R_{2n+1}AR_{2n}) | \text{ran } R_{2n}$$

and therefore

$$\|K_\varepsilon^*K_\varepsilon\| \le \sup\{\|R_{2n}A^*R_{2n+1}\| \cdot \|R_{2n+1}AR_{2n}\| : n \ge 1\} < \varepsilon^2.$$

Hence, $||K_\varepsilon|| < \varepsilon$.

To see that $T_\varepsilon = A - K_\varepsilon$ is actually triangular, observe that (with $P_n = \sum_{k=1}^n R_k$) $(1-P_{2n})K_\varepsilon P_{2n} = R_{2n+1}AR_{2n} = (1-P_{2n})AP_{2n}$, so that

$$(1-P_{2n})T\ P_{2n} = 0. \qquad \square$$

The results of Section 6.2.1 and Proposition 6.9 imply that (i) => (vi) and therefore, that any two of the conditions (i) through (vi) of Theorem 6.4 are equivalent. This suffices, in particular (as observed in Remark 6.6), to complete the proof of Corollary 6.5.

Given $T \in L(H)$, define the *minimum modulus* $m(T)$ (the *essential minimum modulus* $m_e(T)$) of T by

$$m(T) = \min\{\lambda \in \sigma((T^*T)^{\frac{1}{2}})\}$$
$$(m_e(T) = \min\{\lambda \in \sigma_e((T^*T)^{\frac{1}{2}})\} = \min\{\lambda \in \sigma((\tilde{T}^*\tilde{T})^{\frac{1}{2}})\}). \tag{6.6}$$

Clearly, $m_e(T+K) = m_e(T)$ for all $K \in K(H)$. By using the spectral decompositions of $\tilde{T}^*\tilde{T}$ and T^*T and the Weyl-von Neumann theorem, it is not difficult to prove the following

PROPOSITION 6.10. *Let* $T \in L(H)$; *then*

(i) If T *is not a semi-Fredholm operator of positive index, then*

$$m_e(T) = \sup\{m(T+K):\ K \in K(H)\} = \max\{m(T+K):\ K \in K(H)\}.$$

(ii) $\liminf(n \to \infty)\ ||Tx_n|| \geq m_e(T)$ *for any ON sequence* $\{x_n\}_{n=1}^\infty$ *of* H *and there exists an ON sequence* $\{y_n\}_{n=1}^\infty$ *such that* $m_e(T) = \liminf(n \to \infty)\ ||Ty_n||$.

(iii) If $E(.)$ *is the spectral measure of* $H = (T^*T)^{\frac{1}{2}}$, *then the smallest non-negative* α *such that* $\operatorname{rank} E((\alpha-\varepsilon, \alpha+\varepsilon))$ *is infinite, for every* $\varepsilon > 0$, *is* $\alpha = m_e(T)$.

(iv) Given $\varepsilon > 0$ *and* $p > 1$ *there exists* $K_\varepsilon \in C^p(H)$, $|K_\varepsilon|_p < \varepsilon$, *such that* $\operatorname{nul}(H+K_\varepsilon-m_e(T)) = \infty$.

(v) $\tilde{T}$ *is left (right) invertible if and only if* $m_e(T) > 0$ *(*$m_e(T^*) > 0$, *resp.).*

(vi) $\operatorname{ran} T$ *is closed and* $\operatorname{nul} T < \infty$ *(*$\operatorname{ran} T^*$ *is closed and* $\operatorname{nul} T^* < \infty$*) if and only if* $m_e(T) > 0$ *(*$m_e(T^*) > 0$, *resp.).*

(vii) T *is Fredholm if and only if* $m_e(T) > 0$ *and* $m_e(T^*) > 0$. *In that case,* $m_e(T) = m_e(T^*)$.

(viii) If T_1 *is another operator, then*

$$|m_e(T) - m_e(T_1)| \leq ||\tilde{T} - \tilde{T}_1|| \leq ||T - T_1||.$$

(ix) $m_e(\lambda - T)$ *is continuous for* $\lambda \in \mathbb{C}$; *moreover,*

$$|m_e(\lambda_1-T) - m_e(\lambda_2 - T)| \leq |\lambda_1-\lambda_2|.$$

We have the following

COROLLARY 6.11. *Let* $A \in L(H)$ *and assume that* $\rho_{s-F}^{-}(A) \neq \emptyset$; *then*

$$\operatorname{dist}[A,(QT)] \geq \max\{m_e(\lambda-A): \lambda \in \rho_{s-F}^{-}(A)\} > 0.$$

PROOF. Let $\lambda \in \rho_{s-F}^{-}(A)$. By Proposition 6.10(i) there exists $K_\lambda \in K(H)$ such that $m_e(A-\lambda) = m(A+K_\lambda-\lambda)$. By Lemma 3.39, the distance from $A+K_\lambda$ to the set of all triangular operators cannot be smaller than $m(A+K_\lambda-\lambda)$. On the other hand, by Corollary 6.5, (QT) is invariant under compact perturbations. Since the triangular operators are dense in (QT) (equivalence between (i) and (iv) of Theorem 6.4) and the above observations hold for all λ in $\rho_{s-F}^{-}(A)$, we conclude that

$$\operatorname{dist}[A,(QT)] \geq \sup\{m_e(\lambda-A): \lambda \in \rho_{s-F}^{-}(A)\}.$$

Since $m_e(\lambda-A) = 0$ for all $\lambda \in \partial\rho_{s-F}^{-}(A) \subset \sigma_{\ell re}(A)$ and (by Proposition 6.10(ix)) $m_e(\lambda-A)$ is a continuous function of λ, the above supremum is actually attained for some $\lambda \in \rho_{s-F}^{-}(A)$. By Proposition 6.10(vi), this maximum is strictly positive. □

6.2.3 Spectral characterization of quasitriangularity

Corollary 6.11 implies, in particular, that $\rho_{s-F}^{-}(A) = \emptyset$ for all $A \in (QT)$. Conversely, if $\rho_{s-F}^{-}(A) = \emptyset$ and $\varepsilon > 0$, then according to Theorem 6.1 there exists a simple model

$$S_\varepsilon = \begin{pmatrix} R & B \\ 0 & S_- \end{pmatrix} \begin{matrix} H_1 \\ H_2 \end{matrix},$$

where R is algebraic and S_- has the form of (6.1)(iv), such that $\|A-S_\varepsilon\| < \varepsilon$. (If $\rho_{s-F}^{+}(A) = \emptyset$, then S_- is absent and $S_\varepsilon = R$.) R is obviously triangular with respect to some ONB of H_1.

On the other hand, if Φ is an analytic Cauchy domain, then either by a direct argument (based on Runge's approximation theorem [102,p. 28]) or by using Vitushkin's theorem [102,p.207], we can easily see that the linear span of the functions $\{(\zeta-\lambda)^{-1}: \lambda \notin \partial\Phi\}$ is uniformly dense in the space of all continuous functions defined on $\partial\Phi$. Since the mapping $\lambda \to (\zeta-\lambda)^{-1}$ is a norm-continuous mapping from $\mathbb{C}\setminus\partial\Phi$ into $L^2(\partial\Phi)$, it is not difficult to deduce that if $\{\lambda_n\}_{n=1}^{\infty}$ is an arbitrary denumerable dense subset of Φ, then $L^2(\partial\Phi)\ominus H^2(\partial\Phi) = \bigvee\{P_-[(\zeta-\lambda)^{-1}]\}_{n=1}^{\infty}$, where P_- denotes the orthogonal projection of $L^2(\partial\Phi)$ onto $L^2(\partial\Omega)\ominus H^2(\partial\Omega)$.

Since $M_-(\partial\Phi)(P_-[(\zeta-\lambda)^{-1}]) = \lambda P_-[(\zeta-\lambda)^{-1}]$ for all $\lambda \in \Phi$, it readily follows that $M_-(\partial\Phi)$ is triangular with respect to the Gram-Schmidt orthonormalization of of the sequence $\{P_-[(\zeta-\lambda_n)^{-1}]\}_{n=1}^{\infty}$.

Now it is easily seen that if $\{\mu_k\}_{k=1}^{\infty}$ is a denumerable dense subset of $\rho^{+}_{s-F}(S_-)$ and $\{g_{kr}\}_{r=1}^{\infty}$ is a denumerable dense subset of $\ker(S_- - \mu_k)$ $(k = 1,2,\ldots)$, then S_- is triangular with respect to the Gram-Schmidt orthonormalization of the family $\{g_{kr}\}_{k,r=1}^{\infty}$.

Hence, $R\oplus S_-$ is triangular. Since $\sigma(R)\cap\sigma(S_-) = \emptyset$, $S_\varepsilon \sim R\oplus S_-$ (Corollary 3.22) and therefore S_ε is also triangular. By taking $\varepsilon = 1, 1/2, \ldots, 1/n, \ldots$, we conclude that A is the limit of a sequence of triangular operators, i.e., A satisfies (iv).

The proof of Theorem 6.4 is now complete. □

6.3 Biquasitriangular operators

Let $(QT)^* = \{T \in L(H): T^* \in (QT)\}$. An operator $A \in L(H)$ is *biquasitriangular* if both A and A* are quasitriangular. Hence, the class (BQT) of all biquasitriangular operators is equal to the set $(QT)\cap(QT)^*$.

6.3.1 Block-diagonal and quasidiagonal operators

An operator $B \in L(H)$ is *block-diagonal* if there exists an increasing sequence $\{P_n\}_{n=1}^{\infty}$ in $PF(H)$ such that $P_n \to 1$ (strongly) and $P_nB = BP_n$ for all $n = 1,2,\ldots$. Clearly, this is equivalent to saying that $B = \oplus_{n=1}^{\infty} B_n$, where $B_n = (P_n - P_{n-1})B|\mathrm{ran}(P_n - P_{n-1}) = B|\mathrm{ran}(P_n - P_{n-1})$ $(P_0 = 0;\ n = 1,2,\ldots)$. (It is obvious that the B_n's act on finite dimensional spaces.)

An operator A is *quasidiagonal* if there exists $\{P_n\}_{n=1}^{\infty}$ as above such that $\|P_nA - AP_n\| \to 0$ $(n \to \infty)$.

Let (QD) and (BD) denote the classes of all quasidiagonal and all block-diagonal operators, respectively. The main properties of the quasidiagonal operators will be summarized in the following three theorems. (Their proofs are either very simple or minor modifications of the proofs of the analogous statements in Theorem 6.4 and will be omitted.)

THEOREM 6.12. *The following conditions are equivalent for* $A \in L(H)$:

(i) A *is quasidiagonal;*

(ii) There exists a (not necessarily increasing) sequence $\{P_n\}_{n=1}^{\infty}$ *in* $PF(H)$ *such that* $P_n \to 1$ *(strongly) and* $\|P_nA - AP_n\| \to 0$ *as* $n \to \infty$;

(iii) The modulus of quasidiagonality

$$qd(A) = \lim\inf(P \to 1)\ \|PA - AP\|, \tag{6.7}$$

where P *runs over* $PF(H)$, *is equal to* 0;

(iv) A *is the limit of a sequence of block-diagonal operators;*

(v) $A = B + K$, *where* B *is block-diagonal and* K *is compact;*

(vi) Given $\varepsilon > 0$, A *can be written as* $A = B_\varepsilon + K_\varepsilon$, *where* B_ε *is*

block-diagonal, K_ε *is compact and* $\|K_\varepsilon\| < \varepsilon$.

THEOREM 6.13. *For each* $A \in L(H)$,

$$q(A) \leq \max\{q(A), q(A^*)\} \leq qd(A) = \mathrm{dist}[A, (QD)].$$

Moreover, there exists $C \in (QD)$ *such that* $\|A - C\| = qd(A)$.

THEOREM 6.14. (QD) *is a closed subset of* $L(H)$, *invariant under unitary equivalence and under compact perturbations. Furthermore, if* $A \in (QD)$, *then* $C^*(A) \subset (QD)$. *In particular,* $Nor(H) + K(H) \subset (QD)$.

But (as proven in [188,p.14,Example 1.3]; see also [149],[150]) the classes (QD) and (BD) are not invariant under similarities and this makes it impossible to give a spectral characterization of (QD) (analogous to Theorem 6.4(vii)). As we shall see immediately, such a characterization is actually possible for the larger class of all biquasitriangular operators. (It is completely apparent, from the definition, that $(QD) \subset (BQT)$.)

6.3.2 Characterizations of biquasitriangularity

THEOREM 6.15. *The following conditions are equivalente for* $A \in L(H)$:

(i) A *is biquasitriangular;*

(ii) There exists a (not necessarily increasing) sequence $\{P_n\}_{n=1}^{\infty}$ *in* $PF(H)$ *such that* $P_n \to 1$ *(strongly),* $\|(1 - P_{2n-1})AP_{2n-1}\| \to 0$ *and* $\|P_{2n}A(1 - P_{2n})\| \to 0$, *as* $n \to \infty$.

(iii) Given $\varepsilon > 0$, *there exists an increasing sequence* $\{P_n\}_{n=1}^{\infty}$ *in* $PF(H)$ *such that* $P_n \to 1$ *(strongly),* $\|(1 - P_{2n-1})AP_{2n-1}\| < \varepsilon/4^n$ *and* $\|P_{2n}A(1 - P_{2n})\| < \varepsilon/4^n$ *for all* $n = 1, 2, \ldots$.

(iv) $\max\{q(A), q(A^*)\} = 0$.

(v) $A \in Alg(H)^-$.

(vi) $A \in Alg(H)^- + K(H)$.

(vii) $A \in \{R \in L(H)$: *R is similar to a normal operator with finite spectrum*$\}^-$.

(viii) $A \in \{R \in L(H)$: $R \sim N$, *N is normal*$\}^-$.

(ix) $A \in \{R \in L(H)$: $R \sim B$, $B \in Nor(H) + K(H)\}^-$.

(x) $A \in \{R \in L(H)$: $\sigma(R)$ *is totally disconnected*$\}^-$.

(xi) $A \in \{R \in L(H)$: interior $\sigma(R) = \emptyset\}^-$.

(xii) $A \in \{R \in L(H)$: $R \sim C$, $C \in (QD)\}^-$.

(xiii) $\rho_{s-F}^{-}(A) = \rho_{s-F}^{+}(A) = \emptyset$.

(xiv) (Staircase representation) Given $\varepsilon > 0$ *there exists* $K_\varepsilon \in K(H)$, $\|K_\varepsilon\| < \varepsilon$, *and a family* $\{R_n\}_{n=1}^{\infty} \subset PF(H)$ *of pairwise orthogonal projections, such that* $A - K_\varepsilon$ *admits a matrix representation of the form*

$$A - K_\varepsilon = \begin{pmatrix}
A_1 & B_1 & & & & & & & & \\
 & C_1 & & & & & & & & \\
 & D_1 & A_2 & B_2 & & & & & 0 & \\
 & & & C_2 & & & & & & \\
 & & & D_2 & A_3 & & & & & \\
 & & & & & \ddots & & & & \\
 & & & & & & A_n & B_n & & \\
 & & & & & & & C_n & & \\
 & 0 & & & & & & D_n & A_{n+1} & \\
 & & & & & & & & & \ddots
\end{pmatrix}
\begin{matrix} \mathcal{R}_1 \\ \mathcal{R}_2 \\ \mathcal{R}_3 \\ \mathcal{R}_4 \\ \mathcal{R}_5 \\ \vdots \\ \mathcal{R}_{2n-1} \\ \mathcal{R}_{2n} \\ \mathcal{R}_{2n+1} \\ \vdots \end{matrix}, \qquad (6.8)$$

where $\mathcal{R}_n = \operatorname{ran} R_n$, $n = 1, 2, \ldots$.

(xv) There exists a decomposition $\mathcal{H} = \mathcal{H}_1 \oplus \mathcal{H}_2$ *of* $\mathcal{H}$ *into two infinite dimensional subspaces such that*

$$A = \begin{pmatrix} D_1 & * \\ K & D_2 \end{pmatrix} \begin{matrix} \mathcal{H}_1 \\ \mathcal{H}_2 \end{matrix}, \qquad (6.9)$$

where D_j *is a quasidiagonal operator acting on* $\mathcal{H}_j$, *such that* $\sigma_e(D_j) = \sigma_e(A)$, $j = 1, 2$, *and* K *is compact.*

(xvi) Given $\varepsilon > 0$ *there exists* $K_\varepsilon \in K(\mathcal{H})$, $\|K_\varepsilon\| < \varepsilon$, *such that*

$$A - K_\varepsilon = \begin{pmatrix} B_1 & * \\ 0 & B_2 \end{pmatrix} \begin{matrix} \mathcal{H}_1 \\ \mathcal{H}_2 \end{matrix}, \qquad (6.10)$$

where B_j *is a block-diagonal operator acting on the infinite dimensional subspace* $\mathcal{H}_j$, *such that* $\sigma_e(B_j) = \sigma_e(A)$, $j = 1, 2$.

(xvii) Given $\varepsilon > 0$ *there exists* $K_\varepsilon \in K(\mathcal{H})$, $\|K_\varepsilon\| < \varepsilon$, *such that*

$$A - K_\varepsilon = \begin{pmatrix} N\oplus D_1 & * & * & * \\ 0 & N\oplus D_2 & * & * \\ 0 & 0 & N\oplus D_3 & * \\ 0 & 0 & 0 & N\oplus D_4 \end{pmatrix} \begin{matrix} \mathcal{M}_1 \\ \mathcal{M}_2 \\ \mathcal{M}_3 \\ \mathcal{M}_4 \end{matrix},$$

where N *is a normal operator and the* D_j*'s are block-diagonal operators such that* $\sigma(N) = \sigma_e(N) = \sigma_e(D_j) = \sigma_e(A)$, $j = 1, 2, 3, 4$.

PROPOSITION 6.16. (BQT) *is a closed subset of* $L(\mathcal{H})$, *invariant under compact perturbations and under similarities. Furthermore, if* $A \in$ (BQT) *and* p *is a polynomial, then* $p(A)$, $p(A^*) \in$ (BQT).

PROOF. Since (BQT) = (QT) $\cap$ (QT)* and (QT) and (QT)* are closed, invariant under compact perturbations and invariant under similarities (Corollary 6.5), (BQT) has the same properties. The second statement follows from Corollary 6.5 as well. □

PROOF OF THEOREM 6.15. The equivalence between any two of the conditions (i), (ii), (iii), (iv) or (xiii) follows immediately from Theorem 6.4.

Let $(BQT)_v$, $(BQT)_{vi}$, ... , $(BQT)_{xii}$ denote the families of operators described in (v), (vi), ... , (xii), respectively. By Corollary 5.10 and the equivalence between (i) and (xiii), we have

$$(BQT)_v = (BQT)_{vi} = (BQT)_{vii} = (BQT).$$

Since

$$\{N \in \mathcal{Nor}(H): \ \sigma(N) \text{ is finite}\} \subset \mathcal{Nor}(H) \subset \mathcal{Nor}+\mathcal{K}(H) \subset (QD) \subset (BQT),$$

we easily see that $(BQT)_{vii} = (BQT)_{viii} = (BQT)_{ix} = (BQT)$.

Similarly, since (by Corollary 1.14(i) and Theorem 1.13)

$$(BQT) = (BQT)_{vii} \subset (BQT)_x \subset (BQT)_{xi} \subset \{B \in L(H): \ \rho_{s-F}^-(B) = \rho_{s-F}^+(B) = \emptyset\} = (BQT),$$

we see that $(BQT)_x = (BQT)_{xi} = (BQT)$.

Hence, any two of the conditions (i) - (xiii) are equivalent.

(xvi) => (vx) => (xiii) The first implication is trivial. On the other hand, it is easily seen that if A satisfies (xiii), then A-K has the form (6.10) for some K in $\mathcal{K}(H)$. Since (BQT) is invariant under compact perturbations (Proposition 6.16), it suffices to show that every operator of the form (6.10) belongs to (BQT).

Since (BD) ⊂ (QD) ⊂ (BQT), it readily follows from the equivalence between (i) and (xiii) that $\sigma_e(B_j) = \sigma_{\ell re}(B_j)$ and $\text{ind}(\lambda - B_j) = 0$ for all $\lambda \in \rho_{s-F}(B_j)$, $j = 1,2$. A fortiori, $\sigma_{\ell re}(A-K_\varepsilon) = \sigma_e(A-K_\varepsilon)$ and $\text{ind}(A-K_\varepsilon-\lambda) = 0$ for all $\lambda \in \rho_{s-F}(A-K_\varepsilon)$.

Hence $A-K_\varepsilon$ satisfies (xiii). Since K_ε is compact, it is completely apparent that A also satisfies (xiii).

(xiv) => (i) Assume that $A-K_\varepsilon$ admits the representation (6.8) (for some $K_\varepsilon \in \mathcal{K}(H)$); then $P_n = \sum_{j=1}^{n^\varepsilon} R_n \in \mathcal{PF}(H)$, $P_n \to 1$ (strongly, as $n \to \infty$), $(1-P_{2n-1})(A-K_\varepsilon)P_{2n-1} = 0$ and $P_{2n}(A-K_\varepsilon)(1-P_{2n}) = 0$, so that $A-K_\varepsilon$ and $(A-K_\varepsilon)^*$ are triangular operators. Hence $A-K_\varepsilon \in (BQT)$. A fortiori (since K_ε is compact), $A \in (BQT)$.

(iii) => (xiv) Define $P_{-1} = P_0 = 0$ and

$$K_\varepsilon = \sum_{n=1}^{\infty}\{(1-P_{2n-1})A(P_{2n-1}-P_{2n-3})+(P_{2n}-P_{2n-2})A(1-P_{2n})\}. \qquad (6.11)$$

Since $(1-P_{2n-1})A(P_{2n-1}-P_{2n-3})$ and $(P_{2n}-P_{2n-2})A(1-P_{2n})$ are finite rank operators and

$$\max\{\|(1-P_{2n-1})A(P_{2n-1}-P_{2n-3})\|,\|(P_{2n}-P_{2n-2})A(1-P_{2n})\|\}$$

$$\leq \max\{\|(1-P_{2n-1})AP_{2n-1}\|,\|P_{2n}A(1-P_{2n})\|\} < \varepsilon/4^n,$$

it is easily seen that the series (6.11) converges in the norm, $K_\varepsilon \in K(H)$ and $\|K_\varepsilon\| < 2\sum_{n=1} \varepsilon/4^n < \varepsilon$.

It is straightforward to check that $A - K_\varepsilon$ admits the matrix representation (6.8.). (Roughly: $A - (1-P_1)A(P_1-P_{-1})$ has 0's everywhere in the first column, except for the term A_1; $A - (1-P_1)A(P_1-P_{-1}) - (P_2-P_0)A(1-P_2)$ has 0's everywhere in the first two rows except for the terms A_1, B_1, C_1, etc.)

(i) => (xvi) Since $A \in$ (BQT), it follows that $\sigma_{\ell re}(A) = \sigma_e(A)$. By Proposition 4.29, there exists $S \simeq_a A$, $A - S \in K(H)$, $\|A - S\| < \varepsilon/2$ and

$$S = A'\oplus\begin{pmatrix} N_\ell & * & * \\ 0 & B & * \\ 0 & 0 & N_r \end{pmatrix}$$

where $A' \simeq A$ and N_ℓ and N_r are unitarily equivalent diagonal normal operators of uniform infinite multiplicity such that $\sigma(N_\ell) = \sigma_e(N_\ell) = \sigma(N_r) = \sigma_e(N_r) = \sigma_e(A)$. Assume that $N_\ell \simeq N_r \simeq (\text{diag}\{\lambda_1,\lambda_2,\ldots\})^{(\infty)}$. Let $\{e_n\}_{n=1}^\infty$ be an ONB of H.

Now we proceed essentially as in the proof of Theorem 6.4(iii) => (i). First we find $P_1 \in PF(H)$ such that ran P_1 contains e_1 and a vector $f_1^1 \in \ker(N_\ell-\lambda_1)$, and $\|(1-P_1)SP_1\| < \varepsilon/5$. Now we find $P_2 \in PF(H)$ such that ran P_2 contains ran P_1, e_2 and $g_1^1, g_1^2 \in \ker(N_r-\lambda_j)$, $g_2^j \perp$ ran P_1 $(j = 1,2)$, and $\|P_2S(1-P_2)\| < \varepsilon/5$.

Assume that $P_1 \leq P_2 \leq \ldots \leq P_{2n}$ have been chosen. Now we choose $P_{2n+1} \in PF(H)$ such that ran P_{2n+1} contains ran P_{2n}, e_{2n+1} and vectors $f_{n+1}^1, f_{n+1}^2, \ldots, f_{n+1}^{n+1}$, where $f_{n+1}^j \in \ker(N -\lambda_j)$ and $f_{n+1}^j \perp$ ran P_{2n} $(j = 1, 2,\ldots,n+1)$, and $\|(1-P_{2n+1})SP_{2n+1}\| < \varepsilon/5^{n+1}$; then we choose $P_{2n+2} \in PF(H)$ such that ran P_{2n+2} contains ran P_{2n+1}, e_{2n+2} and vectors $g_{n+1}^1, g_{n+1}^2,\ldots,g_{n+1}^{n+1}$, where $g_{n+1}^j \in \ker(N_r-\lambda_j)$ and $g_{n+1}^j \perp$ ran P_{2n+1} $(j=1,2,\ldots,n+1)$, and $\|P_{2n+2}S(1-P_{2n+2})\| < \varepsilon/5^{n+1}$.

By induction, we have constructed a sequence $\{P_n\}_{n=1}^\infty$ satisfying (iii) with A replaced by S and ε replaced by $\varepsilon/2$. Define $P_1 = P_0 = 0$ and let

$$C_\varepsilon = \sum_{n=1}^\infty\{(1-P_{2n-1})S(P_{2n-1}-P_{2n-3})+(P_{2n}-P_{2n-2})S(1-P_{2n})\}.$$

It follows from the previous step (see (6.11)) that $C_\varepsilon \in K(H)$, $\|C_\varepsilon\| < \varepsilon/2$, $K_\varepsilon = (S-A)+C_\varepsilon \in K(H)$, $\|K_\varepsilon\| < \varepsilon$ and $A - K_\varepsilon$ admits a staircase representation of the form (6.8). Furthermore, by construction,

λ_n is an eigenvalue of A_k and an eigenvalue of C_k for infinitely many values of k, for each $n = 1,2,\ldots$. More precisely, $\lambda_n \in \sigma_p(A_k)$ and $\lambda_n \in \sigma_p(C_k)$ for all $k \geq n$. Hence

$$\{\lambda_n\}^- = \sigma_{\ell re}(A) \subset \sigma_{\ell re}(\oplus_{k=1}^{\infty} A_k) \cap \sigma_{\ell re}(\oplus_{k=1}^{\infty} C_k). \tag{6.12}$$

Now observe that

$$A - K_\varepsilon = \left(\begin{array}{ccc|c|cc|c|cccc} A_1 & 0 & B_1 & 0 & 0 & 0 & 0 & \cdot & \cdot & \cdot \\ & A_2 & D_1 & 0 & B_2 & 0 & 0 & \cdot & \cdot & \cdot \\ & & C_1 & 0 & 0 & 0 & 0 & \cdot & \cdot & \cdot \\ & & & A_3 & D_2 & 0 & B_3 & \cdot & \cdot & \cdot \\ & & & & C_2 & 0 & 0 & \cdot & \cdot & \cdot \\ & & & & & A_4 & D_3 & \cdot & \cdot & \cdot \\ & & & & & & C_3 & \cdot & \cdot & \cdot \\ & & 0 & & & & & \cdot & & \\ & & & & & & & & \cdot & \\ & & & & & & & & & \cdot \end{array}\right) \begin{array}{l} R_1 \\ R_3 \\ R_2 \\ R_5 \\ R_4 \\ R_7 \\ R_6 \\ \cdot \\ \cdot \\ \cdot \end{array} \tag{6.13}$$

Since the column of A_n contains A_n in the main diagonal and 0's in any other entry, it is easily seen that

$$A-K_\varepsilon = \left(\begin{array}{cccc|cccc} A_1 & & & & B_1 & & & \\ & A_2 & & & D_1 & B_2 & & \\ & & A_3 & 0 & & D_2 & B_3 & 0 \\ & & & \cdot & & & \cdot & \cdot \\ & 0 & & & & & & \cdot \\ & & & \cdot & & & & \cdot \quad \cdot \\ \hline & & & & C_1 & & & \\ & & & & & C_2 & & \\ & & & & & & C_3 & 0 \\ & & 0 & & & 0 & & \cdot \\ & & & & & & & \cdot \\ & & & & & & & \cdot \end{array}\right) \begin{array}{l} R_1 \\ R_3 \\ R_5 \\ \cdot \\ \cdot \\ \cdot \\ R_2 \\ R_4 \\ R_6 \\ \cdot \\ \cdot \\ \cdot \end{array} = \begin{pmatrix} \oplus_{n=1}^{\infty} A_n & B \\ 0 & \oplus_{n=1}^{\infty} C_n \end{pmatrix} \begin{array}{l} H_1 \\ H_2 \end{array},$$

where $H_1 = \oplus_{n=1}^{\infty} R_{2n-1}$ and $H_2 = \oplus_{n=1}^{\infty} R_{2n}$.

Clearly, $X_1 = \oplus_{n=1}^{\infty} A_n$ and $X_2 = \oplus_{n=1}^{\infty} C_n$ are block-diagonal operators (acting on H_1 and H_2, respectivley) and $\sigma_e(X_1) \cup \sigma_e(X_2) \subset \sigma_e(A) = \sigma_{\ell re}(A)$. Combining this observations with (6.12) we conclude that $\sigma_e(X_1) = \sigma_e(X_2) = \sigma_e(A)$.

(i) => (xvii) Let K_ε, $X_1 = \oplus_{n=1}^{\infty} A_n$ and $X_2 = \oplus_{n=1}^{\infty} C_n$ be the operators constructed in the previous step. Recall that $\lambda_n \in \sigma_p(A_k) \cap \sigma_p(C_k)$ for all $k \geq n$. Thus, if $k \geq n$ we have

$$A_{2k-1}\oplus A_{2k} = \begin{pmatrix} \lambda_n & * & 0 \\ 0 & E_{n,k} & 0 \\ 0 & 0 & A_{2k} \end{pmatrix} \begin{matrix} M_{n,k} \\ R_{2k-1}\ominus M_{n,k} \\ R_{2k} \end{matrix} = \begin{pmatrix} \lambda_n & * \\ 0 & F_{n,k} \end{pmatrix} \begin{matrix} M_{n,k} \\ N_{n,k} \end{matrix},$$

where $M_{n,k}$ is one-dimensional, $F_{n,k} = E_{n,k}\oplus A_{2k} \in L(N_{n,k})$, $N_{n,k} = (R_{2k-1}\oplus R_{2k})\ominus M_{n,k}$ and $\lambda_n \in \sigma_p(F_{n,k})$.

The same argument, applied to $(A_{2k-1}\oplus A_{2k})^*$ shows that

$$A_{2k-1}\oplus A_{2k} = \begin{pmatrix} G_{n,k} & * \\ 0 & \lambda_n \end{pmatrix} \begin{matrix} N'_{n,k} \\ M'_{n,k} \end{matrix},$$

where $M'_{n,k}$ is one-dimensional, $G_{n,k} \in L(N'_{n,k})$, $N'_{n,k} = (R_{2k-1}\oplus R_{2k})\ominus M'_{n,k}$ and $\lambda_n \in \sigma_p(G_{n,k})$ $(k \geq n)$.

Since we have infinitely many possible choices of k for each n, we can easily see that

$$X_1 = \oplus_{n=1}^{\infty} A_n = [\oplus_{k=1}^{\infty}(A_{4k-3}\oplus A_{4k-1})]\oplus[\oplus_{k=1}^{\infty}(A_{4k-2}\oplus A_{4k})]$$

$$= \begin{pmatrix} \oplus_{n=1}^{\infty} G_n & * \\ 0 & N \end{pmatrix} \oplus \begin{pmatrix} N & * \\ 0 & \oplus_{m=1}^{\infty} F_m \end{pmatrix},$$

where $N = (\mathrm{diag}\{\lambda_1,\lambda_2,\ldots\})^{(\infty)}$, the operators G_n have the form G_{n,k_n} for a suitable subsequence $\{k_n\}_{n=1}^{\infty}$ and the operators F_m have the form F_{m,k_m} for a suitable subsequence $\{k_m\}_{m=1}^{\infty}$, so that $\oplus_{n=1}^{\infty} G_n$ and $\oplus_{m=1}^{\infty} F_m$ are block-diagonal operators. Moreover, $\lambda_r \in \sigma_p(G_n)\cap\sigma_p(F_m)$ for all n and m large enough (depending on r), whence we easily conclude that $\sigma_e(\oplus_{n=1}^{\infty} G_n) = \sigma_e(\oplus_{m=1}^{\infty} F_m) = \sigma(N) = \sigma_e(N) = \sigma_e(A)$.

Define $D_1 = \oplus_{n=1}^{\infty} G_n$, $D_2 = \oplus_{m=1}^{\infty} F_m$; then

$$X_1 = \begin{pmatrix} D_1 & X'_1 \\ 0 & N \end{pmatrix} \oplus \begin{pmatrix} N & X''_1 \\ 0 & D_2 \end{pmatrix} = \begin{pmatrix} N & 0 & 0 & X''_1 \\ 0 & D_1 & X'_1 & 0 \\ 0 & 0 & N & 0 \\ 0 & 0 & 0 & D_2 \end{pmatrix} = \begin{pmatrix} N\oplus D_1 & * \\ 0 & N\oplus D_2 \end{pmatrix}.$$

Similarly, we can obtain

$$X_2 = \begin{pmatrix} N\oplus D_3 & * \\ 0 & N\oplus D_4 \end{pmatrix}$$

and, a fortiori,

$$A - K_\varepsilon = \begin{pmatrix} X_1 & * \\ 0 & X_2 \end{pmatrix} = \begin{pmatrix} N\oplus D_1 & * & * & * \\ 0 & N\oplus D_2 & * & * \\ 0 & 0 & N\oplus D_3 & * \\ 0 & 0 & 0 & N\oplus D_4 \end{pmatrix}.$$

(xvii) => (xiii) Observe that $\sigma_e(N\oplus D_j) = \sigma_{\ell re}(N\oplus D_j) = \sigma_e(A-K_\varepsilon) = \sigma_{\ell re}(A-K_\varepsilon) = \sigma_e(A) = \sigma_{\ell re}(A)$, j = 1,2,3,4. Now the result follows exactly as in (xv) => (xiii).

The proof of Theorem 6.15 is complete now. □

6.4 On the relative size of the sets (QT), (QT)*, (BQT), $[N(H)+K(H)]^-$ and $N(H)^-$

THEOREM 6.17. *Each of the following inclusions is proper:*

$$N(H)^- \subset [N(H)+K(H)]^- \subset (BQT) \subset (QT) \subset L(H).$$

Moreover, each of these subsets (except $L(H)$) is nowhere dense in the next one.

Furthermore, the same is true for the chain of inclusions

$$N(H)^- \subset [N(H)+K(H)]^- \subset (BQT) \subset (QT)^* \subset L(H).$$

and $(QT) \cap (QT)^*$ *is nowhere dense in* $L(H)$.

PROOF. Consider the first chain of inclusions. Since each set in the chain is closed, in order to prove that a set is nowhere dense in the next one it suffices to show that its complement is dense in the containing set.

(i) Observe that "$T \notin (QT)$" is a "bad property" (as defined in Section 3.5.2) and the unilateral shift is a concrete example of an non-quasitriangular operator. By Theorem 3.51 $L(H)\setminus(QT)$ is dense in $L(H)$.

(ii) Let $T \in (BQT)$ and $\varepsilon > 0$. By Proposition 4.29, there exists $T_1 \in (BQT)$, $||T - T_1|| < \varepsilon/2$ such that

$$T_1 = \begin{pmatrix} \lambda & * \\ 0 & T_2 \end{pmatrix}$$

for some $\lambda \in \partial\sigma_e(T) \cap \partial\rho(T)$, where $\sigma(T_2) = \sigma(T)$. (Take $\Gamma_\ell = \{\lambda\}$, $\Gamma_r = \emptyset$.) Proceeding as in the proof of Theorem 3.51, we can find

$$T_\varepsilon = \begin{pmatrix} \lambda+\eta+\delta S^* & * \\ 0 & T_2 \end{pmatrix},$$

where $\varepsilon > 0$ and S^* is a backward shift of multiplicity one, such that $\sigma(\lambda+\eta+\delta S^*) \cap \sigma(T_2) = \emptyset$, T_2 is biquasitriangular and $||T - T_\varepsilon|| < \varepsilon$. Clearly, $T_\varepsilon \in (QT)\setminus(BQT)$. Hence, $(QT)\setminus(BQT)$ is dense in (QT).

(iii) Let $T \in [N(H)+K(H)]^-$ and $\varepsilon > 0$. If $\mu \in \sigma_e(T)_{\varepsilon/2}\setminus\sigma_e(T)$, then $\mu \neq 0$ and it follows from Theorem 5.34 that there exists $Q \in N(H)+K(H)$ such that $||T - (\mu+Q)|| < \varepsilon$. Clearly, $\mu+Q \in (BQT)\setminus[N(H)+K(H)]^-$ and therefore $(BQT)\setminus[N(H)+K(H)]^-$ is dense in (BQT).

(iv) If $T \in N(H)^-$, $\varepsilon > 0$ and $\mu \in \partial\sigma(T) \cap \sigma_{\ell re}(T)$, then we can use the argument of the first part of the proof of Theorem 3.49 in order to obtain a finite rank operator F_ε, $\|F_\varepsilon\| < \varepsilon$, such that $\sigma_0(T+F_\varepsilon)\cap D(\mu,\varepsilon) \neq \emptyset$. It is easily seen that $T+F_\varepsilon \in [N(H)+K(H)]^- \setminus N(H)^-$, so that $[N(H)+K(H)]^- \setminus N(H)^-$ is dense in $[N(H)+K(H)]^-$.

By taking adjoints, we conclude that (BQT) is nowhere dense in (QT)* and (QT)* is nowhere dense in $L(H)$. Hence, both (QT) and (QT)* are nowhere dense in $L(H)$. A fortiori, so is their union. □

REMARK 6.18. Theorem 6.17 indicates that $N(H)^-$, $[N(H)+K(H)]^-$, (BQT), (QT), (QT)* and (QT)∪(QT)* are "very small" subsets of $L(H)$. However (as proved in Theorem 5.15), even $N(H)^-$ is large enough to guarantee that $L(H) = N(H)^- + N(H)^-$.

On the other hand, we also have

$$\{L(H)\setminus[(QT)\cup(QT)^*]\}+\{L(H)\setminus[(QT)\cup(QT)^*]\} = L(H).$$

Indeed, by Proposition 4.21(ii), given $T \in L(H)$ there exists $K \in K(H)$ such that $T-K = A\oplus B$ (A, B acting on infinite dimensional spaces). If S is the unilateral shift of multiplicity one and $r > 2\|T\|$, then $T = T_1+T_2$, where

$$T_1 = (A+2r+rS)\oplus(B-2r-rS^*)+K \quad\text{and}\quad T_2 = (-2r-rS)\oplus(2r+rS^*).$$

It can be easily checked (use Theorem 1.13(iii) and (v)) that T_1+2r and T_2-2r are Fredholm operators of index 1, and T_1-2r and T_2+2r are Fredholm operators of index -1. Hence, $T_1, T_2 \in L(H)\setminus[(QT)\cup(QT)^*]$.

6.5 A Riesz decomposition theorem for operators with disconnected essential spectrum

The classical Riesz decomposition theorem asserts that if $\sigma(T)$ is the disjoint union of two nonempty clopen subsets σ_1 and σ_2, then the space can be written as the algebraic direct sum $H = H_1\dot{+}H_2$ of two invariant subspaces of T, H_1 and H_2, such that $\sigma(T|H_j) = \sigma_j$, $j=1,2$. (See Chapter I.) This result is false, in general, if we merely assume that the essential spectrum splits into two clopen subsets. Namely, if $T = M_-(\partial\Omega)$, where Ω is the annulus $\{\lambda \in \mathbb{C}: 1 < |\lambda| < 3\}$, then such a decomposition is impossible: If $H = H_1\dot{+}H_2$, where H_1 and H_2 are invariant under T, $\sigma_e(T|H_1) = \{\lambda: |\lambda| = 1\}$ and $\sigma_e(T|H_2) = \{\lambda: |\lambda| = 3\}$, then ind $(\lambda-T|H_1) = 0$ for all λ in Ω and therefore $\mathrm{ind}(\lambda-T|H_2) = -1$ for all λ in Ω. Since $\sigma_e(T|H_2)\cap D(0,1)^- = \emptyset$, we conclude that $\mathrm{ind}(\lambda-T|H_2) = -1$ for all λ in $D(0,3)$. Therefore, $\mathrm{ind}(\lambda-T|H_1) = \mathrm{ind}(\lambda-T) - \mathrm{ind}(\lambda-T|H_2) = +1$ for all $\lambda \in D(0,1)$ and this implies that $D(0,1) \subset \sigma_p(T)$, a contradiction.

A more careful analysis indicates that $\sigma_e(M_+(\partial\Omega)+A)$ splits into two nonempty clopen subsets σ_1 and σ_2, but $M_+(\partial\Omega)+A$ does not have two complementary invariant subspaces, H_1 and H_2, such that $\sigma_e(M_+(\partial\Omega)+A|H_1) = \sigma_1$ and $\sigma_e(M_+(\partial\Omega)+A|H_2) = \sigma_2$, for any $A \in L(H)$ such that $||A|| < 1$. However, it readily follows from the BDF theorem that $M_+(\partial\Omega)$ is unitarily equivalent to a compact perturbation of $3S\oplus S^*$, where S denotes the unilateral shift of multiplicity one. Clearly, $\sigma_e(3S) = \{\lambda: \ |\lambda| = 3\}$ and $\sigma_e(S^*) = \{\lambda: \ |\lambda| = 1\}$.

Consider the bilateral shift B defined by $Be_n = e_{n+1}$ for $n \le 0$ and $Be_n = 2^{-n}e_{n+1}$ for $n > 0$ with respect to an ONB $\{e_n\}_{-\infty<n<\infty}$; then $\sigma(B) = D(0,1)^-$, $\sigma_e(B) = \{0\}\cup\partial D(0,1)$ and $\text{ind}(\lambda-B) = \text{nul}(\lambda-B) = 1$ for all $\lambda \in D(0,1)\setminus\{0\}$. Either by general results about the structure of bilateral weighted shifts (see, e.g., [103], [131], [184]) or by a direct argument, it can be shown that B does not have any pair of complementary invariant subspaces. However, if B_m is the result of replacing the m-th weight by a 0 (i.e., $B_m = B-2^{-m}e_{m+1}\otimes e_m$) for some $m > 0$, then B_m admits two complementary reducing subspaces $H_1 = \vee\{e_n\}_{n\le m}$ and $H_2 = \vee\{e_n\}_{n=m+1}^{\infty}$ such that $\sigma_e(B_m|H_2) = \{0\}$ and $\sigma_e(B_m|H_1) = \partial D(0,1)$. (Indeed, $B_m|H_2$ is a compact quasinilpotent and $B_m|H_1 \sim S^*$!)

These two examples illustrate both the possibility of restoring the validity of Riesz decomposition theorem for some small compact perturbation of T (where $T \in L(H)$ has a disconnected essential spectrum) and the "index obstructions" that make it an impossible attempt.

PROPOSITION 6.19. *Let* $T \in L(H)$ *and let* Γ *be a clopen nonempty subset of* $\sigma_{\ell re}(T)$ *contained in a unique component* Ω *of* $\rho_{s-F}(T)\cup\Gamma$ *such that*

$$\infty > \inf\{\text{ind}(\lambda - T): \ \lambda \in \Omega\cap\rho_{s-F}(T)\} = n > 0 \qquad (6.14)$$

and $\text{ind}(\lambda - T) = n$ *for every* λ *in a component* Φ *of* $\Omega\setminus\Gamma$ *whose boundary* $\partial\Phi$ *intersects* $\partial\Omega$.

Given $\varepsilon > 0$, *there exists* $K \in K(H)$, $||K|| < \varepsilon$, *such that*

$$T - K \sim A\oplus R,$$

where R *is a smooth operator,* $\sigma_{\ell re}(R) = \Gamma$, $\sigma(R) = \Gamma\cup\{\lambda \in \Omega\cap\rho_{s-F}(T): \text{ind}(\lambda - T) > n\}$, $\sigma_e(A)\cap\Omega = \emptyset$ *and* $\text{nul}(\lambda - A) = n$ *and* $\text{nul}(\lambda-A)^* = 0$ *for all* $\lambda \in \Omega$.

PROOF. Proceeding as in the proof of Theorem 4.11, we can find $C \in K(H)$ such that

$$T - C = \begin{pmatrix} V & * \\ 0 & B \end{pmatrix},$$

where $V \simeq N_+(\Omega_o)^{*(n)}$, $\Omega_o = (\text{interior } \Omega^-)^*$, $\sigma(B) = \Gamma\cup\sigma_o$, $\sigma_{\ell re}(B) = \Gamma\cup\sigma_1$,

$\sigma_1 \cap \Omega = \emptyset$ and $\sigma_0 \cap \Omega = \{\lambda \in \Omega \cap \rho_{s-F}(T): \text{ind}(\lambda - T) > n\}$ (obviously, $\text{ind}(\lambda-B) = \text{ind}(\lambda-T) - n$, for all $\lambda \in \sigma_0 \cap \Omega$).

It readily follows that B is the algebraic (not necessarily orthogonal) direct sum of an operator G whose spectrum does not intersect Ω and an operator J' such that $\sigma_{\ell re}(J') = \Gamma$ and $\sigma(J') = \Gamma \cup (\sigma_0 \cap \Omega)$. Furthermore, $J' \in (QT)$ because $\text{ind}(\lambda - J') \geq 0$ for all $\lambda \in \rho_{s-F}(J')$. (Use Theorem 6.4.) Hence,

$$T - C = \begin{pmatrix} V & * & * \\ 0 & G & * \\ 0 & 0 & J \end{pmatrix} \begin{matrix} H_1 \\ H_2 \\ H_3 \end{matrix} = \begin{pmatrix} W & * \\ 0 & J \end{pmatrix} \begin{matrix} H_1 \oplus H_2 \\ H_3 \end{matrix},$$

where $J \sim J'$ and

$$W = \begin{pmatrix} V & * \\ 0 & G \end{pmatrix} \in L(H_0),\ H_0 = H_1 \oplus H_2.$$

Let

$$C = \begin{pmatrix} C_{00} & C_{03} \\ C_{30} & C_{33} \end{pmatrix} \begin{matrix} H_0 \\ H_3 \end{matrix};$$

then

$$T = \begin{pmatrix} W + C_{00} & * \\ C_{30} & J + C_{33} \end{pmatrix} \begin{matrix} H_0 \\ H_3 \end{matrix}.$$

Since $J \in (QT)$ and C_{33} is compact, $J+C_{33}$ also belongs to (QT) and therefore (by Theorem 6.4) $J+C_{33} = J_0 + C_0$, where $(1-P_n)J_0P_n = 0$ for an increasing sequence $\{P_n\}_{n=1}^{\infty}$ in $PF(H_3)$ such that $P_n \to 1$ (strongly, as $n \to \infty$) and C_0 is compact. Furthermore, by Proposition 3.45 and its proof, C_0 can be chosen so that $\sigma_0(J_0) = \emptyset$. Since C_{30} is compact, the operator

$$K_1(n) = \begin{pmatrix} 0 & 0 \\ (1-P_n)C_{30} & (1-P_n)(J_0+C_0)P_n \end{pmatrix} \begin{matrix} H_0 \\ H_3 \end{matrix}$$

is compact and $\|K_1(n)\| \to 0$ $(n \to \infty)$.

Let $A_4(n)$ be the restriction of $T - K_1(n)$ to its invariant subspace $H_{4(n)} = H_0 \oplus \text{ran } P_n$ and let $M(n)$ be the compression of J_0+C_0 to the subspace $H_{5(n)} = \text{ran}(1-P_n)$.

Clearly, there exists n_0 such that if $K_1 = K_1(n_0)$, $A_4 = A_4(n_0)$ and $M = M(n_0)$, then $\|K_1\| < \varepsilon/2$ and $\sigma_0(M) \subset \Omega \cap \Gamma_{\varepsilon/2}$. By Proposition 3.45 and Theorem 3.48 there exist compact operators K_4 and K_5 such that $\|K_4\| < \varepsilon/2$, $\|K_5\| < \varepsilon/2$, $A = A_4 - K_4$ has the desired properties and $\sigma_0(M-K_5) = \emptyset$, so that $R = M - K_5$ also has the desired properties.

Thus, if $K_2 = K_4 \oplus K_5$ and $K = K_1 + K_2$, then $K \in K(H)$, $\|K\| < \varepsilon$ and

$$T - K = \begin{pmatrix} A & * \\ 0 & R \end{pmatrix}.$$

Since $\sigma(R)$ is contained in Ω and $\Omega \subset \rho_r(A)$, it follows from Corollary 3.22 that $T-K \sim A\oplus R$. □

REMARKS 6.20. (i) By taking adjoints, it is easily seen that Proposition 6.19 remains true if (6.14) is replaced by

$$-\infty < \sup\{\text{ind}(\lambda-T): \lambda \in \Omega\cap\rho_{s-F}(T)\} = n < 0 \qquad (6.15)$$

and "$\sigma(R) = \Gamma\cup\{\lambda \in \Omega\cap\rho_{s-F}(T): \text{ind}(\lambda-T) > n\}$, $\sigma_e(A)\cap\Omega = \emptyset$ and $\text{nul}(\lambda-A) = n$ and $\text{nul}(\lambda-A)^* = 0$ for all $\lambda \in \Omega$" is replaced by "$\sigma(R) = \Gamma\cup\{\lambda \in \Omega\cap\rho_{s-F}(T): \text{ind}(\lambda-T) < n\}$, $\sigma_e(A)\cap\Omega = \emptyset$ and $\text{nul}(\lambda-A) = 0$ and $\text{nul}(\lambda-A)^* = -n$ for all $\lambda \in \Omega$".

(ii) An especially important case of this proposition is the one corresponding to the case when Γ is a singleton (or, more generally, a finite subset of $\sigma_{\ell re}(T)$). This case will be analyzed with detail in Chapter VIII.

6.6 Notes and remarks

The concept of quasitriangularity plays a central role in the proofs of the Aronszajn-Smith theorem [36] on the existence of invariant subspaces for compact operators, and in the proofs of several generalizations [39], [51], [118]. However, this important concept stayed hidden for many years until P. R. Halmos explicitly stated it (and some of its related properties) in his article [121]. The notion of quasitriangularity immediately drew the attention of many operator theorists who used it to extend the previously known results on the existence of invariant subspaces for operators "somehow related to compact operators" (see, e.g., [9], [70], [158], [161], [192], [193]) or worked on the structure of the quasitriangular operators and its implications. The most important result in this line is the spectral characterization of quasitriangularity (i.e., the equivalence between (i) and (vii) in Theorem 6.4).

The equivalence between any two of the properties (i) - (vi) in Theorem 6.4 is due to P. R. Halmos [121] (see also [123,Problem 4]), who also proved that (QT) is closed, invariant under compact perturbations and closed under the formation of polynomials. (The simplified proof of the step (ii) => (iii) given here is due to G. R. Luecke [154].) That (QT) is also invariant under similarities is an observation of W. B. Arveson [121,p.291]; the first proof of this fact (very different from the present one) appears in [75].

In [121,Theorem 3], Halmos proved that the unilateral shift of multiplicity one is not quasitriangular. The argument of the proof was used by R. G. Douglas and C. M. Pearcy in [75,Theorem 1] to show that (i) => (vii).

The final step, (vii) => (i), proved to be the most difficult one: In [75], Douglas and Pearcy proved that every operator with a finite spectrum is quasitriangular. Halmos extended this result to operators T with the property that $\sigma(T)$ has analytic capacity zero [124]. (The reader is referred to [102] for the definition of analytic capacity. In particular, if $\sigma(T)$ has analytic capacity zero, then $\sigma(T)$ is totally disconnected and has planar Lebesgue measure zero.) C. Apostol proved in [4] that every decomposable operator (in the sense of C. Foiaş; see [67]) is quasitriangular, thus extending the previous results, and introduced the notion of "modulus of quasitriangularity" (q(T)). In [19] (see also [21]), C. Apostol, C. Foiaş and L. Szidó obtained several partial results about non-quasitriangular operators. Finally, the goal was attained by C. Apostol, C. Foiaş and D. Voiculescu in a series of papers published in the Revue Roumaine de Mathématiques Pures et Appliqueés: [21], [22], [23], [24], [25]. (The main result was previously announced in [20].) In fact, a first proof of the implication (vii) => (i) was obtained in [23] and a second one, including a stronger statement, was obtained in [25].

Simultaneously (and independently), L. A. Fialkow obtained several relevant results on the subject, including a proof of the equality q(T) = dist[T,(QT)].

A simplified proof of the main theorem of C. Apostol, C. Foiaş and D. Voiculescu was given by R. G. Douglas and C. M. Pearcy in [74]. The four proofs of this result (i.e., the two obtained in [23] and [25], Douglas and Pearcy's version and the one given here) strongly depend on constructive techniques of approximation. The present proof (largely based on Voiculescu's theorem, which was unknown at the time of publication of the other three proofs) provides a more natural approach.

The relevance of the achievement of C. Apostol, C. Foiaş and D. Voiculescu goes far from the significance of their main result. It is worth mentioning, at least, two other consequences: a) The introduction of a large number of constructive techniques that have become standard tools since then (including Theorem 3.49 of Chapter III). b) From a heuristic viewpoint, the result strongly suggested that the only real obstructions for many problems of approximation are those derived from the stability of the semi-Fredholm operators and the upper semicontinuity of separate parts of the spectrum. (This is, precisely, the aspect of the Approximation of Operators that has been continuously emphasized throughout this monograph.)

Finally, it is convenient to recall that the notion of quasitriangularity also admits several other alternative definitions; namely, R. G. Douglas and C. M .Pearcy [75] have also shown

THEOREM 6.21. (Addendum to Theorem 6.4) *The following conditions are equivalent for* $A \in L(H)$

(i) A *is quasitriangular.*

(viii) There exists $\varepsilon > 0$ *and a sequence* $\{F_n\}_{n=1}^{\infty}$ *of finite rank operators converging strongly to* 1 *such that each* F_n *is bounded below by* ε *on* $(\ker F_n)^{\perp}$ *and* $\|(1-F_n)AF_n\| \to 0$ $(n \to \infty)$.

(ix) There exists a sequence $\{K_n\}_{n=1}^{\infty}$ *of positive semi-definite compact operators that converges strongly to* 1 *and satisfies* $\|(1-K_n)AK_n\| \to 0$ $(n \to \infty)$.

Furthermore, if $(FC_+,\leq)$ denotes the directed set of all positive semi-definite finite rank contractions acting on H (partially ordered by the relation: $A_1 \leq A_2$ if and only if $\langle A_1x,x\rangle \leq \langle A_2x,x\rangle$ for all $x \in H$, as usual) and

$$q_a(T) = \lim\inf(A \to 1)\ \|(1-A)^{\frac{1}{2}}TA^{\frac{1}{2}}\|$$

$(A \in FC_+)$, then we have [21,Section 3]

THEOREM 6.22. *If* $T \in L(H)$, *then*

(i) $q_a(T) = q_a(T+K) = q_a(T\oplus 0) = q(T\oplus 0)$ *for all* $K \in K(H)$.

(ii) $\frac{1}{2}q_a(T) \leq q_a(T+\lambda) \leq q(T)$ *for all* $\lambda \in \mathbb{C}$.

(iii) $q(T) = \lim(\lambda \to \infty)\ q_a(T+\lambda)$.

In [192],[193], D. Voiculescu analyzed certain notions of "extended quasitriangularity" in Hilbert spaces (where the role of $K(H)$ is played by a Schatten p-class $C^p(H)$). On the other hand, P. Meyer-Nieberg defined quasitriangularity in arbitrary Banach spaces (the definition coincides with the usual one in case the underlying Banach space is a Hilbert space) and proved several invariant subspace theorems in this new setting [158]. C. Apostol [7], [8] and C. Apostol and D. Voiculescu [33] analyzed the properties of this newly defined quasitriangularity. Among many other results, the last paper extends the Apostol-Foiaş-Voiculescu's spectral characterization of quasitriangularity to the case when the Banach space is either c_o or ℓ^p, $1 \leq p < \infty$.

The introductory result of this chapter (Apostol-Morrel simple models) is one of the most important "tools" of approximation. Theorem 6.1 has been proved by C. Apostol and B. B. Morrel in [27] where it has been applied to obtain (among other results) a spectral characterization of the closure of the set of all operators with spectrum contained in a fixed nonempty subset of $\mathbb{C}$. (This result will be proved and analyzed in the second monograph.)

The existence of quasitriangular approximants (an operator T in (QT) is a quasitriangular approximant for $A \in L(H)$ if $||A-T|| = \text{dist}[A, (QT)]$) is due to C. Apostol, C. Foiaş and D. Voiculescu [21,Theorem 2.2]. The proof presented here (Lemma 6.7 and Propositions 6.8 and 6.9) is a refinement obtained by R. A. Smucker in his doctoral dissertation [188]. (Proposition 6.9 refines Halmos' result in [123,p.905].) The essential minimum modulus $m_e(.)$ has been analyzed by several authors; namely, in [95] the main theorem states that seven conditions are equivalent, and each of these conditions is equivalent to $m_e(T) = 0$. Proposition 6.10 (essentially equal to [53,Theorem 2]) sums up the most relevant properties of the essential minimum modulus and Corollary 6.11 is just a quantitative version of [75,Theorem 1].

Theorems 6.12 and 6.13 are due to P. R. Halmos [123,Problem 4]. Halmos' article and the doctoral dissertation of R. A. Smucker are the basic references for quasidiagonality from the viewpoint of approximation (The interested reader can also examine [149], [150], [155] and [197,Chapter XIV].) In particular, Theorem 6.14 is also due to R. A. Smucker [188,Theorem 5.5].

The equivalence between any two of the conditions (i) - (xiii) of Theorem 6.15 is either an immediate consequence of the Apostol-Foiaş-Voiculescu result on spectral characterization of quasitriangularity or a consequence of Voiculescu's theorem on algebraic limits of algebraic operators (Theorem 5.9). The equivalences between (i) and any of the conditions (xv) or (xvi) are due to C. Apostol and C. Foiaş [17]. The staircase representation (xiv) is a result of C. Foiaş, C. M. Pearcy and D. Voiculescu [98] and the representation of (xvii) is one of the steps of Voiculescu's proof of Theorem 5.9 [194,Proposition 3.2].

C. Apostol, C. Foiaş and D. Voiculescu [21,Theorem 2.8] proved that (QT) is nowhere dense in $L(H)$ (and, a fortiori, that $(QT) \cup (QT)^*$ is nowhere dense). In [155], G. R. Luecke proved that each of the sets of the chain

$$Nor(H) \subset Nor(H)+K(H) \subset (QD) \subset (QT)$$

(except (QT)) is nowhere dense in the containing set and D. A. Herrero [149,Theorem 3.1] proved that the same is true for the chain

$$(QD) \subset (BQT) \subset (QT).$$

Hence, Theorem 6.17 admits the following companion result.

THEOREM 6.23. *Each of the following inclusions is proper*

$$Nor(H) \subset Nor(H)+K(H) \subset (QD) \subset (BQT) \subset (QT) \subset (QT) \cup (QT)^* \subset L(H).$$

Moreover, each of these subsets (except $L(H)$*) is nowhere dense in the next one.*

Finally, the results of Section 6.5 are due to J. Barría and D. A. Herrero [42].

7 The structure of a polynomially compact operator

An operator T is called *polynomially compact* if $p(T) \in K(H)$ for some monic polynomial p. Every algebraic operator is polynomially compact ($p(T) = 0$) and so is the sum of an algebraic operator and a compact one. Indeed, this is the most general example of a polynomially compact operator (Theorem 7.2 below).

The structure of these operators did not play any special role in the previous chapters. The intrinsic reason was that in those chapters we considered the set of *all* algebraic operators (or *all* nilpotent operators) and the structure of the whole set "hides" the structure of many interesting subsets. In order to analyze several more specific approximation problems, we shall need a finer analysis of these operators. This is the content of the present chapter, as a preparation for the results of Chapter VIII on closures of similarity orbits.

7.1 Reduction to the (essentially) nilpotent case

The following auxiliary result will help to reduce many problems about polynomially compact operators to the case when the essential spectrum is the singleton $\{0\}$, i.e., to *essentially nilpotent* operators. (T is essentially nilpotent if $\tilde{T}$ is a nilpotent element of the Calkin algebra.)

LEMMA 7.1. *Let* $T \in L(H)$ *and assume that* $\sigma(T)$ *is the disjoint union of finitely many clopen subsets* $\sigma_1, \sigma_2, \ldots, \sigma_m$. *Let* $H_j = H(\sigma_j;T)$ *(*$j = 1,2,\ldots,m$*); then* $H = H_1 \dot{+} H_2 \dot{+} \ldots \dot{+} H_m$ *and*

(i) $S(T) = \{A \in L(H): A \sim \oplus_{j=1}^{m} A_j, A_j \sim T|H_j, j = 1,2,\ldots,m\}$;

(ii) If $A \in S(T)^-$ *and* $\sigma(A)$ *is the disjoint union of finitely many clopen subsets* $\sigma_1', \sigma_2', \ldots, \sigma_m'$ *such that* $\sigma_j' \supset \sigma_j$ *for each* $j = 1,2,\ldots,m$, *then* $A \sim \oplus_{j=1}^{m} A_j$, *where* $A_j \in S(T|H_j)^-$ *(*$j = 1,2,\ldots,m$*)*.

(iii) If T *is algebraic with minimal monic polynomial* p, $p(\lambda) = \Pi_{j=1}^{m} (\lambda - \lambda_j)^{k_j}$ *(*$\lambda_i \neq \lambda_j$ *if* $i \neq j$*), then* $\sigma(T)$ *is the disjoint union of* m *clopen (finite or denumerable) subsets* $\sigma_1, \sigma_2, \ldots, \sigma_m$ *such that* $\sigma_j \cap$

$\sigma_e(T) = \{\lambda_j\}$ *for each* $j = 1,2,\ldots,m$. *Furthermore, if* $A \in S(T)^-$ *then* $\sigma(A) = \sigma(T)$ *and* $\sigma_e(A) = \sigma_e(T) = \{\lambda_1,\lambda_2,\ldots,\lambda_m\}$.

(iv) If $\sigma(A) = \sigma(T)$ *for all* A *in* $S(T)^-$, *then*

$$S(T)^- = \{A \in L(H): \ A \sim \oplus_{j=1}^m A_j,\ A_j \in S(T|H_j)^-,\ j = 1,2,\ldots,m\}.$$

(v) If T *is polynomially compact,* $\tilde{T}$ *has minimal monic polynomial* p *(as in (iii)) and* $T|H_j = L_j+K_j$, *where* $L_j \in L(H_j)$, $(\lambda_j - L_j)^{k_j} = 0$ *and* $K_j \in L(H_j)$ *(for each* $j = 1,2,\ldots,m$*), then* $T = L+K$, *where* $L \in L(H)$, $p(L) = 0$ *and* $K \in K(H)$.

PROOF. (i) Since $\{\sigma_j\}_{j=1}^m$ is a finite family of pairwise disjoint compact sets, it follows from Corollary 3.22 that $T \sim \oplus_{j=1}^m T_j$, where $T_j \in L(M_j)$, $T_j \simeq T|H_j$, $j = 1,2,\ldots,m$. (Observe that $\sigma(A) = \sigma(T) = \cup_{j=1}^m \sigma_j$.) Assume that $W = (W_{ij})_{i,j=1}^m$, where $W_{ij}: M_j \to N_i$ $(i,j = 1,2,\ldots,m)$, is an invertible operator such that $\oplus_{j=1}^m A_j = W(\oplus_{j=1}^m T_j)W^{-1}$; then $(\oplus_{j=1}^m A_j) W = W(\oplus_{j=1}^m T_j)$. Proceeding exactly as in the proof of Corollary 3.22, we conclude that $W_{ij} = 0$ for all $i \neq j$ and W_{jj} is invertible for all $j = 1,2,\ldots,m$. Therefore $M_j \simeq N_j$ and $A_j = W_{jj}TW_{jj}^{-1} \sim T_j \simeq T|H_j$ for each $j = 1,2,\ldots,m$.

(ii) Proceeding as in (i), we can write $T \sim \oplus_{j=1}^m T_j$, $T_j \in L(M_j)$, and $A \sim \oplus_{j=1}^m A_j$, $A_j \in L(N_j)$, where $\sigma(T_j) = \sigma_j \subset \sigma_j' = \sigma(A_j)$, $j = 1,2,\ldots,m$. Clearly, it suffices to show that A_j is unitarily equivalent to an operator in $S(T_j)^-$ (for each $j = 1,2,\ldots,m$).

If $W_n = (W_{ij,n})_{i,j=1}^m (W_{ij,n}: M_j \to N_i,\ i,j = 1,2,\ldots,m;\ n = 1,2,\ldots)$ and $\|\oplus_{j=1}^m A_j - W_n(\oplus_{j=1}^m T_j)W_n^{-1}\| \to 0$ $(n \to \infty)$, then we can combine the argument of the proof of Corollary 3.22 (as in (i)) with a limit argument in order to show that $\|W_{ij,n}\| \to 0$ $(n \to \infty)$ for all $i \neq j$, $M_j \simeq N_j$ (so that we can directly assume that $M_j = N_j$) and $\|A_j - W_{jj,n}T_jW_{jj,n}^{-1}\| \to 0$ $(n \to \infty)$.

(iii) Clearly, $\sigma_e(T) = \sigma(T) = \{\lambda_1,\lambda_2,\ldots,\lambda_m\}$ and $\sigma(T)\setminus\sigma_e(T)$ can only accumulate at $\sigma_e(T)$, whence we immediate deduce that $\sigma(T)$ is the disjoint union of m clopen subsets $\sigma_1, \sigma_2,\ldots, \sigma_m$ such that $\sigma_j \cap \sigma_e(T) = \{\lambda_j\}$ $(j = 1,2,\ldots,m)$.

If $A \in S(T)^-$, then $\tilde{A} \in S(\tilde{T})^-$ and therefore (by Proposition 1.7) $p(\tilde{A}) = 0$. Hence $\sigma_e(A) \subset \sigma_e(T) = \{\lambda_1,\lambda_2,\ldots,\lambda_m\}$. It follows that $\sigma(A)$ is a finite or denumerable set and therefore it is a totally disconnected set. By Corollary 1.6, $\sigma_e(A) = \sigma_e(T)$ and $\sigma(A) = \sigma(T)$.

(iv) This is a corollary of (ii).

(v) Let W be an invertible operator such that $WTW^{-1} = \oplus_{j=1}^m T_j$, where $T_j \simeq T|H_j$. Then there exists operators $L_j' \simeq L_j$ and $K_j' \simeq K_j$ such that $T_j = L_j'+K_j'$ $(j = 1,2,\ldots,m)$. Then $T = L+K$, where $L = W^{-1}(\oplus_{j=1}^m L_j')W$ satisfies the polynomial equation $p(L) = 0$ and $K = W^{-1}(\oplus_{j=1}^m K_j')W \in K(H)$.

□

THEOREM 7.2. *Let* $T \in L(H)$ *and assume that* $p(\tilde{T}) = 0$ *for some polynomial* p, $p(\lambda) = \Pi_{j=1}^{m} (\lambda - \lambda_j)^{k_j}$. *Then there exists a compact operator* K *such that* $p(T-K) = 0$.

LEMMA 7.3. *Let* P, $R \in L(H)$ *be two orthogonal projections with compact product* PR; *then for every* $\varepsilon > 0$ *there exists a finite rank projection* $F \le R$ *such that* $\|P(R-F)\| < \varepsilon$.

PROOF. Let $\{F_n\}_{n=1}^{\infty}$ be an increasing sequence of finite rank projections converging strongly to R. Since $PR \in K(H)$, $\|PR(R-F_n)\| \to 0$ $(n \to \infty)$. Thus, for a sufficiently large n, $\|P(R-F_n)\| = \|PR(R-F_n)\| < \varepsilon$.

□

LEMMA 7.4. *Let* H *be a positive operator,* $0 \le H \le C$ *(where* C *is a constant); then there exist an invertible operator* W *and a sequence* $\{P_n\}_{n=1}^{\infty}$ *of pairwise orthogonal projections such that*

(i) $WH = HW = \sum_{n=1}^{\infty} 2^{-n} P_n$;

(ii) $P_n W = WP_n$ *and* $P_n H = HP_n$ *for all* $n = 1,2,\ldots$;

(iii) $\operatorname{ran} P_m H = \operatorname{ran} P_m \subset \operatorname{ran} H \subset \oplus_{n=1}^{\infty} \operatorname{ran} P_n = (\operatorname{ran} H)^{-}$ $(m = 1,2,\ldots)$;

(iv) $\|W\| \le 1/C$, $\|W^{-1}\| \le 2C$.

PROOF. Let $H = \int \lambda \, dE(\lambda)$ be the spectral decomposition of H and set $P_0 = P_{\ker H}$, $P_n = E((2^{-n}C, 2^{-(n-1)}C])$ $(n = 1,2,\ldots)$. The sequence $\{P_n\}_{n=1}^{\infty}$ is pairwise orthogonal, $HP_n = P_n H$ for all n, and it is also clear that the P_n's satisfy (iii). Define $W = \int_{[0,C]} f(\lambda)\, dE(\lambda)$, where

$$f(\lambda) = \begin{cases} 1, & \text{if } \lambda = 0, \\ 1/(2^n \lambda), & \text{if } \lambda \in (2^{-n}C, 2^{-(n-1)}C], \\ & n = 1,2,\ldots . \end{cases}$$

Then $WH = HW = \int_{[0,C]} \lambda f(\lambda)\, dE(\lambda) = \sum_{n=1}^{\infty} 2^{-n} P_n$, $WP_n = P_n W$ for all $n = 1,2,\ldots$, $\|W\| = \sup(\lambda \in [0,C])\, f(\lambda) \le 1/C$, W is bounded below by $\inf(\lambda \in [0,C])\, f(\lambda) = (2C)^{-1}$ (so that W is invertible) and $W^{-1} = \int_{[0,C]} f(\lambda)^{-1} dE(\lambda)$. Hence, $\|W^{-1}\| \le \sup(\lambda \in [0,C])\, f(\lambda)^{-1} = 2C$. □

COROLLARY 7.5. *Let* A, $B \in L(H)$ *and assume that* $AB \in K(H)$. *Then there exists a projection* $P \in L(H)$ *such that* AP *and* $(1-P)B$ *are both compact.*

PROOF. Let $A = UH_1$ and $B^* = VH_2$ be the polar decompositions of A and B^*. Then $H_1H_2 = U^*(AB)V$ is compact and this reduces our problem to show that there exists a projection P such that H_1P and $(1-P)H_2$ are both compact. Hence we can directly assume that $A = H_1$ and $B = H_2$ are positive semi-definite hermitian operators.

Two projections M and N will be constructed so that AM, NB, and (1-M)(1-N) are compact. Then (1-M)B will also be compact, since

$$(1-M)B = (1-M)NB + (1-M)(1-N)B.$$

By applying Lemma 7.4 to A and B, we obtain invertible operators W, V and two families of pairwise orthogonal projections $\{M_j\}_{j=1}^{\infty}$ (commuting with A) and $\{N_k\}_{k=1}^{\infty}$ (commuting with B) such that

$$WA = \sum_{j=1}^{\infty} 2^{-j} M_j,\ \sum_{j=1}^{\infty} M_j = 1 - P_{\ker A},\ BV = \sum_{k=1}^{\infty} 2^{-k} N_k \text{ and } \sum_{k=1}^{\infty} N_k = 1 - P_{\ker B}.$$

Thus, we can write

$$M_j WABVN_k = M_j \left(\sum_{m,n=1}^{\infty} 2^{-m-n} M_m N_n\right) N_k = 2^{-j-k} M_j N_k.$$

Therefore $M_j N_k$ is compact for each $j,k = 1,2,\ldots$. Thus the product $(\sum_{j=1}^{k} M_j)N_k$ is also compact. By applying Lemma 7.3 to this product, we obtain a finite rank projection $F_k \le N_k$ such that

$$\|(\sum_{j=1}^{k} M_j)(N_k - F_k)\| < 2^{-2k},\ k = 1,2,\ldots\ .$$

In particular, $\|M_j(N_k - F_k)\| < 2^{-k-j}$, $j \le k$, $k = 1,2,\ldots$.

Similarly, there exist finite rank projections $G_j \le M_j$ such that $\|(M_j - G_j)N_k\| < 2^{-k-j}$, $k \le j$, $j = 1,2,\ldots$. Thus for all positive integers k, j, we have

$$\|(M_j - G_j)(N_k - F_k)\| < 2^{-k-j}. \tag{7.1}$$

Define the projections M, N $\in L(H)$ by

$$M = (\sum_{j=1}^{\infty} G_j) + P_{\ker A},\ N = (\sum_{k=1}^{\infty} F_k) + P_{\ker B}.$$

Then

$$AM = (W^{-1}\sum_{j=1}^{\infty} 2^{-j} M_j)[(\sum_{k=1}^{\infty} G_k) + P_{\ker A}] = W^{-1}\sum_{j=1}^{\infty} 2^{-j} G_j,$$

which is compact because $2^{-j} \to 0$ $(j \to \infty)$ and $G_j G_i = 0$ if $i \ne j$.

Similarly, NB is compact. Observe that

$$(1-M) = [(\sum_{j=1}^{\infty} M_j) + P_{\ker A}] - [(\sum_{j=1}^{\infty} G_j) + P_{\ker A}] = \sum_{j=1}^{\infty} (M_j - G_j)$$

and

$$(1-N) = \sum_{k=1}^{\infty} (N_k - F_k).$$

Thus

$$(1-M)(1-N) = \sum_{j,k=1}^{\infty} (M_j - G_j)(N_k - F_k). \tag{7.2}$$

Now

$$(M_j - G_j)(N_k - F_k) = (M_j - G_j)M_j N_k (N_k - F_k)$$

is compact because $M_j N_k \in K(H)$. It follows from (7.1) that (1-M)(1-N) is the norm limit of the *compact* partial sums of (7.2). Hence,

$$(1-M)(1-N) \in K(H).$$ □

PROOF OF THEOREM 7.2. By Lemma 7.1, we can restrict ourselves to the case when $\tilde{T}^k = 0$ for some $k \geq 1$. We proceed by induction over k. If $\tilde{T}^1 = 0$, then T is compact. Suppose that $T^k \in K(H)$ for some $k \geq 2$ and the result is true for k-1.

By Corollary 7.5, there exists a projection P such that $T^{k-1}P$, $(1-P)T \in K(H)$; then

$$T = \begin{pmatrix} A & X \\ K_{21} & K_{22} \end{pmatrix} \begin{matrix} \text{ran } P \\ \text{ker } P \end{matrix},$$

where K_{21} and K_{22} are compact operators and $A = T_{\text{ran } P}$.

Observe that

$$\left(T - \begin{pmatrix} 0 & 0 \\ K_{21} & K_{22} \end{pmatrix}\right)^{k-1} P = \begin{pmatrix} A^{k-1} & * \\ 0 & 0 \end{pmatrix} P \in K(H).$$

Hence, $A^{k-1} \in K(PH)$. By our inductive hypothesis, there exists a compact operators $K_{11} \in L(\text{ran } P)$ such that $(A-K_{11})^{k-1} = 0$.

Define

$$K = \begin{pmatrix} K_{11} & 0 \\ K_{21} & K_{22} \end{pmatrix};$$

then

$$(T-K)^k = \begin{pmatrix} A-K_{11} & X \\ 0 & 0 \end{pmatrix}^{k-1} \begin{pmatrix} A-K_{11} & X \\ 0 & 0 \end{pmatrix} = \begin{pmatrix} 0 & * \\ 0 & 0 \end{pmatrix} \begin{pmatrix} A-K_{11} & X \\ 0 & 0 \end{pmatrix} = \begin{pmatrix} 0 & 0 \\ 0 & 0 \end{pmatrix} = 0. \quad \square$$

COROLLARY 7.6. *Let Q be an essentially nilpotent operator of essential order* k; *then there exists a sequence* $\{K_n\}_{n=1}^{\infty}$ *of compact operators such that*

(i) $\|K_n\| \to 0$ $(n \to \infty)$;

(ii) $\sigma(Q-K_n) = \{0\} \cup [\sigma(Q) \setminus D(0,1/n)]$, $n = 1,2,\ldots$;

(iii) $Q-K_n \sim R_n \oplus F_n$, *where* $F_n \simeq Q|H([\sigma_0(Q) \setminus D(0,1/n)];Q)$ *and* R_n *is nilpotent.*

Furthermore, if Q *is quasinilpotent, then the* K_n*'s can be chosen so that* rank $R_n^j \leq$ rank $Q^j \leq \infty$- *and* rank R_n^j *is finite for all* j *such that* $Q^j \in K(H)$.

PROOF. Let $\sigma_n = \sigma_0(Q) \setminus D(0,1/n)$ $(n = 1,2,\ldots)$. Clearly, σ_n is a clopen subset of $\sigma(Q)$ and

$$Q = \begin{pmatrix} F_n & X_n \\ 0 & A_n \end{pmatrix} \begin{matrix} M_n \\ N_n \end{matrix},$$

where $M_n = H(\sigma_n;Q)$, $F_n = Q|M_n$ and $A_n = Q_{N_n}$, so that $\sigma(A_n) = \sigma(Q) \setminus \sigma_n$.

By Proposition 3.45 (and its proof) there exists a normal compact operator $N_n \in K(N_n)$ such that $\|N_n\| = \text{sp}(A_n) < 1/n$ and $B_n = A_n - N_n$ is quasinilpotent. By hypothesis, $\tilde{Q}^k = 0$; it readily follows that $(\tilde{A}_n)^k = (\tilde{B}_n)^k = 0$. By Theorem 7.2, $B_n = L_n + C_n$, where $L_n^k = 0$ and $C_n \in K(N_n)$. Since

$\sigma(B_n)=\{0\}$, it follows from the upper semicontinuity of the spectrum (Corollary 1.6) and Proposition 3.45 that there exists a finite rank operator G_n such that $\|C_n - G_n\| < 1/n$ and $\sigma(L_n+G_n)=\{0\}$.

Define $R_n = L_n+G_n$ and

$$K_n = Q - \begin{pmatrix} F_n & X_n \\ 0 & R_n \end{pmatrix} = \begin{pmatrix} 0 & 0 \\ 0 & N_n+(C_n-G_n) \end{pmatrix}$$

Then $K_n \in K(H)$, $\|K_n\| < 2/n \to 0$ $(n\to\infty)$, i.e., $\{K_n\}_{n=1}^{\infty}$ satisfies (i). On the other hand, (ii) and (iii) are immediate consequences of the above constructions: Observe that $R_n^{\,k} = (L_n+G_n)^k$ is a finite rank quasinilpotent; therefore, $R_n^{\,k}$ is a nilpotent. A fortiori, so is R_n.

In the case when $\sigma(Q) = 0$, the R_n's can be chosen so as to satisfy the condition "rank $R_n^{\,j} \le$ rank $Q^j \le \infty$- and rank $R_n^{\,j}$ is finite for all j such that $Q^j \in K(H)$", by an "ad hoc" modification of the previous argument. □

Corollary 7.5 yields the following

COROLLARY 7.7. *Let* $\{A_j\}_{j=1}^{m}$ *be a finite family of operators acting on* H *and assume that the product* $A_1A_2\ldots A_m$ *is compact. Then there exist compact operators* $K_1, K_2,\ldots, K_m$ *such that*

$$(A_1-K_1)(A_2-K_2)\ldots\ldots(A_m-K_m) = 0.$$

PROOF. By Corollary 7.5 and an obvious inductive argument, we can find orthogonal projections $P_1, P_2,\ldots, P_{m-1}$ such that $K_1 = A_1P_1$, $K_j = (1-P_{j-1})A_jP_j$ $(j=2,3,\ldots,m-1)$ and $K_m = (1-P_{m-1})A_m$ are compact operators. Then,

$$\begin{aligned}
&(A_1-K_1)(A_2-K_2)(A_3-K_3)\ldots(A_{m-1}-K_{m-1})(A_m-K_m)\\
&= [A_1-A_1P_1][A_2-(1-P_1)A_2P_2][A_3-(1-P_2)A_3P_3]\ldots[A_{m-1}-(1-P_{m-2})A_{m-1}P_{m-1}]\\
&\times [A_m-(1-P_{m-1})A_m] = [A_1(1-P_1)][A_2(1-P_2)+P_1A_2P_2][A_3(1-P_3)+P_2A_3(1-P_3)]\\
&\ldots[A_{m-1}(1-P_{m-1})+P_{m-2}A_{m-1}P_{m-1}][P_{m-1}A_m]\\
&= [A_1(1-P_1)A_2(1-P_2)][A_3(1-P_3)+P_2A_3P_3]\ldots[A_{m-1}(1-P_{m-1})\\
&+P_{m-2}A_{m-1}P_{m-1}][P_{m-1}A_m] = \ldots = [A_1(1-P_1)A_2(1-P_2)A_3(1-P_3)\\
&\ldots A_{r-1}(1-P_{r-1})][A_r(1-P_r)+P_{r-1}A_rP_r][A_{r+1}(1-P_{r+1})+P_rA_{r+1}P_{r+1}]\\
&\ldots[A_{m-1}(1-P_{m-1})+P_{m-2}A_{m-1}P_{m-1}][P_{m-1}A_m] = [A_1(1-P_1)A_2(1-P_2)\\
&\ldots A_{m-1}(1-P_{m-1})][P_{m-1}A_m] = 0.
\end{aligned}$$

□

7.3 Restrictions of nilpotent operators

Recall that q_k is the $k \times k$ Jordan nilpotent cell ($k = 1,2,\ldots$) defined by (2.1).

LEMMA 7.8. *If* Q *is a nilpotent of order* $k > 1$ *and* M *is an invariant subspace of* Q *such that* $Q|M \sim q_k^{(\alpha)}$ *for some cardinal* α, $1 \le \alpha \le \infty$, *then there exists an invariant subspace* N *for* Q *such that* $M \cap N = \{0\}$, $M+N$ *is closed and* $H = M \dot{+} N$.

PROOF. Let $M_j = M \cap \ker Q^j$, $j = 1,2,\ldots,k$. By induction on j we shall prove that there exists an invariant subspace $N_j \subset \ker Q^j$ such that

1) $N_j \cap M_j = \{0\}$, and
2) $N_j + M_j = \ker Q^j$.

For $j = 1$ we can take $N_1 = \ker Q \ominus M_1$. Suppose N_j has been constructed for some $j < k$. We have $Q^{-1}(N_j) \cap M = M \cap \ker Q = M_1$. Define

$$N_{j+1} = [Q^{-1}(N_j) \ominus (N_j + M_1)] \oplus N_j.$$

It is completely apparent that $N_j + N_1$ is closed and that $N_{j+1} \cap M = \{0\}$. If $x \in \ker Q^{j+1}$, then $Qx = y_j + z_j$, where $y_j \subset M_j$ and $z_j \in N_j$. Since $Q|M \sim q_k^{(\alpha)}$, it follows that $y_j = Qy_{j+1}$ for some $y_{j+1} \in M_{j+1}$. Hence $x - y_{j+1} \in Q^{-1}(N_j) = N_{j+1} + M_1$, and therefore $x \in M_{j+1} + N_{j+1} + M_1 = M_{j+1} + N_{j+1}$.

Since N_j is invariant under Q ($j = 1,2,\ldots,k$), if we define $N = N_k$ we are done. □

LEMMA 7.9. *(i) Let* Q *be a nilpotent of order* k *and let* (5.1) *be the canonical representation of* Q*; then* $Q_{r,r+1}Q_{r+1,r+2}\cdots Q_{p-1,p}$ *maps* $\ker Q^p \ominus \ker Q^{p-1}$ *injectively into* $\ker Q^r \ominus \ker Q^{r-1}$ *for* $1 \le r < p$ *and* $p = 2,3,\ldots,k$.

(ii) Assume that $\operatorname{ran} Q^{k-1}$ *contains a subspace* M *of dimension* α. *Then there exist a subspace* $R_o \subset (\ker Q^{k-1})^\perp$ *of dimension* α *such that* $M = Q^{k-1}R_o$ *and a constant* $\delta > 0$ *such that*

$$\|Q_{r,r+1}Q_{r+1,r+2}\cdots Q_{p-1,p}(Q_{p,p+1}\cdots Q_{k-1,k}x)\| \ge \delta\|Q_{p,p+1}\cdots Q_{k-1,k}x\| \tag{7.3}$$

for all $x \in R_o$ *and for* $1 \le r \le p \le k-1$. *The subspace* $R = \bigvee_{j=0}^{k-1} Q^j R_o$ *is invariant under* Q *and is actually equal to the algebraic direct sum* $R_o \dot{+} QR_o \dot{+} Q^2R_o \dot{+} \ldots \dot{+} Q^{k-1}R_o$. *Furthermore,* $Q_1 = Q|R \sim q_k^{(\alpha)}$.

(iii) If, moreover, $M = \operatorname{ran} Q^{k-1}$ *and* $Q_2 = Q_{M^\perp}$, *then* $Q_2^{k-1} = 0$.

PROOF. (i) For each $p = 2,3,\ldots,k-1$, we have

$$Q^{p-1} = \begin{pmatrix} 0 & 0 & \cdots & 0 & Q_{12}Q_{23}\cdots Q_{p-1,p} & \\ & & \cdot & & 0 & \\ & 0 & & & \cdot & * \\ & & & & \cdot & \\ & & & & 0 & \end{pmatrix} \begin{matrix} \ker Q^1 \ominus \ker Q^0 \\ \ker Q^2 \ominus \ker Q^1 \\ \cdot \\ \cdot \\ \ker Q^k \ominus \ker Q^{k-1} \end{matrix}$$

(The first p-1 columns of the matrix of Q^{p-1} are 0's and the (1,p)-entry is equal to $Q_{12}Q_{23}\cdots Q_{p-1,p}$.)

By definition of the representation (5.1), it follows that $Q_{12}Q_{23}\cdots Q_{p-1,p}$ maps $\ker Q^p \ominus \ker Q^{p-1}$ into $\ker Q^1 \ominus \ker Q^0$ and $\ker(Q_{12}Q_{23}\cdots Q_{p-1,-p}) \subset \ker Q^{p-1} \cap [\ker Q^p \ominus \ker Q^{p-1}] = \{0\}$.

Therefore, $Q_{12}Q_{23}\cdots Q_{p-1,p}$ is injective. A fortiori, $Q_{r,r+1}Q_{r+1,r+2}\cdots Q_{p-1,p}$ is also injective for $1 \le r < p$.

(ii) Thus, if $M \subset \operatorname{ran} Q^{k-1} = Q_{12}Q_{23}\cdots Q_{k-1,k}[(\ker Q^{k-1})]^{\perp}$ is a subspace (hence, closed) of dimension α, then there exists a subspace $R_o \subset (\ker Q^{k-1})^{\perp}$ such that $Q_{12}Q_{23}\cdots Q_{k-1,k}: R_o \to M$ is an isomorphism; this shows, in particular, that $\dim R_o = \alpha$ and $Q_{12}Q_{23}\cdots Q_{k-1,k}|R_o$ is bounded below.

Now it is not difficult to show, by induction, that $Q_{r,r+1}$ maps $Q_{r+1,r+2}\cdots Q_{k-1,k}R_o$ isomorphically onto $Q_{r,r+1}Q_{r+1,r+2}\cdots Q_{k-1,k}R_o$ and therefore there exists a positive constant δ such that (7.3) is satisfied for all $x \in R_o$ and for $1 \le r \le p \le k-1$, whence it readily follows that $Q^j R_o$ must be closed (i.e., a subspace) for all $j = 0,1,2,\ldots,k-1$.

Let $y = Q^p x$ for some p, $0 \le p \le k-1$ and some $x \in R_o$, and let $z = Q^{p+1}x_{p+1} + Q^{p+2}x_{p+2} + \ldots + Q^{k-1}x_{k-1}$, where $x_{p+1}, x_{p+2}, \ldots, x_{k-1} \in R_o$; then

$$\begin{aligned}\|y-z\| &\ge (1/\|Q^{k-1-p}\|)\,\|Q^{k-1-p}(y-z)\| = (1/\|Q^{k-1-p}\|)\,\|Q^{k-1}x\| \\ &= (1/\|Q^{k-1-p}\|)\,\|Q_{12}Q_{23}\cdots Q_{k-1,k}x\| \ge (\delta/\|Q^{k-1-p}\|)\|x\| \\ &\ge (\delta/\|Q^{k-1-p}\|.\|Q^p\|)\,\|y\|,\end{aligned}$$

whence it follows that the subspaces $\vee\{Q^j R_o\}_{j=p+1}^{k-1}$ and $Q^p R_o$ form a positive angle. Therefore the algebraic direct sum $R = \sum_{j=0}^{k-1} Q^j R_o$ is closed and direct, i.e., $R = R_o \dot{+} QR_o \dot{+} Q^2R_o \dot{+} \ldots \dot{+} Q^{k-1}R_o$ is a subspace of H containing $M = Q^{k-1}R_o$.

It is completely apparent that R is invariant under Q and that the action of $Q|R$ can be described by: $Q: Q^j R_o \to Q^{j+1}R_o$ isomorphically for $j = 0,1,2,\ldots,k-1$, and $Q: Q^{k-1}R_o \to \{0\}$, whence it immediately follows that $Q_1 = Q|R \sim q_k^{(\alpha)}$.

(iii) Let

$$Q = \begin{pmatrix} Q_1 & * \\ 0 & Q_2 \end{pmatrix} \begin{matrix} R \\ R^{\perp} \end{matrix}$$

and assume that $M = \operatorname{ran} Q^{k-1} (\subset R)$. Observe that

$$Q^{k-1} = \begin{pmatrix} Q_1^{k-1} & * \\ 0 & Q_2^{k-1} \end{pmatrix}$$

and $Q^{k-1}H = M$. It follows that $\operatorname{ran} Q_2^{k-1} = \{0\}$, i.e., $Q_2^{k-1} = 0$. □

Corollary 7.6 implies, in particular, that if $\tilde{Q}^k = 0$, $\sigma(Q) = \{0\}$

and rank $Q^j = \infty-$ (i.e., Q^j is a compact operator, not of finite rank; see Section 1.3) for all $j \geq k$, then Q is the limit of a sequence $\{R_n\}_{n=1}^{\infty}$ of nilpotent operators such that rank $R_n^{\,j}$ < rank Q^j for all n and j (n, j = 1,2,...) and $Q-R_n \in K(H)$ for all n = 1,2,... . By using Corollary 7.7 we can obtain an analogous result for the case when $Q^k = 0$.

PROPOSITION 7.10. *Let* Q *be a nilpotent of order* k *and assume that* $Q^{k-1}, Q^{k-2}, \ldots, Q^r$ *are non-zero finite rank operators, and* rank $Q^j = \infty-$ *for* $j = s, s+1, \ldots, r-1$, *but* $Q^{s-1} \notin K(H)$ $(1 \leq s \leq r \leq k)$. *Then there exists a sequence* $\{R_n\}_{n=1}^{\infty}$ *in* $N_k(H)$ *such that* $Q - R_n \in K(H)$ *for all* n, rank $R_n^{\,j}$ = rank Q^j *for* $j = r, r+1, \ldots, k-1$, $\{\text{rank } R_n^{\,j}\}_{n=1}^{\infty}$ *is strictly increasing for each* j, $s \leq j \leq r-1$, *and* $\|Q - R_n\| \to 0$ $(n \to \infty)$.

PROOF. By Lemmas 7.9 and 7.8 and an obvious inductive argument, we have

$$Q = \begin{pmatrix} F & * \\ 0 & T \end{pmatrix} = W(F \oplus T)W^{-1},$$

where W is invertible, $F \sim \oplus_{j=r+1}^{k} q_j^{(\alpha_j)}$, the α_j's are non-negative integers, $\alpha_k = \text{rank } Q^{k-1} \neq 0$, $T^r = 0$ and rank $T^j = \infty-$ for $j = s, s+1, \ldots, r-1$, but T^{s-1} is not compact.

If $s = r$, there is nothing to prove. Assume that $1 \leq s < r \leq k$ and let

$$T = \begin{pmatrix} 0 & T_{12} & T_{13} & \cdot & \cdot & \cdot & T_{1,r-1} & T_{1r} \\ & 0 & T_{23} & \cdot & \cdot & \cdot & T_{2,r-1} & T_{2r} \\ & & 0 & \cdot & \cdot & \cdot & T_{3,r-1} & T_{3r} \\ & & & \cdot & & \cdot & \cdot & \cdot \\ & & & & \cdot & \cdot & \cdot & \cdot \\ & 0 & & & & \cdot & \cdot & \cdot \\ & & & & & & 0 & T_{r-1,r} \\ & & & & & & & 0 \end{pmatrix} \begin{matrix} M_1 \\ M_2 \\ M_3 \\ \cdot \\ \cdot \\ \cdot \\ M_{r-1} \\ M_r \end{matrix}$$

be the canonical matrix of $T \in L(M)$ with respect to the decomposition $M = \oplus_{j=1}^{r} M_j$, where $M_j = \ker T^j \ominus \ker T^{j-1}$, $j = 1,2,\ldots,r$.

Observe that $T_{12}T_{23}\ldots T_{r-1,r}$ is a compact operator with rank $\infty-$. Thus, by Corollary 7.7 there exists compact operators $K_{12}, K_{23}, \ldots, K_{r-1,r}$ such that $(T_{12}-K_{12})(T_{23}-K_{23})\ldots(T_{r-1,r}-K_{r-1,r}) = 0$.

Given $n \geq 1$, we can find compact operators $C_{j-1,j,n}$ $(j = 1,2,\ldots, r-1)$ such that $\|C_{j-1,j,n}\| < (rn)^{-1}$ and

$$n < \text{rank}(T_{12}-C_{12,n})(T_{23}-C_{23,n})\ldots(T_{r-1,r}-C_{r-1,r,n}) = d_n < \infty-.$$

Let T_n' be the operator obtained from T by replacing $T_{j-1,j}$ by $T_{j-1,j}-C_{j-1,j,n}$, $j = 2,3,\ldots,r$. It is easily seen that $T - T_n' \in K(H)$,

$\|T-T_n'\| < (rn)^{-1}$, $(T_n')^r = 0$, $\operatorname{rank}(T_n')^j \leq \infty-$ for $j = s,s+1,\ldots,r-1$, $(T_n')^{s-1}$ is not compact and $\operatorname{rank}(T_n')^{r-1} = d_n$ is finite.

Now we can repeat the same argument with T_n' in order to obtain $T_n'' \in N_r(M)$ such that $T - T_n'' \in K(M)$, $\|T - T_n''\| < 2/rn$, $\operatorname{rank}(T_n'')^j \leq \infty-$ for $j = s,s+1,\ldots,r-1$, $(T_n'')^{s-1}$ is not compact and $\operatorname{rank}(T_n'')^j$ is finite for $j = r-1,r-2$. After $r-s$ steps, we obtain an operator $T_n \in N_r(M)$ such that $T - T_n \in K(M)$, $\|T - T_n\| < 1/n$, rank T_n^j is finite and larger than n for $j = s,s+1,\ldots,r-1$ and T_n^{s-1} is not compact.

Passing, if necessary, to a subsequence we can assume that rank $\{T_n^j\}_{n=1}^{\infty}$ is strictly increasing for each j, $s \leq j \leq r-1$.

Finally, define $R_n = W(F \oplus T_n)W^{-1}$. It is completely apparent that the sequence $\{R_n\}_{n=1}^{\infty}$ satisfies all our requirements. □

7.4 Operators similar to Jordan operators

A Jordan operator J in $L(H)$ is an operator unitarily equivalent to

$$\oplus_{j=1}^{m} [\lambda_j + \oplus_{k=1}^{m_j} q_k^{(\alpha_{kj})}], \tag{7.4}$$

where $1 \leq m < \infty$, $\lambda_1,\ldots,\lambda_m$ are distinct complex numbers, $0 \leq \alpha_{kj} \leq \infty$ and m_j is finite for all $j = 1,2,\ldots,m$.

THEOREM 7.11. *If* $Q \in N(H)$, *then* Q *is similar to a Jordan operator if and only if* ran Q^j *is closed for all* $j = 1,2,\ldots$.

PROOF. If J is a Jordan nilpotent operator, then it is completely apparent from the definition (7.4) that ran T^j is closed for all $j = 1, 2,\ldots$. It is also clear that if $Q = WJW^{-1}$ for some invertible W, then ran $Q^j = W(\text{ran } J^j)$ is closed for all $j = 1,2,\ldots$.

Assume that $Q \in N_k(H)$ and ran Q^j is closed for all $j = 1,2,\ldots,k-1$. Since ran Q^{k-1} is closed, it follows from Lemmas 7.9 and 7.8 that Q has an invariant subspace M_k with invariant complement R_k such that $Q|M_k \sim q_k^{(\alpha_k)}$ $(0 \leq \alpha_k = (1/k)\dim M_k \leq \infty)$ and $Q|R_k \in N_{k-1}(R_k)$. Now we consider $Q|R_k$. By induction, we conclude that Q is similar to an operator of the form $\oplus_{j=1}^{k} q_j^{(\alpha_j)}$. □

It is easily seen that

$$\alpha_j = \alpha_j(Q) = \dim\,(\ker Q \cap \text{ran } Q^{j-1}) \ominus (\ker Q \cap \text{ran } Q^j)\ , \ j = 1,2,\ldots,k,$$

and that this set of cardinals is a similarity invariant. Theorem 7.11 says that this is, in fact, the *only* similarity invariant of Q. More precisely, we have the following.

COROLLARY 7.12. *(i) If* Q_1, $Q_2 \in N_k(H)$ *and* ran $Q_1{}^j$ *and* ran $Q_2{}^j$ *are closed for all* $j = 1,2,..,k-1$, *then* $Q_1 \sim Q_2$ *if and only if* $\alpha_j(Q_1) = \alpha_j(Q_2)$ *for all* $j = 1,2,...,k-1$.

(ii) If $Q \in N_k(H)$ *and* ran Q^j *is closed for all* $j = 1,2,...,k-1$, *then* $Q \sim Q^*$ *and* $Q \sim \lambda Q$ *for all* $\lambda \in \mathbb{C}\backslash\{0\}$.

By Lemma 7.1, we have

COROLLARY 7.13. *If* $T \in \mathcal{A}\ell g(H)$, *then* T *is similar to a Jordan operator if and only if* ran $q(T)$ *is closed for all polynomials* $q|p$, *where* p *is the minimal (monic) polynomial of* T.

Operators similar to Jordan operators are actually dense in $\mathcal{A}\ell g(H)$. We shall need an auxiliary result.

LEMMA 7.14. *Let* $Q \in N_k(H)$ *with canonical representation* (5.1). *Then* Q^j *has closed range for all* j $(j = 1,2,...,k-1)$ *if and only if* $Q_{j,j+1}$ *has closed range for all* j $(j = 1,2,...,k-1)$.

PROOF. Observe that

$$Q^j = \begin{pmatrix} 0 & \cdots & 0 & Q^{(j)}_{1,j+1} & & & & \\ 0 & \cdots & 0 & & Q^{(j)}_{2,j+2} & & * & \\ \vdots & & \vdots & & & \ddots & & \\ \vdots & & \vdots & & 0 & & \ddots & \\ 0 & \cdots & 0 & & & & & Q^{(j)}_{k-j,k} \\ 0 & \cdots & 0 & 0 & \cdot & \cdot & \cdot & 0 \\ \vdots & & \vdots & \vdots & & & & \vdots \\ 0 & \cdots & 0 & 0 & \cdot & \cdot & \cdot & 0 \end{pmatrix},$$

where $Q^{(j)}_{m,m+j} = Q_{m,m+1}Q_{m+1,m+2}\cdots Q_{m+j-1,m+j}$. If $Q_{i,i+1}$ has closed range (for all $i = 1,2,...,k-1$), then $Q_{i,i+1}$ is bounded below by $\delta > 0$ for all $i = 1,2,...,k-1$, and $Q^{(j)}_{m,m+j}$ is bounded below by δ^j, $m = 1,2,...,k-j$. Assume that $x^n = \sum_{i=j+1}^{k} x_i^n \in (\ker Q^j)^\perp$, $x_i^n \in \ker Q^i \ominus \ker Q^{i-1}$, $i = j+1, j+2, ..., k$, $\|x^n\| = 1$ for all $n = 1,2,...$, and $\|Q^j x^n\| \to 0$ $(n \to \infty)$. It is not difficult to see that

$$\delta^j \|x^n_{j+i}\| \le \|Q^{(j)}_{i,j+i}\, x^n_{j+i}\| \to 0 \ (n \to \infty),\ i = 1,2,...,k-j,$$

and, a fortiori, that $\|x^n\| \le \sum_{i=1}^{k-j} \|x^n_{j+i}\| \to 0$ $(n \to \infty)$, a contradiction.

Hence, ran $Q^j = Q^j[(\ker Q^j)]^\perp$ is closed, $j = 1,2,...,k-1$.

Conversely, if Q^j has closed range, then $Q^j|(\ker Q^j)^\perp$ is bounded below by some positive constant η. Then $Q^{(j)}_{1,j+1} = Q_{12}Q_{23}\cdots Q_{j,j+1}$ is also bounded below by η and therefore it is left invertible $(j = 1,2,...,k-1)$.

This clearly implies that ran $Q_{j,j+1}$ is closed for $j=1,2,\dots,k-1$.

□

THEOREM 7.15. *Let* $Q \in N_k(H)$. *Then there exists a sequence* $\{Q_n\}_{n=1}^{\infty}$ *in* $N_k(H)$ *such that* $\|Q - Q_n\| \to 0$ $(n \to \infty)$ *and* $(\text{ran } Q^j)^- = \text{ran } Q_n{}^j$ *for all* $j=1,2,\dots,k-1$ *and for all* $n=1,2,\dots$.

PROOF. Assume that Q has the canonical representation (5.1) and let $Q_{j-1,j} = V_j H_j$ be the polar decomposition of $Q_{j-1,j}$, $j=2,3,\dots,k$. Since $Q_{j-1,j}$ is injective, $\ker H_j = \{0\}$. Define Q_n by (5.1) with $Q_{j-1,j}$ replaced by $Q_{j-1,j,n} = V_j(H_j + 1/n)$, $j=2,3,\dots,k$, $n=1,2,\dots$.

Then $Q_n \in N_k(H)$ and (by Lemma 7.14) ran $Q_n{}^j$ is closed for all $j = 1,2,\dots,k-1$ $(n=1,2,\dots)$. On the other hand, it readily follows from the definition that ran $Q^j \subset$ ran $Q_n{}^j \subset (\text{ran } Q^j)^-$ for all $j=1,2,\dots,k-1$ and all $n=1,2,\dots$. □

By using the arguments of Lemma 7.1, we obtain

COROLLARY 7.16. *Let* T *be an algebraic operator with minimal monic polynomial* p. *Then there exists a sequence* $\{T_n\}_{n=1}^{\infty}$ *in* $A\ell g(H)$ *such that* $p(T_n) = 0$ *for all* $n=1,2,\dots$, $\|T - T_n\| \to 0$ $(n \to \infty)$ *and* $[\text{ran } q(T)]^- = \text{ran } q(T_n)$ *for all* $q|p$ *and all* $n=1,2,\dots$.

7.5 A similarity invariant for polynomially compact operators

PROPOSITION 7.17. *If* $p(T) \in K(H)$ *for some (monic) polynomial* p, *then the following are equivalent:*

(i) $T \sim J+K$, *where* J *is a Jordan operator and* K *is compact.*

(ii) $\tilde{T} \sim \tilde{J}$ *for some Jordan operator* J.

(iii) ran q(T) *is the algebraic sum of a subspace* H_q *and the range* R_q *of a compact operator* R_q, *for each polynomial* $q|p$.

(iv) 0 *is an isolated point of* $\sigma_e[q(T)^*q(T)]$ *for each polynomial* $q|p$.

(v) $\rho(\tilde{T})$ *is similar to a Jordan operator for every unital *-representation* ρ *of* $C^*(\tilde{T})$.

(vi) $\rho(\tilde{T})$ *is similar to a Jordan operator for some faithful unital *-representation* ρ *of* $C^*(\tilde{T})$.

(vii) $\rho(\tilde{T})$ *is similar to a Jordan operator for every unital *-representation* ρ *of* $A(H)$.

The equivalence between (iv) and any of the statements (v), (vi) or (vii) follows easily from Corollary 7.13 and the well-known fact that ran A is closed if and only if 0 is an isolated point of $\sigma(A^*A)$. The equivalence between (iv) and (iii) follows from the analysis of the possible structures of an operator range (or Julia manifold; see [71], [96]) and (i) => (ii) => (iv) are trivial implications. In order to complete the proof (By showing that (vi) => (i)), we shall need the following auxiliary result.

LEMMA 7.18. *If* $T \in L(H)$, $\sigma(T) = \sigma_1 \cup \sigma_2$, *where* σ_1 *and* σ_2 *are two disjoint compact nonempty subsets,* H_1 *is the Riesz spectral subspace corresponding to* σ_1, $H_2 = H \ominus H_1$,

$$T = \begin{pmatrix} A & C \\ 0 & B \end{pmatrix} \begin{matrix} H_1 \\ H_2 \end{matrix},$$

and P_j *is the orthogonal projection of* H *onto* H_j, $j = 1,2$, *then* P_1 *and* P_2 *belong to* $C^*(T)$.

Moreover, there also exists a unique operator $X: H_2 \to H_1$ *such that*

$$\begin{pmatrix} 1 & X \\ 0 & 1 \end{pmatrix} \begin{pmatrix} A & C \\ 0 & B \end{pmatrix} \begin{pmatrix} 1 & X \\ 0 & 1 \end{pmatrix}^{-1} = \begin{pmatrix} A & 0 \\ 0 & B \end{pmatrix} = A \oplus B,$$

$$W = \begin{pmatrix} 1 & X \\ 0 & 1 \end{pmatrix}, \quad W^{-1} = \begin{pmatrix} 1 & X \\ 0 & 1 \end{pmatrix}^{-1} = \begin{pmatrix} 1 & -X \\ 0 & 1 \end{pmatrix}, \quad A \oplus B \in C^*(T)$$

and $C^*(T) = C^*(\{W, A \oplus B\})$.

PROOF. It is well-known that a C*-algebra with identity is inverse closed. If the idempotent

$$E = \begin{pmatrix} 1 & Y \\ 0 & 0 \end{pmatrix} \begin{matrix} H_1 \\ H_2 \end{matrix}$$

is the projection of H onto $H_1 = H(\sigma_1; T)$ along $H(\sigma_2; T)$, then E is the limit of rational functions in T and therefore

$$E, E^* = \begin{pmatrix} 1 & 1 \\ Y^* & 1 \end{pmatrix} \begin{matrix} H_1 \\ H_2 \end{matrix}, \quad EE^* = \begin{pmatrix} 1+Y^*Y & 0 \\ 0 & 0 \end{pmatrix} \text{ and } P_1 = \text{(norm) } \lim(n \to \infty)\ p_n(EE^*)$$

belong to $C^*(T)$, where $\{p_n\}_{n=1}^{\infty}$ is a sequence of polynomials converging uniformly to 0 on the real segment $[-1/3, 1/3]$ and converging uniformly to 1 on the real segment $[2/3, 1+\|Y\|^2]$. It is clear that $P_2 = 1 - P_1$, $A \oplus 0 = P_1 T P_1$, $0 \oplus B = P_2 T P_2$, $A \oplus B = A \oplus 0 + 0 \oplus B$ and $A \oplus B - T$ also belong to $C^*(T)$.

Since $\sigma(A) = \sigma_1$ is disjoint from $\sigma(B) = \sigma_2$, it follows from Corollary 3.20 that τ_{AB} is invertible and

$$X \equiv \tau_{AB}^{-1}(C) = \frac{-1}{2\pi i} \int_\Gamma (\lambda - A)^{-1} C (\lambda - B)^{-1}\, d\lambda,$$

where Γ is the (suitably oriented) boundary of a Cauchy domain Ω containing $\sigma(A)$ such that $\sigma(B) \cap \Omega^- = \emptyset$.

Given $\lambda \in \Gamma$,

$$X(\lambda) = ((\lambda - A) \oplus 1)^{-1}(A \oplus B - T)(1 \oplus (\lambda - B))^{-1} = \begin{pmatrix} 0 & (\lambda-A)^{-1}(-C)(\lambda-B)^{-1} \\ 0 & 0 \end{pmatrix}$$

belongs to $C^*(T)$ and (by taking limits of suitable Riemann sums) we conclude that

$$W = \begin{pmatrix} 1 & X \\ 0 & 1 \end{pmatrix} = 1 + \frac{1}{2\pi i} \int_\Gamma X(\lambda)\, d\lambda \in C^*(T).$$

A fortiori, W, W^{-1} and $A \oplus B$ belong to $C^*(T)$. Since $T = W^{-1}(A \oplus B)W \in C^*(\{W, A \oplus B\})$, it immediately follows that $C^*(T) = C^*(\{W, A \oplus B\})$. □

Now we are in a position to complete the proof of Proposition 7. 17. Assume that there exists a faithful unital *-representation ρ of $C^*(T)$ on $\mathcal{H}_\rho$ such that $A = \rho(\tilde{T})$ is similar to a Jordan operator with minimal polynomial p, $p(\lambda) = \Pi_{i=1}^m (\lambda - \lambda_i)^{k_i}$. Since ρ is faithful, it follows that $p(\tilde{T}) = 0$. (Indeed, p is also the minimal polynomial of $\tilde{T}$.) Hence, by Theorem 7.2, T (or some compact perturbation of T) is an algebraic operator with minimal polynomial p. Let

$$T = \begin{pmatrix} T_1 & & & & * \\ & T_2 & & & \\ & & \cdot & & \\ & & & \cdot & \\ & 0 & & & \cdot \\ & & & & T_m \end{pmatrix} \qquad (7.5)$$

be the matrix of T with respect to the decomposition $\mathcal{H} = \oplus_{i=1}^m \mathcal{R}_i$, where the $\mathcal{R}_i$'s are inductively defined so that $\oplus_{i=1}^n \mathcal{R}_i = \mathcal{H}(\{\lambda_1, \lambda_2, \ldots, \lambda_n\}; T)$ $(n = 1, 2, \ldots, m)$.

Applying Lemma 7.18 (and an obvious inductive argument) to T, we can find a(unique!) invertible operator W in $C^*(T)$ such that $T = W(\oplus_{i=1}^m T_i)W^{-1}$ and $C^*(T) = C^*(\{W, \oplus_{i=1}^m T_i\})$. A fortiori,

$$A = \rho(T) = \begin{pmatrix} A_1 & & & & * \\ & A_2 & & & \\ & & \cdot & & \\ & & & \cdot & \\ & 0 & & & \cdot \\ & & & & A_m \end{pmatrix} = V(\oplus_{i=1}^m A_i)V^{-1}$$

with respect to a decomposition $\mathcal{H}_\rho = \oplus_{i=1}^m \operatorname{ran} \rho(\tilde{R}_i)$, where R_i denotes the orthogonal projection of $\mathcal{H}$ onto $\mathcal{R}_i$ $(i = 1, 2, \ldots, m)$, $V = \rho(\tilde{W})$, $\oplus_{i=1}^m A_i = \rho(\oplus_{i=1}^m \tilde{T}_i)$, and A_i is similar to a Jordan operator with minimal polynomial $(\lambda - \lambda_i)^{k_i}$.

Clearly, this reduces our problem to the case when $\sigma(\tilde{T})$ is a singleton; moreover, replacing (if necessary) T by $T-\lambda$, we can (and shall) directly assume that $\tilde{T}^k = 0$, but $\tilde{T}^{k-1} \neq 0$ for some $k \geq 1$. Then $A^k = 0$, $A^{k-1} \neq 0$ and

$$A = \begin{pmatrix} 0 & A_{12} & & & & & \\ & 0 & A_{23} & & & * & \\ & & 0 & \cdot & & & \\ & & & \cdot & \cdot & & \\ & & & & \cdot & \cdot & \\ & & 0 & & & \cdot & \\ & & & & & 0 & A_{k-1,k} \\ & & & & & & 0 \end{pmatrix} \begin{matrix} H_{\rho,1} \\ H_{\rho,2} \\ H_{\rho,3} \\ \cdot \\ \cdot \\ \cdot \\ H_{\rho,k-1} \\ H_{\rho,k} \end{matrix}, \qquad (7.6)$$

where $H_{\rho,j} = \ker A^j \ominus \ker A^{j-1}$, $j = 1,2,\ldots,k$.

Since A is similar to a Jordan operator, it follows from Lemma 7.14 that the operators $A_{j,j+1}: H_{\rho,j+1} \to H_{\rho,j}$ are bounded below and 0 is an isolated point of the spectrum of $A^{*j}A^j$ for all $j = 1,2,\ldots,k-1$; therefore there exists $\eta > 0$ such that $(0,\eta] \cap [\cup_{j=1}^{k-1} \sigma(A^{*j}A^j)] = (0,\eta] \cap [\cup_{j=1}^{k-1} \sigma(\tilde{T}^{*j}\tilde{T}^j)] = \emptyset$. Thus, if $E_j(.)$ denotes the spectral measure of $T^{*j}T^j$, then rank $E_j([\eta/2,\eta]) = \beta_j$ is finite for all $j = 1,2,\ldots,k-1$. It follows that 0 is an isolated point of $\sigma_e(T^{*j}T^j)$ $(j = 1,2,\ldots,k-1)$ and $[\eta/2,\eta] \cap [\cup_{j=0}^{k} \sigma(T^{*j}T^j)]$ is finite. Hence, $[\eta/2,\eta]$ contains a closed subinterval $[\alpha,\beta]$ disjoint from $\cup_{j=0}^{k} \sigma(T^{*j}T^j)$ and therefore (by the Stone-Weierstrass theorem) there exists a sequence $\{r_n\}_{n=1}^{\infty}$ of polynomials converging uniformly to 1 on $[0,\alpha]$ and converging uniformly to 0 on $[\beta,\infty) \cap [\cup_{j=0}^{k} \sigma(T^{*j}T^j)]$. It readily follows that $\{r_n(T^{*j}T^j)\}$ converges in the norm to the orthogonal projection $L_j = E_j([0,\alpha]) \in C^*(T)$ $(j = 0, 1,2,\ldots,k)$, and $\rho(\tilde{L}_j - \tilde{L}_{j-1}) = P_j = (\text{norm}) \lim(n \to \infty)\{r_n(A^{*j}A^j) - r_n(A^{*j-1}A^{j-1})\} \in C^*(A)$ is the orthogonal projection of H_ρ onto $H_{\rho,j}$, for all $j = 1,2,\ldots,k$.

Since ρ is isometric, there exist compact perturbations $R_j \in C^*(T)+K(H)$ of $L_j - L_{j-1}$, $j = 1,2,\ldots,k$, such that $R_1, R_2, \ldots, R_k$ are pairwise orthogonal projections, $\sum_{j=1}^{k} R_j = 1$, and $\rho(\tilde{R}_j) = P_j$, for all $j = 1,2,\ldots,k$.

Let $T = (T_{ij})_{i,j=1}^{k}$ be the matrix of T with respect to the decomposition $H = \oplus_{j=1}^{k} H_j$, where $H_j = \operatorname{ran} R_j$, $j = 1,2,\ldots,k$. Since $\rho(\tilde{R}_i\tilde{T}\tilde{R}_j) = P_i A P_j$ and ρ is isometric, it follows that T_{ij} must be compact for all (i,j) such that $1 \leq j \leq i \leq k$. Thus, up **to a** compact perturbation, we can directly assume that T has an upper triangular matrix of the form (7.6) (with the A_{ij}'s replaced by T_{ij}'s) with respect to the above decomposition.

Now observe that

$$\rho(1\oplus \oplus_{j=2}^{k} T_{j,j+1}^* T_{j,j+1}) = \rho(R_1 + \sum_{j=2}^{k}(R_j T R_{j+1})^*(R_j T R_{j+1}))$$

$$= P_1 + \sum_{j=2}^{k}(P_j A P_{j+1})^*(P_j A P_{j+1}) = 1\oplus \oplus_{j=2}^{k} A_{j,j+1}^* A_{j,j+1}$$

is invertible in $C^*(A)$, whence we immediately obtain that $T_{j,j+1}: H_{j+1} \to \operatorname{ran} T_{j,j+1}$ ($\subset H_j$; $j = 1,2,\ldots,k-1$) is a Fredholm operator with $0 \leq \operatorname{ind} T_{j,j+1} < \infty$.

If $k = 1$, then $T = 0$ is a lifting of $\tilde{T}$, and we are done. Assume that $k > 1$; then, by using the above construction and an inductive argument, we can assume that $(1-R_k)(T+K_{k-1})|\ker R_k$ ($\ker R_k = \oplus_{j=1}^{k-1} H_j$) is equal to

$$\begin{pmatrix} 0 & T_{12}' & & & & \\ & 0 & T_{23}' & & & \\ & & 0 & \cdot & * & \\ & & & \cdot & \cdot & \\ & & & & \cdot & \cdot \\ & 0 & & & \cdot & \cdot \\ & & & & 0 & T_{k-2,k-1}' \\ & & & & & 0 \end{pmatrix} \oplus 0_{d_{k-1}},$$

with respect to a decomposition $\ker R_k = (\oplus_{j=1}^{k-1} H_j') \oplus \mathbb{C}^{d_{k-1}}$ ($0 \leq d_j < \infty$ for all j), where K_{k-1} is a finite rank operator, H_j' is a subspace of finite codimension of H_j and $T_{j,j+1}': H_{j+1}' \to H_j'$ is bounded below for all $j = 1,2,\ldots,k-1$, and $0_{d_{k-1}}$ denotes the zero operator acting on the finite dimensional space $\mathbb{C}^{d_{k-1}}$.

Let R_{k-1}' be the orthogonal projection of H_{k-1} onto H_{k-1}'. Since $T_{k-1,k}: H_k \to \operatorname{ran} T_{k-1,k}$ is a Fredholm operator with negative index, $R_{k-1}' T_{k-1,k}: H_k \to \operatorname{ran} R_{k-1}' T_{k-1,k}$ is also a Fredholm operator with ind $R_{k-1}' T_{k-1,k} = d_k' \geq 0$ and $T_{k-1,k}' = R_{k-1,k}' T_{k-1,k} | [\ker R_{k-1}' T_{k-1,k}]^\perp$ is an isomorphism of Hilbert spaces between $H_k' = [\ker R_{k-1}' T_{k-1,k}]^\perp$ and $\operatorname{ran} T_{k-1,k}'$. Thus, if $\mathbb{C}^{d_k} = \mathbb{C}^{d_{k-1}} \oplus \ker R_{k-1}' T_{k-1,k}$ ($\dim \mathbb{C}^{d_k} = d_k = d_{k-1} + d_k' < \infty$), then there exists a finite rank operator K_k' such that if $K_k = K_{k-1} + K_k'$, then $T+K_k$ is unitarily equivalent to the orthogonal direct sum of an operator matrix of the form (7.6), with the A_{ij}'s replaced by the T_{ij}''s, and the operator 0_{d_k} acting on the finite dimensional space $\mathbb{C}^{d_k}$.

Since $T_{j,j+1}': H_{j+1}' \to \operatorname{ran} T_{j,j+1}'$ is an isomorphism of Hilbert spaces for all $j = 1,2,\ldots,k-1$, it follows from Lemma 7.14 and Theorem 7.11 that $T+K_k$ is similar to a Jordan operator, and this completes the proof of Proposition 7.17. □

REMARK 7.19. It is convenient to observe that in the above proof (i.e., (vi) => (i) in Proposition 7.17) we did not use the hypothesis "p(T) is compact". Indeed, this hypothesis is an easy consequence of any of the hypotheses (i), (ii), (v), (vi) or (vii), but false, in general, if we merely assume (iii) or (iv) (see [30,p.868]).

7.6 Nice Jordan operators

A Jordan operator J, given by (7.4) will be called *nice* if, for each j ($j=1,2,\ldots,m$), α_{kj} is an infinite cardinal for at most one value of k, $k=1,2,\ldots,m_j$, and *very nice* if, for each j, α_{kj} is an infinite cardinal for exactly one value k_j of k and $\alpha_{kj}=0$ for $k \neq k_j$. Namely, Q is a nice Jordan nilpotent if and only if

$$Q \simeq q_s^{(\infty)} \oplus \oplus_{j=1,j\neq s}^{k} q_j^{(\alpha_j)}, \quad \sum_{j=1,j\neq s}^{k} \alpha_j < \infty, \tag{7.7}$$

and a very nice Jordan nilpotent if and only if $Q \simeq q_s^{(\infty)}$ for some s.

LEMMA 7.20. *Let* C *be a* C*-*algebra with identity* 1 *and let* $t \in N_k(C)$ $(k \geq 2)$. *Then the following are equivalent*

(i) $t^{k-j}+t^{*j}$ *is invertible for all* $j=1,2,\ldots,k-1$.

(ii) $t^{k-j}+t^{*j}$ *is invertible for some* j, $1 \leq j \leq k-1$.

*(iii) There exists a faithful unital *-homomorphism* $\tau: L(\mathbb{C}^k) \to C$ *and an invertible element* $v \in C$ *such that* $v\tau(q_k)v^{-1}=t$.

*(iv) There exists a faithful *-representation* $\rho: C \to L(H_\rho)$ *and* j, $1 \leq j \leq k-1$, *such that the sequence*

$$H_\rho \overset{\rho(t^j)}{\to} H_\rho \overset{\rho(t^{k-j})}{\to} H_\rho \overset{\rho(t^j)}{\to} H_\rho$$

is exact (i.e., $\ker \rho(t^{k-j}) = \operatorname{ran} \rho(t^j)$ *and* $\ker \rho(t^j) = \operatorname{ran} \rho(t^{k-j})$*).*

*(v) For every faithful unital *-representation* $\rho: C \to L(H_\rho)$, *the sequences*

$$H_\rho \overset{\rho(t^j)}{\to} H_\rho \overset{\rho(t^{k-j})}{\to} H_\rho \overset{\rho(t^j)}{\to} H_\rho, \quad j=1,2,\ldots,k-1,$$

are exact.

(vi) $\rho(t) \sim q_k \otimes 1$ *for every faithful unital *-representation* $\rho: C \to L(H)$.

PROOF. (i) => (ii) and (iii) => (vi) => (v) => (iv) are trivial implications.

(v) => (i) and (iv) => (ii). Since $\operatorname{ran} \rho(t^j)$ and $\operatorname{ran} \rho(t^{k-j})$ are closed, so are $\operatorname{ran} \rho(t^j)^* = \operatorname{ran} \rho(t^{*j})$ and $\operatorname{ran} \rho(t^{k-j})^* = \operatorname{ran} \rho(t^{*k-j})$. Observe that

$$[\operatorname{ran} \rho(t^j)^*]^\perp = \ker \rho(t^j) = \operatorname{ran} \rho(t^{k-j}) = [\ker \rho(t^{k-j})^*]^\perp$$

and

$$[\ker \rho(t^j)^*]^\perp = \operatorname{ran} \rho(t^j) = \ker \rho(t^{k-j}) = [\operatorname{ran} \rho(t^{k-j})^*]^\perp,$$

so that

$$\operatorname{ran} \rho(t^{k-j}+t^{*j}) = \operatorname{ran} \rho(t^{k-j}) + \operatorname{ran} \rho(t^{j})^* = H_\rho$$

and

$$\operatorname{ran} \rho(t^{k-j}+t^{*j})^* = \operatorname{ran} \rho(t^{k-j})^* + \operatorname{ran} \rho(t^{j}) = H_\rho .$$

Hence $\rho(t^{k-j}+t^{*j})$ is invertible in $L(H_\rho)$. Since ρ is faithful and unital, it readily follows that $t^{k-j}+t^{*j}$ must be invertible in C.

(ii) => (iii) Assume that $t^{k-h}+t^{*h}$ is invertible (for some h, $1 \leq h \leq k-1$) and let $\rho: C^*(t) \to L(H_\rho)$ be a faithful unital *-representation. If $\rho(t) = T$, then $T \in N_k(H_\rho)$ and $T^{k-h}+T^{*h}$ is invertible. Let

$$T = \begin{pmatrix} 0 & T_{12} & & & & & \\ & 0 & T_{23} & & * & & \\ & & 0 & \cdot & & & \\ & & & \cdot & \cdot & & \\ & & & & \cdot & \cdot & \\ & & 0 & & & \cdot & \cdot \\ & & & & & 0 & T_{k-1,k} \\ & & & & & & 0 \end{pmatrix} \begin{matrix} H_1 \\ H_2 \\ H_3 \\ \cdot \\ \cdot \\ \cdot \\ H_{k-1} \\ H_k \end{matrix}$$

(where $H_j = \ker T^j \ominus \ker T^{j-1}$, $j = 1,2,\ldots,k-1$) be the canonical representation of T. Now

$$T^{k-h}+T^{*h} = \begin{pmatrix} 0 & T(k,h) \\ T(k^*,h^*) & 0 \end{pmatrix},$$

where

$$T(k,h) = \begin{pmatrix} T^{(h)}_{1,k-h+1} & & & & * \\ & T^{(h)}_{2,k-h+2} & & & \\ & & \cdot & & \\ & & & \cdot & \\ & 0 & & & \cdot \\ & & & & T^{(h)}_{h,k} \end{pmatrix},$$

$$T(k^*,h^*) = \begin{pmatrix} (T^{(h)}_{k-h,k})^* & & & & 0 \\ & (T^{(h)}_{k-h-1,k-1})^* & & & \\ & & \cdot & & \\ & & & \cdot & \\ & * & & & \cdot \\ & & & & (T^{(h)}_{1,h+1})^* \end{pmatrix}$$

and $T^{(h)}_{m,m+h} = T_{m,m+1}T_{m+1,m+2}\ldots T_{m+h-1,m+h}$. Since $T^{k-h}+T^{*h}$ is invertible, it is not difficult to conclude that the operators

$$T^{(h)}_{m,m+h}: H_{m+h} \to H_m, \quad m = 1,2,\ldots,k-h,$$

are invertible. Since $T_{j,j+1}$ is injective for all $j = 1,2,\ldots,k-1$, an elementary inductive argument shows that $T_{j,j+1}: H_{j+1} \to H_j$ is invertible for all $j=1,2,\dots,k-1$. In particular, this means that all the subspaces H_j $(j=1,2,\dots,k)$ have the same dimension.

Thus, up to a suitable identification of H_{j+1} with H_j (via a fixed unitary mapping form H_{j+1} onto H_j) we can directly assume that $T_{j,j+1}$ is a positive invertible operator for each $j=1,2,\dots,k-1$.

Clearly, $T^{(j)}_{m,m+j}$ $(m=1,2,\dots,k-j)$ and $T^{k-j}+T^{*j}$ are also invertible operators (on their respective spaces) for all $j=1,2,\dots,k-1$. If $A_j = T^{k-j}T^{*k-j}$ and $B_j = T^{*j}T^j$, then

$$A_j+B_j = (T^{k-j}+T^{*j}).(T^{k-j}+T^{*j})^* = (T^{k-j}+T^{*j})(T^{*k-j}+T^j)$$

is invertible and $A_jB_j = T^{k-j}T^{*k}T^j = 0$. Since A_j and B_j are positive, $R_j = A_j(A_j+B_j)^{-1}$ $(j=0,1,2,\dots,k)$ is an hermitian idempotent; it is easily seen that $P_j = R_j - R_{j-1}$ is the orthogonal projection of H onto H_j, $j=1,2,\dots,k$.

Let 1_j denote the identity on H_j, $j=1,2,\dots,k$. Now it is easily seen that for each i, $1 \le i < k$, $C^*(T)$ contains the operator

$$P_iTP_{i+1}[1_1\oplus 1_2\oplus\dots\oplus 1_{i-1}\oplus(T^*_{i,i+1}T_{i,i+1})\oplus 1_{i+1}\oplus\dots\oplus 1_k]^{-\frac{1}{2}}$$

$$= P_iTP_{i+1}[1_1\oplus 1_2\oplus\dots\oplus 1_{i-1}\oplus T_{i,i+1}\oplus 1_{i+1}\oplus\dots\oplus 1_k]^{-1}$$

$$= \begin{pmatrix} 0 &&&&&&&& \\ & 0 &&&&&&& \\ && \cdot &&&&&& \\ &&& \cdot &&&&& \\ &&&& \cdot &&&& \\ &&&&& 0 & 1 && \\ &&&&&& 0 && \\ &&&&&&& \cdot & \\ &&&&&&&& \cdot \\ &&&&&&&& \cdot \\ &&&&&&&& 0 \end{pmatrix} \begin{matrix} H_1 \\ H_2 \\ \cdot \\ \cdot \\ \cdot \\ H_i \\ H_{i+1} \\ \cdot \\ \cdot \\ \cdot \\ H_k \end{matrix}$$

(the empty entries are 0's), where $1 = 1_{i,i+1}$ denotes the identification between H_{i+1} and H_i (i.e., the operator $U_{i,i+1}$).

Since this is true for all i, $1 \le i < k$, it is not difficult to conclude that all $k \times k$ operator matrices (with respect to the decomposition $H = \oplus_{j=1}^k H_j$) with "constant entries" belong to $C^*(T)$. Hence,

$$Q = \begin{pmatrix} 0 & 1 &&&&& \\ & 0 & 1 &&& 0 & \\ && 0 & \cdot &&& \\ &&& \cdot & \cdot && \\ & 0 &&& \cdot & \cdot & \\ &&&&& 0 & 1 \\ &&&&&& 0 \end{pmatrix} \begin{matrix} H_1 \\ H_2 \\ H_3 \\ \cdot \\ \cdot \\ H_{k-1} \\ H_k \end{matrix} \simeq q_k^{(\infty)} \in C^*(T).$$

If $W = T_{1,k}^{(k-1)} \oplus T_{2,k}^{(k-2)} \oplus \dots \oplus T_{k-2,k}^{(2)} \oplus T_{k-1,k} \oplus 1_k$, then

$$B = W^{-1}TW = \begin{pmatrix} 0 & 1 & B_{13} & \cdot & \cdot & \cdot B_{1,k-1} & B_{1k} \\ & 0 & 1 & \cdot & \cdot & \cdot B_{2,k-1} & B_{2k} \\ & & 0 & \cdot & \cdot & \cdot B_{3,k-1} & B_{3k} \\ & & & \cdot & & \cdot & \cdot \\ & & & & \cdot & \cdot & \cdot \\ & & 0 & & & \cdot & \cdot \\ & & & & & 0 & 1 \\ & & & & & & 0 \end{pmatrix} \begin{matrix} H_1 \\ H_2 \\ H_3 \\ \cdot \\ \cdot \\ \cdot \\ H_{k-1} \\ H_k \end{matrix} .$$

Let

$$W_1 = \begin{pmatrix} 1 & B_{13} & \cdot & \cdot & \cdot & B_{1k} \\ & 1 & \cdot & \cdot & \cdot & B_{2k} \\ & & \cdot & & & \cdot \\ & & & \cdot & & \cdot \\ & 0 & & & \cdot & \cdot \\ & & & & & 1 \end{pmatrix} \begin{matrix} H_1 \\ H_2 \\ \cdot \\ \cdot \\ \cdot \\ H_{k-1} \end{matrix} \quad \oplus 1_k;$$

then

$$C = W_1^{-1} B W_1 = \begin{pmatrix} 0 & 1 & B'_{24} & B'_{25} & \cdot & \cdot & \cdot B'_{2,k-1} & B'_{2k} & 0 \\ & 0 & 1 & B'_{35} & \cdot & \cdot & \cdot B'_{3,k-1} & B'_{3k} & 0 \\ & & 0 & 1 & \cdot & \cdot & \cdot B'_{4,k-1} & B'_{4k} & 0 \\ & & & 0 & \cdot & \cdot & \cdot B'_{5,k-1} & B'_{5k} & 0 \\ & & & & \cdot & & \cdot & \cdot & \cdot \\ & & & & & \cdot & \cdot & \cdot & \cdot \\ & & & & & & \cdot & \cdot & \cdot \\ & & 0 & & & & 0 & 1 & 0 \\ & & & & & & & 0 & 1 \\ & & & & & & & & 0 \end{pmatrix} \begin{matrix} H_1 \\ H_2 \\ H_3 \\ H_4 \\ \cdot \\ \cdot \\ \cdot \\ H_{k-2} \\ H_{k-1} \\ H_k \end{matrix} .$$

Similarly, if

$$W_2 = \begin{pmatrix} 1 & B'_{24} & B'_{25} & \cdot & \cdot & \cdot B'_{2,k-1} & B'_{2k} \\ & 1 & B'_{35} & \cdot & \cdot & \cdot B'_{3,k-1} & B'_{3k} \\ & & 1 & \cdot & \cdot & \cdot B'_{4,k-1} & B'_{4k} \\ & & & \cdot & & \cdot & \cdot \\ & & & & \cdot & \cdot & \cdot \\ & 0 & & & & \cdot & \cdot \\ & & & & & 1 & B'_{k-2,k} \\ & & & & & & 1 \end{pmatrix} \begin{matrix} H_1 \\ H_2 \\ H_3 \\ \cdot \\ \cdot \\ \cdot \\ H_{k-3} \\ H_{k-2} \end{matrix} \quad \oplus 1_{k-1} \oplus 1_k,$$

then

$$D = W_2^{-1} C W_2 = \begin{pmatrix} 0 & 1 & B''_{35} & B''_{36} & \cdot & \cdot & \cdot & B''_{3k} & 0 & 0 \\ & 0 & 1 & B''_{46} & \cdot & \cdot & \cdot & B''_{4k} & 0 & 0 \\ & & 0 & 1 & \cdot & \cdot & \cdot & B''_{5k} & 0 & 0 \\ & & & 0 & \cdot & \cdot & \cdot & B''_{6k} & 0 & 0 \\ & & & & \cdot & & & \cdot & \cdot & \cdot \\ & & & & & \cdot & & \cdot & \cdot & \cdot \\ & & & & & & \cdot & \cdot & \cdot & \cdot \\ & & & 0 & & & & 0 & 1 & 0 \\ & & & & & & & & 0 & 1 \\ & & & & & & & & & 0 \end{pmatrix} \begin{matrix} H_1 \\ H_2 \\ H_3 \\ H_4 \\ \cdot \\ \cdot \\ \cdot \\ H_{k-2} \\ H_{k-1} \\ H_k \end{matrix} .$$

By an obvious inductive argument, after k-1 steps we obtain $Q = V^{-1}TV$, $T = VQV^{-1}$, where $V = WW_1W_2 \ldots W_{k-1} \in C^*(T)$.

Since ρ is faithful and unital, we deduce that there exist (unique) $v, q \in C^*(t) \subset C$ such that $t = vqv^{-1}$, $\rho(v) = V$ and $\rho(q) = Q$. If $M = (\lambda_{ij})_{i,j=1}^{k}$ is a matrix with constant entries in $C^*(T)$, then there exists a *unique* $m \in C^*(t)$ such that $\rho(m) = M$.

Define $\tau: L(\mathbb{C}^k) \to C^*(t) \subset C$ as follows: If $\{e_j\}_{j=1}^{k}$ is the canonical ONB of $\mathbb{C}^k$ and $A = (\lambda_{ij})_{i,j=1}^{k}$ is the matrix of $A \in L(\mathbb{C}^k)$ with respect to this basis, then $\tau(A) = m$.

The properties of ρ make it clear that τ is a faithful unital *-representation. Since $\tau(q_k) = q$, we are done. □

Observe that $q_k^{k-j} + q_k^{*j}$ is a unitary operator. Hence, so is $\tau(q_k^{k-j} + q_k^{*j})$, $j = 1,2,\ldots,k-1$. Let $n = [k/2]$ (= the integral part of k/2) and for $a \in C$ assume that $\delta = \|a - \tau(q_k)\| < \beta_k \equiv (3/2)^{1/(k-n)} - 1$; then it follows from the proof of Proposition 1.10 that

$$\|(a^{k-n} + a^{*n}) - \tau(q_k^{k-n} + q_k^{*n})\| \le \|a^{k-n} - \tau(q_k)^{k-n}\| + \|a^{*n} - \tau(q_k)^{*n}\|$$
$$\le [(1+\delta)^{k-n} - 1] + [(1+\delta)^n - 1] \le 2(1+\delta)^{k-n} - 2 < 1$$

and therefore $a^{k-n} + a^{*n}$ is invertible in C.

By Lemma 7.20, we conclude that if $a \in N_k(C)$ and $\|a - \tau(q_k)\| < \beta_k$, then $a \sim \tau(q_k)$. (Compare this result with those of Section 2.3.2.) Morevoer, if $a = v\tau(q_k)v^{-1}$ and $\|a - b\| < \beta_k/(\|v\| . \|v^{-1}\|)$, then

$$\|v^{-1}bv - \tau(q_k)\| = \|v^{-1}(b-a)v\| < \beta_k$$

and therefore $b \sim \tau(q_k)$. Hence, we have

COROLLARY 7.21. *Let C be a C*-algebra with identity* 1. *The set* $\{a \in C: \ a \sim \tau(q_k)\}$ *(where* τ *is some faithful unital *-homomorphism of* $L(\mathbb{C}^k)$ *into C) is open in* $N_k(C)$ $(k = 1,2,\ldots)$.

Thus, the subset

$$\tilde{N}_k^+(A(H)) = \{\tilde{T} \in N_k(A(H)): \quad \tilde{T}^{k-1}+\tilde{T}^* \text{ is invertible}\} \quad (7.8)$$

is open in $N_k(A(H))$ and

$$N_k^+(H) = \pi^{-1}[\tilde{N}_k^+(A(H))] = \{T \in L(H): \quad \tilde{T}^k = 0, \ \tilde{T}^{k-1}+\tilde{T}^* \text{ is invertible}\} \quad (7.9)$$

intersects $N_k(H)$ in an open subset of $N_k(H)$ $(k = 1,2,\ldots)$.

PROPOSITION 7.22. *If* $T \in L(H)$ *and* T^k *is compact, but* $\tilde{T}^{k-1} \neq 0$ *(for some* $k \geq 2$*), then the following are equivalent*

(i) $T \sim J+K$, *where* J *is a nice Jordan nilpotent and* K *is compact.*

(ii) $T = WJV+K$, *where* W, V *are Fredholm operators such that* $\tilde{W}\tilde{V} = \tilde{V}\tilde{W} = \tilde{I}$, J *is a nice Jordan nilpotent and* K *is compact.*

(iii) $T = WJV+K$, *where* W, V *are Fredholm operators such that* $\tilde{W}\tilde{V} = \tilde{V}\tilde{W} = \tilde{I}$, J *is a very nice Jordan nilpotent and* K *is compact.*

(iv) $\tilde{T}^{k-j}+\tilde{T}^{*j}$ *is invertible for all* $j = 1,2,\ldots,k-1$.

(v) $\tilde{T}^{k-j}+\tilde{T}^{*j}$ *is invertible for some* j, $1 \leq j \leq k-1$.

*(vi) There exists a faithful unital *-homomorphism* $\tau: L(\mathbb{C}^k) \to C^*(\tilde{T})$ *and an invertible element* $\tilde{V} \in C^*(\tilde{T})$ *such that* $\tilde{V}\tau(q_k)\tilde{V}^{-1} = \tilde{T}$.

*(vii) There exists a faithful *-representation* $\rho: C^*(\tilde{T}) \to L(H_\rho)$ *and* j, $1 \leq j \leq k-1$, *such that the sequence*

$$H_\rho \overset{\rho(\tilde{T}^j)}{\to} H_\rho \overset{\rho(\tilde{T}^{k-j})}{\to} H_\rho \overset{\rho(\tilde{T}^j)}{\to} H_\rho$$

is exact.

*(viii) For every faithful unital *-representation* $\rho: A(H) \to L(H_\rho)$ *the sequences*

$$H_\rho \overset{\rho(\tilde{T}^j)}{\to} H_\rho \overset{\rho(\tilde{T}^{k-j})}{\to} H_\rho \overset{\rho(\tilde{T}^j)}{\to} H_\rho \quad (7.10)$$

are exact for all $j = 1,2,\ldots,k-1$.

(ix) $\rho(\tilde{T}) \sim q_k \otimes 1$ *for every faithful unital *-representation* $\rho: C^*(\tilde{T}) \to L(H_\rho)$.

We shall need an auxiliary result.

LEMMA 7.23. *Suppose* $H = H_1 \oplus H_2$ *with* $H_1 \simeq H$ *and* $0 < \dim H_2 < \infty$ *and let* $T = T_1 \oplus T_2$ *be in* $L(H)$; $T_j \in L(H_j)$, $j = 1,2$. *Then, under the identification of* H *with* H_1, $\tilde{T}$ *is unitarily equivalent to* $\tilde{T}_1$ *in* $A(H)$.

PROOF. Let $U: H_1 \to H_1 \oplus H_2$ be a unitary mapping from H_1 onto $H_1 \oplus H_2$ and let $S = UT_1U^*$ in $L(H)$; it suffices to prove that $\tilde{S} \simeq \tilde{T}$. Define $V: H_1 \to H_1 \oplus H_2$ by $Vx = x \oplus 0$. Now $UV^*TVU^* = S$, $VU^* \in L(H)$ is an isometry and co-rank $VU^* = \text{nul}(VU^*)^* = \text{nul } V^* = \dim H_2$ is finite; thus $\tilde{V}\tilde{U}^*$ is unitary in $A(H)$ and it follows that $\tilde{T} \simeq \tilde{S}$. □

PROOF OF PROPOSITION 7.22. (i) => (ii) and (iii) => (ii) are triv ial implications, (ii) => (iii) follows immediately from Lemma 7.23, and the equivalence between any two of the conditions (iv) - (ix) follows from Lemma 7.20 and its proof. (Take $t = \tilde{T}$, and $C = C^*(\tilde{T})$ and $C = A(H)$.)

If T has the form of (iii), then $J \simeq q_k^{(\infty)}$. Since $U(q_k^{(\infty)})$ is closed (Proposition 4.22), it easily follows from Proposition 4.21 that $\rho(J) \simeq q_k^{(\infty)}$ for any faithful unital *-representation ρ of $A(H)$ on a Hilbert space H_ρ of dimension α. By Lemma 7.20, (iii) => (ix).

Assume that $C^*(T)$ admits a faithful unital *-representation ρ_o such that $\rho_o(T) \simeq q_k^{(\infty)}$. By Proposition 7.17 and its proof, $T = W(J+K)W^{-1}$, where $W \in G(H)$, $K \in K(H)$ and $J \simeq \oplus_{j=1}^{k} q_j^{(\alpha_j)}$ with $\alpha_k = \infty$. If $\sum_{j=1}^{k-1} \alpha_j = \infty$, then $J \simeq q_k^{(\infty)} \oplus q_s^{(\infty)} \oplus R$, where $1 \leq s < k$ and $R^{k-1} = 0$.

Let $\rho: A(H) \to L(H_\rho)$ be a faithful *-representation. Since $U(J)$ is closed (Proposition 4.22), it is easily seen from Proposition 4.21 that $\rho(\tilde{J}) \simeq q_k^{(\alpha)} \oplus q_s^{(\beta)} \oplus L$, $\alpha \neq 0$, $\beta \neq 0$, $L^{k-1} = 0$ and therefore the sequence

$$H_\rho \xrightarrow{\rho(\tilde{J}^s)} H_\rho \xrightarrow{\rho(\tilde{J}^{k-s})} H_\rho \xrightarrow{\rho(\tilde{J}^s)} H_\rho$$

is not exact. Since $\tilde{T} = \tilde{W}\tilde{J}\tilde{W}^{-1}$, it readily follows that the sequence (7.10) cannot be exact for $j = s$, contradicting the equivalence between (ix) and (viii).

We conclude that $\sum_{j=1}^{k-1} \alpha_{j-1} < \infty$, i.e., $J \simeq q_k^{(\infty)} \oplus F$, where F acts on a finite dimensional space. Since J is a nice Jordan operator, we see that (ix) => (i).

The proof of Proposition 7.22 is complete now. □

COROLLARY 7.24. *Assume that* $Q \in L(H)$ *is a quasinilpotent operator such that* $\tilde{Q}^k = 0$, $\tilde{Q}^{k-1} \neq 0$ *and* $\tilde{Q}^{k-1} + \tilde{Q}^*$ *is invertible (for some* $k \geq 2$*). Then Q is similar to a nice nilpotent operator if and only if* rank Q^k *is finite.*

PROOF. The necessity of the condition "rank Q^k is finite" is clear, since ran Q^k must be closed if Q is similar to a nice nilpotent.

Assume that rank Q^k is finite. Then rank Q^j is finite for all $j \geq k$ and rank $Q^{j+1} <$ rank Q^j for all $j \geq k$ such that rank $Q^j \neq 0$. Hence, Q is a nilpotent and, by Lemmas 7.9 and 7.8, $Q \sim R \oplus F$, where $F = \oplus_{j=k}^{m} q_j^{(\alpha_j)}$, $\sum_{j=k}^{m} j\alpha_j < \infty$ and R is a nilpotent of order k acting on an infinite dimensional space H_o such that $R^{k-1} + R^*$ is invertible. It readily follows from (5.1) that $R^*: (\ker R^*)^{\perp} \to \text{ran}\, R^*$ is a Fredholm operator and therefore ran R^{*j} is closed for all $j = 1,2,\ldots,k-1$. A fortiori, ran R^j is also closed for all $j = 1,2,\ldots,k-1$.

By Theorem 7.11, R is similar to a Jordan nilpotent J of order k. Since $\tilde{R}^{k-1} + \tilde{R}^*$ is invertible, so is $\tilde{J}^{k-1} + \tilde{J}^*$ (Proposition 7.22), but this is impossible unless J is nice.

Hence, $Q \sim J \oplus F$ and $J \oplus F$ is a nice Jordan nilpotent. □

COROLLARY 7.25. *If* $Q \in N_k(H)$, $Q^{k-1} \neq 0$ *and the canonical decomposition of* Q *is given by* (5.1) *with* $\ker Q^j \ominus \ker Q^{j-1} = H_j$, $j = 1, 2, \ldots, k$, *then the following are equivalent:*

(i) If $Q_{j,j+1} = V_j H_j$ *(polar decomposition), where* $H_j = (Q_{j,j+1}{}^* Q_{j,j+1})^{\frac{1}{2}} : H_{j+1} \to H_{j+1}$ *and* $V_j : H_{j+1} \to H_j$ *is a (necessarily injective) partial isometry, then* nul $V_j^* < \infty$ *for all* $j = 1, 2, \ldots, k-1$.

(ii) If $Q^{k-j} + Q^{*j} = W_j R_j$ *(polar decomposition), then* nul $W_j^* < \infty$ *for all* $j = 1, 2, \ldots, k-1$.

(iii) nul $W_j^* < \infty$ *for some* j, $1 \le j \le k-1$.

(iv) nul $W_j < \infty$ *for all* $j = 1, 2, \ldots, k-1$.

(v) nul $W_j < \infty$ *for some* j, $1 \le j \le k-1$.

PROOF. Given $\varepsilon > 0$, let Q_ε be the operator obtained from Q by replacing $Q_{j,j+1}$ by $V_j(H_j + \varepsilon)$. By Theorem 7.15 and its proof, Q_ε is similar to a Jordan nilpotent of order k.

(i) => (ii),(iv). If Q satisfies (i), then Q_ε is similar to a nice Jordan nilpotent. (Use Theorem 7.11 and its proof.) By Proposition 7.22, $Q_\varepsilon^{k-j} + Q_\varepsilon^{*j}$ is a Fredholm operator. On the other hand, it is easily seen from our construction and (5.1) that $\ker(Q^{k-j} + Q^{*j}) = \ker(Q_\varepsilon^{k-j} + Q_\varepsilon^{*j})$ and $[\operatorname{ran}(Q^{k-j} + Q^{*j})]^- = \operatorname{ran}(Q_\varepsilon^{k-j} + Q_\varepsilon^{*j})$ for all $\varepsilon > 0$ and all $j = 1, 2, \ldots, k-1$, whence the results follow.

(iii) => (i) Conversely, if nul $V_j^* = \infty$ for some j, $j = 1, 2, \ldots, k-1$, then $M = \bigvee_{\varepsilon > 0} \operatorname{ran}(Q_\varepsilon^{k-j} + Q_\varepsilon^{*j})$ has infinite codimension in H. Since $(\operatorname{ran} W_j R_j)^- = \operatorname{ran} W_j \subset M$, it follows that nul $W_j^* = \infty$, a contradiction.

(iv) => (i) Similarly, if nul $V_j^* = \infty$ for some j, $1 \le j \le k-1$, then $\ker(Q^{k-j} + Q^{*j}) = \bigvee_{\varepsilon > 0} \ker(Q_\varepsilon^{k-j} + Q_\varepsilon^{*j})$ is infinite dimensional and therefore nul $W_j = \infty$, a contradiction.

Since (ii) => (iii) and (iv) => (v) are trivial implications, we are done. □

COROLLARY 7.26. *Assume that* Q *has the form of Corollary 7.25 and* nul V_j^* *is finite for all* $j = 1, 2, \ldots, k-1$; *then the following are equivalent:*

(i) $\tilde{Q}^{k-1} + \tilde{Q}^*$ *is not invertible.*

(ii) If $H_j = \int \lambda \, dE_j$ *(spectral decomposition), then* rank $E_j([0, \varepsilon]) = \infty$ *(for all* $\varepsilon > 0$*) for some* j, $1 \le j \le k-1$.

(iii) If $R_j = \int \lambda \, dF_j$ *(spectral decomposition), then* rank $F_j([0, \varepsilon]) = \infty$ *(for all* $\varepsilon > 0$*) for some* j, $1 \le j \le k-1$.

(iv) rank $F_j([0, \varepsilon]) = \infty$ *(for all* $\varepsilon > 0$*) for all* $j = 1, 2, \ldots, k-1$.

PROOF. Since nul V_j^* is finite for all $j = 1,2,\ldots,k-1$, it is not difficult to conclude from (5.1), Corollary 7.25 and Proposition 7.22 that the following four conditions are equivalent

(a) $\tilde{Q}^{k-1}+\tilde{Q}^*$ is invertible;

(b) $Q_{j,j+1}$ is a Fredholm operator for all $j = 1,2,\ldots,k-1$;

(c) $Q^{k-j}+Q^{*j}$ is a Fredholm operator for all $j = 1,2,\ldots,k-1$;

(d) $Q^{k-j}+Q^{*j}$ is a Fredholm operator for some j, $1 \le j \le k-1$.

The contrapositive of this equivalence yields the result. □

COROLLARY 7.27. *Assume that* $Q \in N_k(H)$, rank Q^j *is finite for* $j = r, r+1,\ldots,k-1$, rank $Q^j = \infty$- *for* $j = s,s+1,\ldots,r-1$, $\tilde{Q}^{s-1} \neq 0$ *and either* $s = 1$ *or* $s \ge 2$ *and* $\tilde{Q}^{s-1}+\tilde{Q}^*$ *is not invertible. Then there exists a sequence* $\{Q_n\}_{n=1}^{\infty}$ *in* $N_k(H)$ *such that* $||Q - Q_n|| \to 0$ $(n \to \infty)$ *and*

(i) $Q_n \sim M_n \oplus [\oplus_{j=s+1}^{k} q_j^{(\alpha_{jn})}]$;

(ii) $\sum_{j=s+1}^{k} \alpha_{jn} < \infty$;

(iii) α_{jn} *is independent of* n *and* rank $Q_n^{j-1} =$ rank Q^{j-1} *for* $j = r+1,r+2,\ldots,k$;

(iv) $\{\text{rank } Q_n^j\}_{n=1}^{\infty}$ *is strictly increasing for* $j = s+1,\ldots,r-1$;

(v) M_n *is a Jordan operator of order* s; *and*

(vi) $\tilde{M}_n^{s-1} \neq 0$ *and either* $s = 1$ *or* $s \ge 2$ *and* $\tilde{M}_n^{s-1}+\tilde{M}_n^*$ *is not invertible.*

PROOF. By Proposition 7.10 there exists a sequence $\{R_n\}_{n=1}^{\infty}$ in $N_k(H)$ such that $Q - R_n \in K(H)$ for all n, rank $R_n^j =$ rank Q^j for $j = r,r+1,\ldots,k-1$, $\{\text{rank } R_n^j\}_{n=1}^{\infty}$ is strictly increasing for each j such that $s \le j \le r-1$, and $||Q - R_n|| \to 0$ $(n \to \infty)$. By Lemmas 7.9 and 7.8, $R_n \sim L_n \oplus F_n$, where $L_n^s = 0$ and $F_n = \oplus_{j=s+1}^{k} q_j^{(\alpha_{jn})}$ acts on a finite dimensional space. (It is easily seen that α_{jn} is independent of n for $j = r+1,r+2,\ldots,k$.)

Since $Q - R_n$ is compact, it follows from Lemma 7.23 that $\tilde{L}_n^{s-1} \neq 0$ and either $s = 1$ or $s \ge 2$ and $\tilde{L}_n^{s-1}+\tilde{L}_n^*$ is not invertible. Thus, in order to complete the proof, it suffices to show that if $T \in N_s(H)$, $\tilde{T}^{s-1} \neq 0$ and $\tilde{T}^{s-1}+\tilde{T}^*$ is not invertible, then T can be approximated arbitrarily by operators similar to Jordan models with the same properties.

Let $\varepsilon > 0$ and let

$$T = \begin{pmatrix} 0 & T_{12} & T_{13} & \cdot & \cdot & \cdot & T_{1,s-1} & T_{1s} \\ & 0 & T_{23} & \cdot & \cdot & \cdot & T_{2,s-1} & T_{2s} \\ & & 0 & \cdot & \cdot & \cdot & T_{3,s-1} & T_{3s} \\ & & & \cdot & & & \cdot & \cdot \\ & & & & \cdot & & \cdot & \cdot \\ & & 0 & & & \cdot & \cdot & \cdot \\ & & & & & & 0 & T_{s-1,s} \\ & & & & & & & 0 \end{pmatrix} \begin{matrix} M_1 \\ M_2 \\ M_3 \\ \cdot \\ \cdot \\ \cdot \\ M_{s-1} \\ M_s \end{matrix} \qquad (7.11)$$

be the canonical decomposition (given by (5.1)) of T. Since $\tilde{T}^{s-1} \neq 0$, it is not difficult to see that $M_j = \ker T^j \ominus \ker T^{j-1}$ is infinite dimensional for all $j = 1,2,\ldots,s$. Since $\tilde{T}^{s-1}+\tilde{T}^*$ is not invertible, it follows from Corollary 7.26 that either $(\operatorname{ran} T_{j,j+1})^-$ has infinite codimension or $\operatorname{ran} T_{j,j+1}$ is not closed, for some j, $1 \leq j \leq k-1$.

Let $T_{j,j+1} = V_j H_j$ be the polar decomposition of $T_{j,j+1}$. If $(\operatorname{ran} T_{h,h+1})^- = \operatorname{ran} V_h$ has infinite codimension for some h, $1 \leq h \leq k-1$, then we choose T_ε as the operator defined by (7.11) with $T_{j,j+1}$ replaced by $V_j(H_j+\varepsilon/2)$, $j = 1,2,\ldots,s-1$. It is easily seen that $\|T - T_\varepsilon\| = \varepsilon/2 < \varepsilon$, (7.11) (with the new $(j,j+1)$-entries, $j = 1,2,\ldots,s-1$) is the canonical decomposition of $T_\varepsilon \in N_s(H)$ and (by Lemma 7.14) $\operatorname{ran} T_\varepsilon^j$ is closed for all $j = 1,2,\ldots,s-1$. By Theorem 7.11, T_ε is similar to a Jordan operator. It is completely apparent that $T_\varepsilon^{s-1} \neq 0$ and, since the $(h,h+1)$-entry of T_ε is $V_h(H_h+\varepsilon/2)$ and $\operatorname{ran} V_h(H_h+\varepsilon/2) = \operatorname{ran} V_h$ has infinite codimension, it follows from Corollary 7.25 that $T_\varepsilon^{s-1}+T_\varepsilon^*$ cannot be invertible.

Assume that $s \geq 2$ and $(\operatorname{ran} T_{j,j+1})^-$ has finite codimension for all $j = 1,2,\ldots,s-1$, and let h be the last index such that $\operatorname{ran} T_{j,j+1}$ is not closed. Since $\tilde{T}^{s-1} \neq 0$, it is not difficult to see that $T_{h,h+1}$ cannot be compact. Hence, there exists η, $0 < \eta < \varepsilon/2$, such that if $H_h = \int \lambda \, dE_h$ (spectral decomposition) then $E_h([0,\eta)) = E_h((0,\eta))$ and $E_h([\eta,\infty))$ are both of infinite rank. In this case, we define T'_ε as the operator defined by (7.10) with $T_{j,j+1}$ replaced by $T'_{j,j+1} = V_j(H_j+\varepsilon/4)$ for $j = 1,2,\ldots,h-1$ and $j = h+1,h+2,\ldots,s-1$ and $T_{h,h+1}$ replaced by $T'_{h,h+1} = V_h(\int_{[\eta,\infty)} \lambda \, dE_h)$. It is easily seen that $\|T - T'_\varepsilon\| = \max\{\varepsilon/4,\eta\} < \varepsilon/2$, $T'^{s}_\varepsilon = 0$ and $\tilde{T}'^{s-1}_\varepsilon \neq 0$. Furthermore, (use (7.10)!)

$$T'^{s-1}+T'^* = \left(\begin{array}{cccc|c} 0 & 0 & \cdots & 0 & T'_{12}T'_{23}\cdots T'_{s-1,s} \\ \hline & & & & 0 \\ & & * & & \cdot \\ & & & & \cdot \\ & & & & 0 \end{array}\right) \begin{array}{c} M_1 \\ M_2 \\ \cdot \\ \cdot \\ M_s \end{array}$$

Since $T'_{j,j+1}$ is a Fredholm operator for all $j \neq h$ and $\operatorname{nul} T_{h,h+1} = \infty$, it is not difficult to check that $\operatorname{nul} T'_{12}T'_{23}\cdots T'_{s-1,s} = \infty$. Thus, if $T'^{s-1}_\varepsilon+T'^* = W_1R_1$ (polar decomposition), then $\operatorname{nul} W_1 = \infty$. Let

$$T'_\varepsilon = \begin{pmatrix} 0 & A_{12} & A_{13} & & & & \\ & 0 & A_{23} & & & * & \\ & & 0 & \cdot & & & \\ & & & \cdot & \cdot & & \\ & & & & \cdot & \cdot & \\ & & 0 & & & \cdot & \cdot \\ & & & & & 0 & A_{s-1,s} \\ & & & & & & 0 \end{pmatrix} \begin{array}{l} \ker T'^1_\varepsilon \ominus \ker T'^0_\varepsilon \\ \ker T'^2_\varepsilon \ominus \ker T'^1_\varepsilon \\ \ker T'^3_\varepsilon \ominus \ker T'^2_\varepsilon \\ \cdot \\ \cdot \\ \cdot \\ \ker T'^{s-1}_\varepsilon \ominus \ker T'^{s-2}_\varepsilon \\ \ker T'^s_\varepsilon \ominus \ker T'^{s-1}_\varepsilon \end{array}$$

be the canonical decomposition of T'_ε. Since nul $W_1 = \infty$, it follows from Corollary 7.25 that $(\text{ran } A_{j,j+1})^-$ has infinite codimension for some i, $1 \leq i \leq s-1$. Thus, we can proceed exactly as in the first part of the proof in order to find an operator T_ε similar to a Jordan nilpotent of order s such that $\|T'_\varepsilon - T_\varepsilon\| < \varepsilon/2$, $\tilde{T}_\varepsilon^{s-1} \neq 0$ and $\tilde{T}_\varepsilon^{s-1} + \tilde{T}_\varepsilon^*$ is not invertible.

It is completely apparent that $\|T - T_\varepsilon\| < \varepsilon$. Since $\varepsilon > 0$ can be chosen arbitrarily small, we are done. □

REMARK 7.28. It is convenient to observe that, by using translations, direct sums and similarities, all the results of this section can be rephrased as results about compact perturbations of similarities of nice (not necessarily nilpotent!) Jordan operators, etc.

7.7 Notes and remarks

The theorem about the structure of a polynomially compact operator (Theorem 7.2) is due to C. L. Olsen [159]. The auxiliary result Lemma 7.1 is essentially contained in [139] (see also [43]). Corollary 7.6 is a trivial consequence of C. L. Olsen's theorem and Corollary 7.7 is a remark of D. A. Herrero. Lemma 7.8 is an unpublished result of C. Apostol and D. Voiculescu [34]. (A particular case of this result, for $\alpha = 1$, can be found in B. Charles' article [65,Theorem 7].) Lemma 7.9 and Proposition 7.10 are unpublished observations of J. Barría and D. A. Herrero [43] and D. A. Herrero, respectively.

Theorem 7.11 has been simultaneously and independently discovered by C. Apostol and J. G. Stampfli [30], L. J. Gray [109] and L. R. Williams [197]. Theorem 7.15 is due to L. J. Gray [110], but the proof given here is based on C. Apostol's Lemma 3.2 of [12] (Lemma 7.14, above).

The results on similarity invariants of Section 7.5 have been taken from L. A. Fialkow and D. A. Herrero's article [93] and the notion of "nice Jordan operator" is also due to these authors [94].

The key result given by Lemma 7.20 is an unpublished result of C. Apostol and D. Voiculescu [34]. The statement given here is a mild improvement due to L. A. Fialkow and D. A. Herrero and Proposition 7.22 and Corollaries 7.25, 7.26 and 7.27 also correspond to these authors (unpublished results). On the other hand, Corollaries 7.21 and 7.24 are contained in [34].

8 Closures of similarity orbits of nilpotent operators

With the help of Lemma 7.1, the problem of characterizing $S(T)^-$ for T a polynomially compact operator can be easily reduced to the case when $\tilde{T}$ is nilpotent.

This chapter is an account of what is known and what is unknown about this problem. The results include some information about quasinilpotent operators, the complete solution of the corresponding problem in the Calkin algebra and several conjectures and examples that illustrate the difficulties of the still open part of the problem.

8.1 Universal operators

Given a class W of operators acting on H, we shall say that $R \in W$ is a *universal operator* for W if $W \subset S(R)^-$. Namely, q_k is a universal operator for $N_k(\mathbb{C}^k)$ $(k = 1,2,\ldots)$, but $N(H)$ (H an infinite dimensional space) and $\{\lambda: \lambda \in \mathbb{C}\} \subset L(H)$ do not contain any universal operator.

8.1.1 Universal quasinilpotent operators

Let $Q \in L(H)$ be a quasinilpotent operator. If $Q^k = 0$ and ran Q^{k-1} contains a subspace of dimension α $(1 \leq \alpha \leq \infty)$, then $Q \sim q_k^{(\alpha)} \oplus R$ (Lemmas 7.9 and 7.8), where $R \in N_k(H_\beta)$ for some space H_β of dimension β $(0 \leq \beta \leq \infty)$. By Rota's corollary (Corollary 3.36(ii)), $R \underset{\text{sim}}{\to} q_1^{(\beta)}$ (= the 0 operator acting on H_β). A fortiori, (by a new application of Rota's corollary) we have

$$Q \underset{\text{sim}}{\to} q_1^{(\beta)} \oplus q_k^{(\alpha)} \underset{\text{sim}}{\to} q_1^{(\infty)} \oplus q_k^{(\alpha)} \underset{\text{sim}}{\to} q_1^{(\infty)} \oplus q_k. \quad (8.1)$$

Assume that $Q^k \neq 0$ for all $k = 1,2,\ldots$, and let $R_k(Q) = \{x \in H: Q^k x = 0\}$. It is easily seen that $R_k(Q)$ is a *closed nowhere dense subset* of H. Hence, by Baire's category theorem, $H \setminus \cup_{k=1}^{\infty} R_k(Q) \neq \emptyset$, i.e. there exists a vector $y \in H$ such that $Q^k y \neq 0$ for all $k = 1,2,\ldots$. Let $\mathcal{M} = \bigvee_{k\geq 0} Q^k y$ and let

$$Q = \begin{pmatrix} M & * \\ 0 & L \end{pmatrix} \begin{matrix} M \\ M^\perp \end{matrix}.$$

Clearly, $\sigma(Q) = \{0\} = \sigma(M) \cup \sigma(L)$. Thus, by Lemma 3.36 and Corollary 3.35(ii), $Q \underset{\text{sim}}{\to} M \oplus L \underset{\text{sim}}{\to} M \oplus 0$, where $M = Q|M$ is a cyclic quasinilpotent operator such that $M^k \neq 0$ for all $k = 1, 2, \ldots$.

LEMMA 8.1. *If* $Q \in L(H)$ *is quasinilpotent and* $Q^k \neq 0$ *for all* $k = 1, 2, \ldots$, *then* $S(Q)^-$ *contains all compact quasinilpotents.*

PROOF. By Lemma 2.5 and Proposition 2.33, it suffices to show that $Q \underset{\text{sim}}{\to} q_k \oplus 0$ for all k large enough and, by our previous observations, we can restrict ourselves to the case when Q has a cyclic vector y.

Let $\{e_n\}_{n=1}^{\infty}$ be the Gram-Schmidt orthonormalization of the sequence $\{Q^{n-1}y\}_{n=1}^{\infty}$. Since y is cyclic, it is straightforward to check that the matrix of Q with respect to this ONB has the form

$$Q = \begin{pmatrix} q_{11} & q_{12} & q_{13} & \cdot & \cdot & \cdot & q_{1k} & q_{1,k+1} & \cdot & \cdot \\ q_{21} & q_{22} & q_{23} & \cdot & \cdot & \cdot & q_{2k} & q_{2,k+1} & \cdot & \cdot \\ & q_{32} & q_{33} & \cdot & \cdot & \cdot & q_{3k} & q_{3,k+1} & \cdot & \cdot \\ & & q_{43} & \cdot & \cdot & \cdot & q_{4k} & q_{4,k+1} & \cdot & \cdot \\ & & & \cdot & \cdot & & \cdot & \cdot & \cdot & \cdot \\ & & & & \cdot & \cdot & \cdot & \cdot & \cdot & \cdot \\ & & & & & \cdot & \cdot & \cdot & \cdot & \cdot \\ & & & & & & q_{k+1,k} & q_{k+1,k+1} & \cdot & \cdot \\ & & & 0 & & & & q_{k+2,k+1} & \cdot & \cdot \\ & & & & & & & & \cdot & \cdot \\ & & & & & & & & & \cdot \end{pmatrix}.$$

Moreover, $q_{k,k-1} \neq 0$ for all $k \geq 2$ and $|q_{21}q_{32}\cdots q_{k,k-1}|^{1/k} \leq \|Q^k\|^{1/k} \to sp(Q) = 0$ $(k \to \infty)$, so that $\inf\{|q_{k,k-1}| : k \geq 2\} = 0$.

Fix $k \geq 2$ and let W be the diagonal normal operator $\operatorname{diag}(\{1, w_1, w_2, \ldots, w_{k-2}, w_{k-1}, w_{k-1}, w_{k-1}, w_{k-1}, \ldots\})$, where $w_j = (\Pi_{i=1}^{j} q_{i+1,i})^{-1}$, $j = 1, 2, \ldots, k-1$. If $N = \bigvee\{e_1, e_2, \ldots, e_k\}$, then we have

$$WQW^{-1} = \begin{pmatrix} A & B \\ C & D \end{pmatrix} \begin{matrix} N \\ N^\perp \end{matrix},$$

where

$$A = \begin{pmatrix} q_{11} & q_{12}\frac{1}{w_1} & q_{13}\frac{1}{w_2} & q_{14}\frac{1}{w_3} & \cdot & \cdot & \cdot\ q_{1,k-1}\frac{1}{w_{k-2}} & q_{1k}\frac{1}{w_{k-1}} \\ 1 & q_{22} & q_{23}\frac{w_1}{w_2} & q_{24}\frac{w_1}{w_3} & \cdot & \cdot & \cdot\ q_{2,k-1}\frac{w_1}{w_{k-2}} & q_{2k}\frac{w_1}{w_{k-1}} \\ & 1 & q_{33} & q_{34}\frac{w_2}{w_3} & \cdot & \cdot & \cdot\ q_{3,k-1}\frac{w_2}{w_{k-2}} & q_{3k}\frac{w_2}{w_{k-1}} \\ & & 1 & q_{44} & \cdot & \cdot & \cdot\ q_{4,k-1}\frac{w_3}{w_{k-2}} & q_{4k}\frac{w_3}{w_{k-1}} \\ & & & \cdot & \cdot & & \cdot & \cdot \\ & & & & \cdot & \cdot & \cdot & \cdot \\ & & & & & \cdot & \cdot & \cdot \\ & & 0 & & & & \cdot\ q_{k-1,k-1} & q_{k-1}\frac{w_{k-2}}{w_{k-1}} \\ & & & & & & 1 & q_{kk} \end{pmatrix},$$

$$B = \frac{1}{w_{k-1}}\begin{pmatrix} q_{1,k+1} & q_{1,k+2} & q_{1,k+3} & \cdot & \cdot & \cdot \\ q_{2,k+1}w_1 & q_{2,k+2}w_1 & q_{2,k+3}w_1 & \cdot & \cdot & \cdot \\ \cdot & \cdot & \cdot \\ \cdot & \cdot & \cdot \\ \cdot & \cdot & \cdot \\ q_{k-1,k+1}w_{k-2} & q_{k-1,k+2}w_{k-2} & q_{k-1,k+3}w_{k-2} & \cdot & \cdot & \cdot \\ q_{k,k+1}w_{k-1} & q_{k,k+2}w_{k-1} & q_{k,k+3}w_{k-1} & \cdot & \cdot & \cdot \end{pmatrix},$$

$$C = \begin{pmatrix} 0 & 0 & \cdot & \cdot & \cdot & 0 & q_{k+1,k} \\ & & & & & & 0 \\ & & & & & & 0 \\ & & 0 & & & & \cdot \\ & & & & & & \cdot \\ & & & & & & \cdot \end{pmatrix} \quad \text{and} \quad D = (q_{ij})_{i,j=k+1}^{\infty}.$$

It follows that $\|A - \sum_{j=1}^{k-1} e_{j+1}\otimes e_j\| \le \sum_{j=1}^{k} \|Q\|^j$, $\|B\| \le \sum_{j=1}^{k} \|Q\|^j$, $\|C\| \le \|Q\|$ and $\|D\| \le \|Q\|$.

Since $\sum_{j=1}^{k-1} e_{j+1}\otimes e_j = q_k^* \simeq q_k$, we conclude that $\operatorname{dist}[q_k\oplus 0, S(Q)] \le$

$$\|WQW^{-1} - q_k^*\oplus 0\| \le 4\sum_{j=1}^{k} \|Q\|^j.$$

By Rota's corollary (Corollary 3.35(ii)), there exists a sequence $\{V_n\}_{n=1}^{\infty}$ such that $\|V_nQV_n^{-1}\| < 1/n$ $(n = 1,2,\ldots)$. It is completely apparent that $V_nQV_n^{-1}$ is also a cyclic quasinilpotent. Thus, we can apply the above argument to $V_nQV_n^{-1}$ in order to obtain a sequence $\{W_n\}_{n=1}^{\infty}$ in $G(H)$ such that

$$\|W_n(V_nQV_n^{-1})W_n^{-1} - q_k^*\oplus 0\| = \|(W_nV_n)Q(W_nV_n)^{-1} - q_k^*\oplus 0\| < 4k/n \to 0$$

(as $n \to \infty$). Hence, $Q \underset{\text{sim}}{\to} q_k^*\oplus 0 \simeq q_k\oplus 0$ for all $k = 2,3,\ldots$. □

From Lemma 2.5, we obtain the following auxiliary results:

COROLLARY 8.2. *(i)* $q_k^{(k-1)} \underset{\text{sim}}{\to} q_{k-1}^{(k)}$ *for all* $k > 1$.

(ii) $q_1\oplus q_k^{(k-2)} \underset{\text{sim}}{\to} q_{k-1}^{(k-1)}$ *for all* $k > 2$.

(iii) $q_j\oplus q_k^{(\infty)} \underset{\text{sim}}{\to} q_1^{(\infty)}\oplus q_k^{(\infty)} \underset{\text{sim}}{\to} q_{k-1}^{(\infty)}$ *for all* $j \geq 1$ *and all* $k > 1$.

(iv) *If* $A = [\oplus_{j=1}^{s-1} q_j^{(\alpha_j)}]\oplus q_s^{(\infty)}\oplus[\oplus_{j=s+1}^{k} q_j^{(\alpha_j)}]$, $\sum_{j=1,j\neq s}^{k} \alpha_j < \infty$, $T = q_s^{(\infty)}\oplus[\oplus_{j=s+1}^{k} q_j^{(\tau_j)}]$, $\sum_{j=s+1}^{k} \tau_j < \infty$ *and* $\operatorname{rank} A^j \leq \operatorname{rank} T^j$ *for all* $j = s,s+1,\ldots,k-1$, *then there exists a non-negative integer* α *such that* $T \underset{\text{sim}}{\to} q_1^{(\alpha)}\oplus A$.

(v) *If* $\sum_{j=1}^{k-1} \tau_j = \infty$, *then* $[\oplus_{j=1}^{k-1} q_j^{(\tau_j)}]\oplus q_k^{(\infty)} \# q_1^{(\infty)}\oplus q_k^{(\infty)}$ *for all* $k > 1$.

(vi) *If* $k > 1$, *then* $S(q_1^{(\infty)}\oplus q_k^{(\infty)})^- = N_k(H)\backslash N_k^+(H) \supset N_{k-1}(H)$ *(where* $N_k^+(H)$ *is defined by* (7.9)*)*.

PROOF. (i) Since $k(k-1) = (k-1)k$, both operators act on the same finite dimensional space. On the other hand,

$$\operatorname{rank}\, [q_{k-1}^{(k)}]^j = k\max\{k-1-j,0\} \leq (k-1)\max\{k-j,0\} = \operatorname{rank}\, [q_k^{(k-1)}]^j$$

$(j = 1,2,\ldots,k-1)$.

Now the result follows from Lemma 2.5.

(ii) This can be proved by using the same arguments as above.

(iii) By (i), (ii), Lemma 2.5 and an obvious inductive argument, we have

$$q_j\oplus q_k^{(\infty)} = q_j\oplus[q_k^{(k-1)}]^{(\infty)}\oplus q_k^{(\infty)} \underset{\text{sim}}{\to} q_1^{(j)}\oplus[q_{k-1}^{(k)}]^{(\infty)}\oplus q_k^{(\infty)} = q_1^{(j)}\oplus q_{k-1}^{(\infty)}\oplus q_k^{(\infty)} = q_1^{(j)}\oplus[q_{k-1}^{(k-2)}]^{(\infty)}\oplus q_k^{(\infty)} \underset{\text{sim}}{\to} q_1^{(j)}\oplus[q_{k-2}^{(k-1)}]^{(\infty)}\oplus q_k^{(\infty)} \underset{\text{sim}}{\to} \cdots \underset{\text{sim}}{\to} q_1^{(j)}\oplus q_1^{(\infty)}\oplus q_k^{(\infty)} = [q_1\oplus q_k^{(k-2)}]^{(\infty)} \underset{\text{sim}}{\to} q_{k-1}^{(k-1)\,(\infty)} = q_{k-1}^{(\infty)}.$$

(iv) If $\operatorname{rank} A^j \leq \operatorname{rank} T^j$ for all $j = s,s+1,\ldots,k-1$, it is not difficult to find non-negative integers α, τ such that $\alpha+\sum_{j=1,j\neq s}^{k} j\alpha_j = s\tau +\sum_{j=s+1}^{k} j\tau_j$. By Lemma 2.5,

$$q_s^{(\tau)}\oplus[\oplus_{j=s+1}^{k} q_j^{(\tau_j)}] \underset{\text{sim}}{\to} q_1^{(\alpha)}\oplus[\oplus_{j=1,j\neq s}^{k} q_j^{(\alpha_j)}].$$

A fortiori, $T \underset{\text{sim}}{\to} q_1^{(\alpha)}\oplus A$.

(v) By (iii), $q_1^{(\infty)}\oplus q_k^{(\infty)} \underset{\text{sim}}{\to} [\oplus_{j=1}^{k} q_j^{(\tau_k)}]\oplus q_k^{(\infty)}$ and, by Lemma 2.

5, $q_j^{(\tau_j)} \underset{sim}{\to} q_1^{(j\tau_j)}$ $(j=1,2,\ldots,k-1)$, so that $[\oplus_{j=1}^{k-1} q_j^{(\tau_j)}]\oplus q_k^{(\infty)} \underset{sim}{\to} q_1^{(\infty)}\oplus q_k^{(\infty)}$. Hence, $[\oplus_{j=1}^{k-1} q_j^{(\tau_j)}]\oplus q_k^{(\infty)} \# q_1^{(\infty)}\oplus q_k^{(\infty)}$.

(vi) By Corollary 7.21, $S(q_1^{(\infty)}\oplus q_k^{(\infty)})^- \subset N_k(H)\backslash N_k^+(H)$. Let $A \in N_k(H)\backslash N_k^+(H)$ and let s $(1 \le s \le k)$ be the largest exponent such that $\tilde{A}^{s-1} \ne 0$.

If $s=k$, then it follows from Corollary 7.27 that $\|A - T_n\| \to 0$ $(n \to \infty)$ for a sequence $\{T_n\}_{n=1}^{\infty}$ in $N_k(H)$ such that

$$T_n \sim [\oplus_{j=1}^{k-1} q_j^{(\tau_{nj})}]\oplus q_k^{(\infty)}$$

with $\sum_{j=1}^{k-1} \tau_{nj} = \infty$. Since (by (v)) $T_n \# q_1^{(\infty)}\oplus q_k^{(\infty)}$, it readily follows that $\{T_n\}_{n=1}^{\infty} \subset S(q_1^{(\infty)}\oplus q_k^{(\infty)})^-$. A fortiori, $A \in S(q_1^{(\infty)}\oplus q_k^{(\infty)})^-$.

If $s < k$, then it follows from Theorem 7.15 and Corollary 7.12 that $\|A - S_n\| \to 0$ $(n \to \infty)$ for a sequence $\{S_n\}_{n=1}^{\infty}$ in $N_k(H)$ such that $S_n \sim \oplus_{j=1}^{k} q_j^{(\sigma_{nj})}$, where $\sigma_{ns}=\infty$ and σ_{nk} is finite. By (v) and (iii),

$$q_1^{(\infty)}\oplus q_k^{(\infty)} \# [\oplus_{j=1}^{k} q_j^{(\sigma_{nj})}]\oplus q_k^{(\infty)} \underset{sim}{\to} [\oplus_{j=1}^{k} q_j^{(\sigma_{nj})}]\oplus q_s^{(\infty)} = \oplus_{j=1}^{k} q_j^{(\sigma_{nj})},$$

which is similar to S_n. Proceeding exactly as in the first case, we conclude that $q_1^{(\infty)}\oplus q_k^{(\infty)} \underset{sim}{\to} A$.

Since the inclusion $N_{k-1}(H) \subset N_k(H)\backslash N_k^+(H)$ is trivial, we are done. □

PROPOSITION 8.3. *The quasinilpotent operator* $Q \in L(H)$ *is a universal quasinilpotent (and therefore*

$$S(Q)^- = N(H)^- = \{A \in (BQT):\ \sigma(A) \text{ and } \sigma_e(A) \text{ are connected and contain the origin}\}) \tag{8.2}$$

if and only if $\tilde{Q}$ *is not a nilpotent.*

Moreover,

$$\begin{aligned} Q_u(H) &= (def)\ \{Q \in L(H):\ Q \text{ is a universal quasinilpotent}\} \\ &= \{Q \in L(H):\ Q \text{ is quasinilpotent}\}\backslash \cup_{k=1}^{\infty} \pi^{-1}[N_k(A(H))] \end{aligned} \tag{8.3}$$

is a G_δ *in* $L(H)$, *dense in the set* $Q(H)$ *of all quasinilpotent operators.*

PROOF. It is completely apparent that if $Q \in \pi^{-1}[N_k(A(H))]$ for some $k \ge 1$, then $S(Q)^-$ is also contained in the (closed) set $\pi^{-1}[N_k(A(H))]$, so that the condition is necessary.

Assume that Q is quasinilpotent and $\tilde{Q}^k \ne 0$ for all $k=1,2,\ldots$. By Voiculescu's theorem (Proposition 4.21(iii)), if $R = \rho(\tilde{Q})$ for some faithful unital *-representation ρ of $C^*(\tilde{Q})$, then $Q\oplus R^{(\infty)} \in U(Q)^-$. A fortiori, $Q \underset{sim}{\to} Q\oplus R^{(\infty)}$ and it is completely apparent that $\sigma(R) = \{0\}$ and $R^k \ne 0$ for all $k=1,2,\ldots$.

By Corollary 3.35(ii) and Lemma 8.1,

$$Q \oplus R^{(\infty)} \underset{\text{sim}}{\to} 0 \oplus (q_k \oplus 0)^{(\infty)} = q_1{}^{(\infty)} \oplus q_k{}^{(\infty)}$$

for each k, k = 2,3,... . Hence, by Corollary 8.2(vi)

$$S(Q)^- \supset \cup_{k=2}^{\infty} S(q_1{}^{(\infty)} \oplus q_k{}^{(\infty)})^- \supset \cup_{k=2}^{\infty} N_{k-1}(H) = N(H).$$

Since $Q \in N(H)^-$ (Theorem 5.1), we see that $S(Q)^- = N(H)^-$, whence we obtain formulas (8.2) and (8.3).

It is completely apparent that $Q_u(H)$ is dense in $Q(H)$. Namely, if V is the Volterra operator, defined by $Vf(x) = \int_0^x f(t)dt$ in $L^2([0,1], dx)$, then V is a compact quasinilpotent and $V^k \neq 0$ for all k = 1,2,.... It readily follows that $S(V^{(\infty)}) \subset Q_u(H)$ and $S(V^{(\infty)})$ is dense in $Q(H)$. By Corollary 1.3(ii), $Q(H)$ is a G_δ in $L(H)$. Since $\pi^{-1}[N_k(A(H))]$ is a closed subset of $L(H)$ for each k = 1,2,..., it follows that $Q_u(H)$ is also a G_δ. □

8.1.2 Universal compact quasinilpotent operators

Let $K \in K(H)$ be a compact quasinilpotent operator such that $K^k \neq 0$ for all k = 1,2,... . By Lemma 8.1, K is a universal operator for the class of all compact quasinilpotents. By using the same arguments as in the last part of the previous proof, we obtain the following

PROPOSITION 8.4. *The compact quasinilpotent operator* $K \in L(H)$ *is a universal compact quasinilpotent (and therefore*

$$S(K)^- = N(H)^- \cap K(H) = \{C \in K(H): C \text{ is quasinilpotent}\})$$

if and only if K *is not a nilpotent.*

The set of all universal compact quasinilpotents is a G_δ *in* $L(H)$*, dense in the set of all compact quasinilpotent operators.*

8.2 Compact perturbations of not nice nilpotents

Assume that $T \in L(H)$ is an essential nilpotent of order k (i.e., $\tilde{T}^k = 0$, but $\tilde{T}^{k-1} \neq 0$) such that $\tilde{T}^{k-1} + \tilde{T}^*$ is not invertible.

If $\|A - W_n T W_n^{-1}\| \to 0$ $(n \to \infty)$, then $\|\tilde{A} - \tilde{W}_n \tilde{T} \tilde{W}_n^{-1}\| \to 0$ $(n \to \infty)$ and it follows from Corollaries 1.8 and 7.21 that A is an essential nilpotent of order at most k such that $\tilde{A}^{k-1} + \tilde{A}^*$ is not invertible; therefore

$$S(T)^- \subset \pi^{-1}[N_k(A(H))] \setminus N_k^+(H). \tag{8.4}$$

This observation reduces the problem of characterizing $S(T)^-$ to that of characterizing those operators in the second member of (8.4) that can be uniformly approximated by similarities of T. Since $\sigma_e(A) = \{0\}$ for all A in $S(T)^-$, it is not difficult to see (use Corollary 1.6) that $\sigma(A) = \sigma(T)$ and $\dim H(\lambda;A) = \dim H(\lambda;T)$ for all $\lambda \in \sigma(T)$; furthermore, $T|H(\lambda;T) \underset{\text{sim}}{\to} A|H(\lambda;A)$ for each λ in $\sigma(T)$ (use Corollaries 1.8 and 2.8).

By using these results and Corollary 3.22, the problem of characterizing $S(T)^-$ can be reduced to consider two cases: a) the case when T is a quasinilpotent, and b) the case when $\sigma(T)$ is the union of $\{0\}$ and a sequence of normal eigenvalues converging to 0. This will be done in two propositions.

PROPOSITION 8.5. *Let $Q \in L(H)$ and assume that Q is quasinilpotent, $\tilde{Q}^k = 0$ and $\tilde{Q}^{k-1} \neq 0$, but $\tilde{Q}^{k-1} + \tilde{Q}^*$ is not invertible; then*

$$S(Q)^- = \{A \in Q(H):\ \tilde{A}^k = 0,\ \tilde{A}^{k-1} + \tilde{A}^* \text{ is not invertible},$$
$$\operatorname{rank} A^j \le \operatorname{rank} Q^j \le \infty-, \text{ for all } j \ge k\}.$$

PROOF. Let $R = (\tilde{Q})$ for some faithful unital *-representation ρ of $C^*(\tilde{Q})$; then we have

$$Q \underset{\text{sim}}{\to} Q \oplus R^{(\infty)} \underset{\text{sim}}{\to} Q \oplus (q_1^{(\infty)} \oplus q_k^{(\infty)}) \underset{\text{sim}}{\to} (F \oplus q_1^{(\infty)}) \oplus (q_1^{(\infty)} \oplus q_k^{(\infty)}) =$$
$$F \oplus q_1^{(\infty)} \oplus q_k^{(\infty)} \underset{\text{sim}}{\to} F \oplus q_{k-1}^{(\infty)} \underset{\text{sim}}{\to} F \oplus q_1^{(\infty)} \oplus q_{k-1}^{(\infty)} \underset{\text{sim}}{\to} F \oplus q_{k-2}^{(\infty)} \underset{\text{sim}}{\to}$$
$$F \oplus q_1^{(\infty)} \oplus q_{k-2}^{(\infty)} \underset{\text{sim}}{\to} \cdots \underset{\text{sim}}{\to} F \oplus q_1^{(\infty)} \underset{\text{sim}}{\to} 0,$$

where F is any Jordan nilpotent acting on a finite dimensional space such that $\operatorname{rank} F^j \le \operatorname{rank} Q^j$, for all $j \ge k$. (To see this, apply: 1) Proposition 4.21(ii); 2) Lemmas 7.9 and 7.8; 3) Lemmas 7.9, 7.8 and 8.1; 4) Corollary 8.2(ii); 5) Corollary 8.2(i), etc.)

Assume that A is a quasinilpotent such that $\tilde{A}^k = 0$, $\tilde{A}^{k-1} + \tilde{A}^*$ is not invertible and $\operatorname{rank} A^j \le \operatorname{rank} Q^j \le \infty-$, for all $j \ge k$, and let t ($1 \le t \le k$) be the exponent such that $\tilde{A}^t = 0$, but $\tilde{A}^{t-1} \neq 0$. If either $t > 1$ and $\tilde{A}^{t-1} + \tilde{A}^*$ is not invertible or $t = 1$ (i.e., A is compact), then it follows from Corollaries 7.6 and 7.27 that A can be uniformly approximated by a sequence $\{A_n\}_{n=1}^{\infty} \subset N(H)$ such that $A_n \sim F_n \oplus q_1^{(\infty)} \oplus q_t^{(\infty)}$ (for suitably chosen operators F_n acting on finite dimensional spaces, such that $\operatorname{rank} F_n^{\,j} \le \operatorname{rank} Q^j$, for all $j \ge k$; $n = 1,2,\ldots$). As in the proof of Proposition 8.3, we conclude that $Q \underset{\text{sim}}{\to} A$.

If $t > 1$ and $\tilde{A}^{t-1} + \tilde{A}^*$ is invertible, then $t < k$, A is similar to a compact perturbation of $q_1^{(h)} \oplus q_t^{(\infty)}$ (for some h, $0 \le h < t$; use Proposition 7.22). By Corollary 7.6, A can be uniformly approximated by a sequence $\{B_n\}_{n=1}^{\infty} \subset N_t^+(H) \cap N(H)$ such that $B_n \sim G_n \oplus q_t^{(\infty)}$ (for suitably chosen operators G_n acting on finite dimensional spaces, such that rank $G_n^j \leq$ rank Q^j, for all $j \geq k$; $n=1,2,\ldots$). We conclude as above that $Q \underset{\text{sim}}{\to} A$.

Conversely, if $Q \underset{\text{sim}}{\to} A$, then it follows from our previous observations (at the beginning of this section) and Corollary 1.8 that A is quasinilpotent, $\tilde{A}^k = 0$ and $\tilde{A}^{k-1}+\tilde{A}^*$ cannot be invertible. □

PROPOSITION 8.6. *Let* $T \in L(H)$ *and assume that* $\tilde{T}^k = 0$, $\tilde{T}^{k-1} \neq 0$, $\tilde{T}^{k-1}+\tilde{T}^*$ *is not invertible and* $\sigma(T) = \{0\}\cup\{\lambda_n\}_{n=1}^{\infty}$ *is an infinite set (where* $\lambda_n \neq 0$ *for all* n, $\lambda_n \neq \lambda_m$ *if* $n \neq m$ *and* $\lambda_n \to 0$, *as* $n \to \infty$*); then* $S(T)^-$ *is the set of all those* $A \in L(H)$ *such that* $\tilde{A}^k = 0$, $\tilde{A}^{k-1}+\tilde{A}^*$ *is not invertible,* $\sigma(A) = \sigma(T)$, $\dim H(\lambda_n;A) = \dim H(\lambda_n;T)$ *and* $T|H(\lambda_n;T) \underset{\text{sim}}{\to} A|H(\lambda_n;A)$ *for all* $n = 1,2,\ldots$.

PROOF. Let ρ be a faithful unital *-representation of $C^*(\tilde{T})$ and let $R = \rho(\tilde{T})$. By Propositions 4.21(ii) and 8.5, $T \underset{\text{sim}}{\to} T\oplus R^{(\infty)} \underset{\text{sim}}{\to} T\oplus q_1^{(\infty)} \oplus q_k^{(\infty)}$. Let $\Lambda_m = \{\lambda_1,\lambda_2,\ldots,\lambda_m\}$. By Corollaries 3.22 and 3.35(ii), $T \sim T_m\oplus[\oplus_{n=1}^m R_n]$, where $R_n = \lambda_n+|\lambda_n|J_n$ and J_n is the Jordan form of the nilpotent operator $T|H(\lambda_n;T)$ $(n=1,2,\ldots)$, $\sigma(T_m) = \sigma(T)\setminus\Lambda_m$ and $\|T_m\| < 2\max\{|\lambda_n|: n > m\}$. By taking suitable limits, we conclude that

$$T \underset{\text{sim}}{\to} T\oplus q_1^{(\infty)}\oplus q_k^{(\infty)} \underset{\text{sim}}{\to} S = q_1^{(\infty)}\oplus q_k^{(\infty)}\oplus[\oplus_{n=1}^{\infty} R_n].$$

Clearly, $S \underset{\text{sim}}{\to} S_m = q_1^{(\infty)}\oplus q_k^{(\infty)}\oplus[\oplus_{n=1}^m R_n]\oplus[\oplus_{n=m+1}^{\infty} \lambda_n 1_{M_n}]$, where 1_{M_n} denotes the identity on a space M_n of dimension equal to $\dim H(\lambda_n;T)$ $(n=m+1,m+2,\ldots)$. Let e_{m+j} be a unit vector in M_{m+j}, $j=1,2,\ldots,p$; then

$$S_m \sim q_1^{(\infty)}\oplus q_k^{(\infty)}\oplus[\oplus_{n=1}^m R_n]$$

$$\oplus\begin{pmatrix} \lambda_{m+1} & 1 & & & & 0 \\ & \lambda_{m+2} & \cdot & & & \\ & & \cdot & \cdot & & \\ & & & \cdot & \cdot & \\ & & & & \cdot & \cdot \\ & & & & \lambda_{m+p-1} & 1 \\ 0 & & & & & \lambda_{m+p} \end{pmatrix}\oplus F_m\oplus[\oplus_{n=m+p+1}^{\infty} \lambda_n 1_{M_n}],$$

where $F_m = \oplus_{j=1}^p \lambda_{m+j} 1_{M_{m+j}\ominus\vee\{e_{m+j}\}}$.

Since $\lambda_n \to 0$ and $\|R_n\| \leq 2|\lambda_n| \to 0$ $(n \to \infty)$, it is easily seen that there exists a sequence $\{S'_m\}_{m=1}^{\infty}$ of operators such that $S'_m \sim S_m$ (for all $m=1,2,\ldots$) and $\|q_p\oplus S - S'_m\| \to 0$ $(m \to \infty)$.

Hence, $S \underset{\text{sim}}{\to} q_p\oplus S \simeq q_p\oplus q_1^{(\infty)}\oplus S$ for all $p=1,2,\ldots$. By Lemma 2.5 and Proposition 2.33, $S \underset{\text{sim}}{\to} K_u\oplus S$, where K_u is a universal compact quasinilpotent.

Assume that $T \underset{\text{sim}}{\to} A$. It follows from our observations at the beginning of this section that: (1) $\sigma_e(A) = \sigma_e(T) = \{0\}$, $\sigma(A) = \sigma(T)$ and $\dim H(\lambda_n;A) = \dim H(\lambda_n;T)$ for all $n = 1,2,\ldots$; (2) $T|H(\lambda_n;T) \underset{\text{sim}}{\to} A|H(\lambda_n;A)$ for each n, $n = 1,2,\ldots$; (3) $\tilde{A}^k = 0$ and $\tilde{A}^{k-1}+\tilde{A}^*$ is not invertible.

Suppose that $A \in L(H)$ satisfies (1), (2) and (3) and let $B = \rho_1(\tilde{A})$ for some faithful unital *-representation ρ_1 of $C^*(\tilde{A})$. Then $B^k = 0$, $B^{k-1}+B^*$ is not invertible and (by Proposition 8.5) $q_1^{(\infty)} \oplus q_k^{(\infty)} \underset{\text{sim}}{\to} B^{(\infty)}$. On the other hand, by Proposition 4.21(ii), $A \oplus B^{(\infty)} \simeq_a A$.

Thus, given $\varepsilon > 0$ there exists W invertible such that $\|W(q_1^{(\infty)} \oplus q_k^{(\infty)})W^{-1} - B^{(\infty)}\| < \varepsilon/2$. By continuity, $\|W(q_1^{(\infty)} \oplus q_k^{(\infty)} + C)W^{-1} - B^{(\infty)}\| < \varepsilon$, provided $\|C\| < \varepsilon(2\|W\|.\|W^{-1}\|)^{-1}$. Let

$$A = \left(\begin{array}{cccccccc|c} A_{11} & A_{12} & \cdot & \cdot & A_{1m} & \cdot & \cdot & & B_1 \\ & A_{22} & \cdot & \cdot & A_{2m} & \cdot & \cdot & & B_2 \\ & & \cdot & & \cdot & & & & \cdot \\ & & & \cdot & \cdot & & & & \cdot \\ & & 0 & & A_{mm} & \cdot & \cdot & & B_m \\ & & & & & \cdot & & & \cdot \\ & & & & & & \cdot & & \cdot \\ \hdashline & & & 0 & & & & & A_\infty \end{array}\right) \begin{array}{c} N_1 \\ N_2 \\ \cdot \\ \cdot \\ N_m \\ \cdot \\ \cdot \\ N_\infty \end{array},$$

where $N_1 = H(\lambda_1;A)$ and $N_m = H(\{\lambda_1,\ldots,\lambda_m\};A) \ominus H(\{\lambda_1,\ldots,\lambda_{m-1}\};A)$, and let

$$C_m = \left(\begin{array}{cccccccccc|c} A_{11} & A_{12} & \cdot & \cdot & A_{1m} & A_{1,m+1} & \cdot & A_{1n} & \cdot & \cdot & B_1 \\ & A_{22} & \cdot & \cdot & A_{2m} & A_{2,m+1} & \cdot & A_{2n} & \cdot & \cdot & B_2 \\ & & \cdot & & \cdot & \cdot & & \cdot & & & \cdot \\ & & & \cdot & \cdot & \cdot & & \cdot & & & \cdot \\ & & & & A_{mm} & A_{m,m+1} & \cdot & A_{mn} & \cdot & \cdot & B_m \\ & & & & & A_{m+1,m+1}-\lambda_{m+1} & \cdot & A_{m+1,n} & & \cdot & B_{m+1} \\ & 0 & & & & & \cdot & & & & \cdot \\ & & & & & & \cdot & \cdot & & & \cdot \\ & & & & & & & A_{nn}-\lambda_n & \cdot & & B_n \\ & & & & & & & & \cdot & & \cdot \\ & & & & & & & & & \cdot & \cdot \\ \hdashline & & & & 0 & & & & & & A_\infty \end{array}\right) \begin{array}{c} N_1 \\ N_2 \\ \cdot \\ \cdot \\ N_m \\ N_{m+1} \\ \cdot \\ \cdot \\ N_n \\ \cdot \\ \cdot \\ N_\infty \end{array}$$

It is easily seen that $A - C_m$ is a normal compact operator, $\|A - C_m\| = \max\{|\lambda_n|: n > m\} \to 0$ $(m \to \infty)$ and

$$C_m \quad \begin{pmatrix} A_{11} & A_{12} & \cdot & \cdot & A_{1m} \\ & A_{22} & \cdot & \cdot & A_{2m} \\ & & \cdot & & \cdot \\ & 0 & & \cdot & \cdot \\ & & & & A_{mm} \end{pmatrix} \oplus D_m,$$

where D_m is quasinilpotent, $\tilde{D}_m^{\,k}=0$ and $\tilde{D}_m^{\,k-1}+\tilde{D}_m^*$ is not invertible. By Proposition 8.5, $K_u\oplus q_k^{(\infty)} \underset{\text{sim}}{\rightarrow} D_m$. A fortiori (use properties (1) and (2)),

$$[\oplus_{n=1}^{m} R_n]\oplus K_u\oplus q_k^{(\infty)} \underset{\text{sim}}{\rightarrow} C_m.$$

Choose m large enough to guarantee that $\|A-C_m\| < \varepsilon(2\|W\|.\|W^{-1}\|)^{-1}$; then $\|A-C_m\| < \varepsilon/2$ and there exist V_1 and V_2 invertible such that $V_2 \simeq W$ and

$$\|(V_1\oplus V_2)(\{[\oplus_{n=1}^{m} R_n]\oplus K_u\oplus q_k^{(\infty)}\}\oplus\{q_1^{(\infty)}\oplus[\oplus_{n=m+1}^{\infty} R_n]\oplus q_k^{(\infty)}\})(V_1\oplus V_2)^{-1}$$

$$- A\oplus B^{(\infty)}\| = \max\{\|V_1\{[\oplus_{n=1}^{m} R_n]\oplus K_u\oplus q_k^{(\infty)}\}V_1^{-1} - A\|,$$

$$\|V_2\{q_1^{(\infty)}\oplus[\oplus_{n=m+1}^{\infty} R_n]\oplus q_k^{(\infty)}\}V_2^{-1} - B^{(\infty)}\|\}$$

$$\le \max\{\|V_1\{[\oplus_{n=1}^{m} R_n]\oplus K_u\oplus q_k^{(\infty)}\}V_1^{-1} - C_m\|+\|C_m - A\|,$$

$$\|W(q_1^{(\infty)}\oplus q_k^{(\infty)})W^{-1} - B^{(\infty)}\| + \varepsilon/2\} < \varepsilon.$$

Since $\{[\oplus_{n=1}^{m} R_n]\oplus K_u\oplus q_k^{(\infty)}\}\oplus\{q_1^{(\infty)}\oplus[\oplus_{n=m+1}^{\infty} R_n]\oplus q_k^{(\infty)}\} \simeq K_u\oplus S$ and ε can be chosen arbitrarily small, we conclude that

$$T \underset{\text{sim}}{\rightarrow} S \underset{\text{sim}}{\rightarrow} K_u\oplus S \underset{\text{sim}}{\rightarrow} A\oplus B^{(\infty)} \underset{\text{sim}}{\rightarrow} A. \qquad \square$$

REMARK 8.7. By Lemma 2.5 and Remark 2.6, the condition "$T|H(\lambda_n;T) \underset{\text{sim}}{\rightarrow} A|H(\lambda_n;A)$" can be replaced by "$\text{nul}\,(\lambda_n-A)^j \ge \text{nul}\,(\lambda_n-T)^j$ for all $j=1,2,\ldots,\dim H(\lambda_n;T)-1$".

8.3 Quasinilpotents in the Calkin algebra

8.3.1 General quasinilpotents

Now assume that $Q \in L(H)$ is an essential nilpotent operator such that $\tilde{Q}^k=0$, $\tilde{Q}^{k-1} \neq 0$ and $\tilde{Q}^{k-1}+\tilde{Q}^*$ is invertible. Proceeding as in the first part of the proof of Proposition 8.5, we can find $R \sim q_k^{(\infty)}$ such that $Q\oplus R \simeq_a Q$. Hence, $Q \simeq_a Q\oplus R \underset{\text{sim}}{\rightarrow} Q\oplus q_1^{(\infty)}$, so that $S(Q)^- \supset S(Q\oplus q_1^{(\infty)})^-$. Combining this fact (7.9) and Propositions 7.22 and 8.5, we conclude that if Q is similar to a compact perturbation of $q_1^{(h)}\oplus q_k^{(\infty)}$, where $k \ge 2$ and $0 \le h < k$, then $S(Q)^-$ is the disjoint union of $S(Q\oplus q_1^{(\infty)})^-$ and

$$S^+(Q) = S(Q)^-\cap N_k^+(H), \tag{8.5}$$

which is open and dense in $S(Q)^{-}$.

As a first step toward a description of the structure of $S^{+}(Q)$, we shall consider similarity orbits in the Calkin algebra. Let us remark from the beginning that we have two kinds of similarities, namely those corresponding to invertible elements in $\pi[G(H)] = \{\tilde{W} \in G(A(H)) : \text{ind } W = 0\}$ and those of arbitrary index. Accordingly, we shall write $\tilde{T} \underset{s\text{-sim}}{\rightarrow} \tilde{A}$ for $\tilde{T}, \tilde{A} \in A(H)$ if there exist invertible elements $\tilde{W}_n$ in $A(H)$ such that ind $W_n = 0$ for all $n = 1,2,\ldots$, and $\|\tilde{A} - \tilde{W}_n\tilde{T}\tilde{W}_n^{-1}\| \to 0$ $(n \to \infty)$ and, respectively, $\tilde{T} \underset{w\text{-sim}}{\rightarrow} \tilde{A}$ if there exist invertible elements $\tilde{W}_n \in G(A(H))$ such that $\|\tilde{A} - \tilde{W}_n\tilde{T}\tilde{W}_n^{-1}\| \to 0$ $(n \to \infty)$.

LEMMA 8.8. *Let* τ *and* τ_1 *be two faithful unital *-homomorphisms of* $L(\mathbb{C}^k)$ *into* $A(H)$; *then*

(i) There exists a unitary element $\tilde{U} \in A(H)$ *such that* $\tilde{U}\tau(X) = \tau_1(X)\tilde{U}$ *for all* $X \in L(\mathbb{C}^k)$.

(ii) There exists an integer h, $0 \le h \le k-1$, *such that*

$$\{\text{ind } W:\ \tilde{W}\tau(q_k) = \tau_1(q_k)\tilde{W},\ \tilde{W} \in G(A(H))\} = h+k\mathbb{Z}.$$

(iii) There exists an integer h, $0 \le h \le k-1$, *such that*

$$\{\text{ind } W:\ \tilde{W}\tau(X) = \tau_1(X)\tilde{W},\ \textit{for all } X \in L(\mathbb{C}^k),\ \tilde{W} \in G(A(H))\} = h+k\mathbb{Z}.$$

PROOF. (i) Let $\tilde{W}$ be a partial isometry (i.e. $\tilde{W}^*\tilde{W}$ is a projection) such that $\tilde{W}^*\tilde{W} = \tau(e_1 \otimes e_1)$, $\tilde{W}\tilde{W}^* = \tau_1(e_1 \otimes e_1)$. Then $\tilde{U} = \sum_{j=1}^{k} \tau_1(e_j \otimes e_1)\tilde{W}\tau(e_1 \otimes e_j)$ is a unitary element of $A(H)$ and $\tilde{U}\tau(X) = \tau_1(X)\tilde{U}$ for all X in $L(\mathbb{C}^k)$.

(ii) It is clearly sufficient to prove that

$$\{\text{ind } V:\ \tilde{V} \in G(A(H)),\ \tilde{V} \text{ commutes with } \tau(q_k)\} = k\mathbb{Z}.$$

Furthermore, it follows from (i) that τ can be replaced by any other faithful unital *-homomorphism of $L(\mathbb{C}^k)$ into $A(H)$, because of the invariance of the index under conjugation by unitary elements of $A(H)$. Thus, we can assume that there exists a unitary operator $U: \mathbb{C}^k \otimes H \to H$ such that $\tau(X) = \pi(U(X \otimes 1)U^*)$. If $\tilde{V}$ commutes with $\tau(q_k)$, it is easily seen from the matrix form of $\tau(q_k)$ that

$$\tilde{V} = \pi[U(\textstyle\sum_{j=0}^{k-1} q_k^{\,j} \otimes A_j)U^*] \qquad (A_j \in L(H))$$

and, since the diagonal part of the triangular operator matrix $\sum_{j=0}^{k-1} q_k^{\,j} \otimes A_j$ is equal to $1_{\mathbb{C}^k} \otimes A_0$, it follows that $1_{\mathbb{C}^k} \otimes A_0$ is Fredholm and ind $V = \text{ind}(1_{\mathbb{C}^k} \otimes A_0) = k(\text{ind } A_o) \equiv 0 \pmod k$.

On the other hand, given $m \in \mathbb{Z}$ and $A \in L(H)$ a Fredholm operator of index m, the element $\pi[U(1_{\mathbb{C}^k} \otimes A)U^*]$ of $G(A(H))$ commutes with $\tau(q_k)$ and ind $U(1_{\mathbb{C}^k} \otimes A)U^* = k(\text{ind } A) = km$.

(iii) Use (i) once again (as in the proof of (ii)). We have to show that {ind V: $\tilde{V} \in G(A(H))$, $\tilde{V}$ commutes with $\tau(X)$ for all $X \in L(\mathbb{C}^k)\} = k\mathbb{Z}$ in case $\tau(X) = \pi(U(X\otimes 1_H)U^*)$. Now it is easily seen that $\tilde{V}$ commutes with $\tau[L(\mathbb{C}^k)]$ if and only if $\tilde{V} = \tau(U(1_{\mathbb{C}^k}\otimes A)U^*)$ and hence, $\tilde{V}$ is invertible if and only if A is Fredholm and ind V = k (ind A). □

Let $H = H_o^h \oplus \oplus_{j=1}^k H_j^h$, where h = dim $H_o^h \leq k-1$ ($k \geq 2$) and $H_j^h \simeq H$ for j = 1,2,...,k. Let

$$T_h = 0_{H_o^h}\oplus \begin{pmatrix} 0 & 1 & & & & & \\ & 0 & 1 & & & 0 & \\ & & 0 & \cdot & & & \\ & & & \cdot & \cdot & & \\ & & & & \cdot & \cdot & \\ & 0 & & & & \cdot & \cdot \\ & & & & & 0 & 1 \\ & & & & & & 0 \end{pmatrix} \begin{matrix} H_1^h \\ H_2^h \\ H_3^h \\ \cdot \\ \cdot \\ \cdot \\ H_{k-1}^h \\ H_k^h \end{matrix} \in L(H); \qquad (8.6)$$

then

$$q_1^{(h)}\oplus q_k^{(\infty)} \simeq T_h = \begin{pmatrix} 0 & S_h & & & & & \\ & 0 & 1 & & & 0 & \\ & & 0 & \cdot & & & \\ & & & \cdot & \cdot & & \\ & & & & \cdot & \cdot & \\ & 0 & & & & \cdot & \cdot \\ & & & & & 0 & 1 \\ & & & & & & 0 \end{pmatrix} \begin{matrix} H_o^h\oplus H_1^h \\ H_2^h \\ H_3^h \\ \cdot \\ \cdot \\ \cdot \\ H_{k-1}^h \\ H_k^h \end{matrix}$$

where $S_h: H_2^h \to H_o^h\oplus H_1^h$ is an isometry of corank h. By Lemma 7.23, $\tilde{T}_0 \simeq \tilde{T}_1 \simeq \tilde{T}_2 \simeq \ldots \simeq \tilde{T}_{k-1}$ in $A(H)$; indeed, straightforward computations show that $(S_h\oplus 1\oplus 1\oplus\ldots\oplus 1)T_0(S_h\oplus 1\oplus 1\oplus\ldots\oplus 1)^* = T_h$, $\pi(S_h\oplus 1\oplus 1\oplus\ldots\oplus 1)$ is unitary in $A(H)$ and ind$(S_h\oplus 1\oplus 1\oplus\ldots\oplus 1) = -h$, h = 0,1,2,...,k-1.

It is easily seen that $\tau_h(q_k) = \tilde{T}_h$ uniquely determines a faithful unital *-homomorphism from $L(\mathbb{C}^k)$ into $A(H)$ for each h, h = 0,1,2,...,k-1. However, by Lemma 8.8, there is no $\tilde{V} \in G(A(H))$, ind V = n such that $\tilde{V}\tau_h(q_k) = \tau_r(q_k)\tilde{V}$, unless $h - r - n \equiv 0 \pmod k$. Furthermore: If $h - r - n \not\equiv 0 \pmod k$, then $\tau_r(q_k)$ is "far" from $\tilde{V}\tau_h(q_k)\tilde{V}^{-1}$! (Corollary 8.11 below)

Given $k \geq 2$, define

$N_{k,h}^+(H) = \{T \in L(H)$: T is similar to a compact perturbation of $T_h\}$ (h = 0,1,2,...,k-1) and

$$\tilde{N}_{k,h}^+(A(H)) = \pi[N_{k,h}^+(H)].$$

By (7.9), (7.8), Proposition 7.22 and Lemma 8.6, $N_k^+(H)$ $(\tilde{N}_k^+(A(H)))$

is the disjoint union of the sets $N^+_{k,h}(H)$ ($\tilde{N}^+_{k,h}(A(H))$, resp), $h=0,1,2,\ldots,k-1$.

LEMMA 8.9. *Let C be a C^*-algebra with identity 1 and let $\tau: L(\mathbb{C}^k) \to C$ be a faithful unital $*$-homomorphism. Assume that $x \in N_k(C)$ and let*

$$s = \sum_{j=1}^{k} \tau(e_j \otimes e_1) x^{j-1}$$

Then $\tau(q_k)s = xs$ *and* $\|s-1\| \leq k\|x-\tau(q_k)\|(1+\|x\|)^{k-1}$.

PROOF. Indeed,

$$\tau(q_k)s = [\sum_{i=1}^{k-1} \tau(e_i \otimes e_{i+1})][\sum_{j=1}^{k} \tau(e_j \otimes e_1) x^{j-1}]$$

$$= \sum_{i=1}^{k-1}\sum_{j=1}^{k} \tau((e_i \otimes e_{i+1})(e_j \otimes e_1)) x^{j-1} = \sum_{i=1}^{k-1} \tau(e_i \otimes e_1) x^j = sx,$$

and, proceeding as in the proof of Proposition 1.10, we have

$$\|s-1\| = \|\sum_{j=1}^{k} \tau(e_j \otimes e_1) x^{j-1} - \sum_{j=1}^{k} \tau(e_j \otimes e_j)\|$$

$$= \|\sum_{j=1}^{k} \tau(e_j \otimes e_1) x^{j-1} - \sum_{j=1}^{k} \tau(e_j \otimes e_1)\tau(q_k)^{j-1}\|$$

$$\leq \sum_{j=2}^{k} \|\tau(e_j \otimes e_1)[x^{j-1} - \tau(q_k)^{j-1}]\| \leq \sum_{j=1}^{k-1} \|x^j - \tau(q_k)^j\|$$

$$\leq \|x-\tau(q_k)\| \sum_{j=1}^{k-1}\sum_{i=0}^{j-1} \|x\|^i \leq k\|x-\tau(q_k)\|(1+\|x\|)^{k-1}. \qquad \square$$

LEMMA 8.10. *Let* $k \geq 2$ *and* $j \in \mathbb{Z}$, *then*

(i) If $j \equiv h \pmod k$, *then*

$\{\tilde{V}\tilde{T}_0\tilde{V}^{-1}: \tilde{V} \in G(A(H))$ *and* $\operatorname{ind} V = -j\} = \tilde{N}^+_{k,h}(A(H))$.

(ii) If $j \not\equiv h \pmod k$, *then*

$\{\tilde{V}\tilde{T}_0\tilde{V}^{-1}: \tilde{V} \in G(A(H))$ *and* $\operatorname{ind} V = -j\} \cap \tilde{N}^+_{k,h}(A(H)) = \emptyset$.

(iii) The subsets $\{\tilde{N}^+_{k,h}(A(H))\}_{h=0}^{k-1}$ *are open in* $\tilde{N}^+_k(A(H))$. *Indeed, these subsets are the components of* $\tilde{N}^+_k(A(H))$.

PROOF. By our previous observations, it only remains to show that $\tilde{N}^+_{k,h}(A(H))$ is connected and open in $\tilde{N}^+_k(A(H))$. The connectedness is an immediate consequence of (i), (ii) and the fact that $\{\tilde{V} \in G(A(H)): \operatorname{ind} V = -h\}$ is connected for each h, $h=0,1,2,\ldots,k-1$.

Let $\tilde{X}_n \in \tilde{N}^+_k(A(H))$ and $\tilde{X} = \tilde{V}\tilde{T}_0\tilde{V}^{-1}$, $\operatorname{ind} V = -h$, be such that $\|\tilde{X}-\tilde{X}_n\| \to 0$ $(n \to \infty)$. Then we have $\lim(n \to \infty) \|\tilde{T}_0 - \tilde{V}^{-1}\tilde{X}_n\tilde{V}\| = 0$ and therefore, by Lemma 8.9, there exist elements $\tilde{S}_n \in G(A(H))$ such that $\tilde{S}_n\tilde{T}_0 = \tilde{V}^{-1}\tilde{X}_n\tilde{V}\tilde{S}_n$ (for all $n=1,2,\ldots$) and $\|\tilde{S}_n - \tilde{1}\| \to 0$ $(n \to \infty)$. Hence, for n large enough, $\tilde{S}_n$ is invertible, $\operatorname{ind} S_n = 0$, and

$$\tilde{X}_n = (\tilde{V}\tilde{S}_n)\tilde{T}_0(\tilde{V}\tilde{S}_n)^{-1},$$

so that $\tilde{X}_n \in \tilde{N}^+_{k,h}(A(H))$.

Therefore, $\tilde{N}^+_{k,h}(A(H))$ is open in $\tilde{N}^+_k(A(H))$. $\square$

From Lemma 8.10, Corollary 7.21 and comments preceeding it, we immediately obtain the following

COROLLARY 8.11. *(i) If* $\tilde{X} = \tilde{V}\tilde{T}_0\tilde{V}^{-1}$, ind $V = -h$ *(*$0 \le h \le k-1$*),* $\tilde{Y} \in N_k(A(H))$ *and* $\|\tilde{X} - \tilde{Y}\| < \beta_k(\|\tilde{V}\|\cdot\|\tilde{V}^{-1}\|)^{-1}$, *then* $\tilde{Y} \in \tilde{N}^+_{k,h}(A(H))$.

(ii) If $\tilde{X} = \tilde{V}\tilde{T}_0\tilde{V}^{-1} \in \tilde{N}^+_{k,h}(A(H))$, $\tilde{Y} = \tilde{W}\tilde{T}_0\tilde{W}^{-1} \in \tilde{N}^+_{k,r}(A(H))$ *and* $h - r \not\equiv 0 \pmod{k}$, *then*

$$\|\tilde{X} - \tilde{Y}\| \ge \beta_k[(\|V\|\cdot\|V^{-1}\|)^{-1} + (\|W\|\cdot\|W^{-1}\|)].$$

PROPOSITION 8.12. *Let* $\tilde{Q}$ *be a quasinilpotent element of* $A(H)$ *and let* $\tilde{A} \in A(H)$*; then*

(i) If $\tilde{Q} \in N_k(A(H))\setminus\tilde{N}^+_k(A(H))$, *and* $\tilde{Q}^{k-1} \ne 0$, *then*

$\tilde{Q} \underset{s\text{-sim}}{\rightarrow} \tilde{A} \iff \tilde{Q} \underset{w\text{-sim}}{\rightarrow} \tilde{A} \iff \tilde{A} \in N_k(A(H))\setminus\tilde{N}^+_k(A(H))$.

(ii) If $\tilde{Q} \in \tilde{N}^+_{k,h}(A(H))$ *for some* $k \ge 2$, *and* h, $0 \le h \le k-1$, *then*

$$\tilde{Q} \underset{s\text{-sim}}{\rightarrow} \tilde{A} \iff \tilde{A} \in N_k(A(H))\setminus\cup_{j=0,j\ne h}^{k-1} \tilde{N}^+_{k,j}(A(H))$$

and

$$\tilde{Q} \underset{w\text{-sim}}{\rightarrow} \tilde{A} \iff \tilde{A} \in N_k(A(H)).$$

(iii) If $\tilde{Q} \notin \cup_{k=1}^{\infty} N_k(A(H))$, *then*

$\tilde{Q} \underset{s\text{-sim}}{\rightarrow} \tilde{A} \iff \tilde{Q} \underset{w\text{-sim}}{\rightarrow} \tilde{A} \iff \tilde{A} \in N(A(H))^-$.

PROOF. (i) We have

$\tilde{Q} \underset{s\text{-sim}}{\rightarrow} \tilde{A} \Rightarrow \tilde{Q} \underset{w\text{-sim}}{\rightarrow} \tilde{A} \Rightarrow \tilde{A} \in N_k(A(H))\setminus\tilde{N}^+_k(A(H))$

(The last implication follows from the fact that the non-invertibility of $\tilde{X}^{k-1}+\tilde{X}^*$ for $\tilde{X} \in N_k(A(H))$) is a similarity invariant; see Lemma 7.20.)

On the other hand, if $\tilde{A} \in N_k(A(H))\setminus\tilde{N}^+_k(A(H))$, then (Theorem 7.2) there exists $Q_0, A_0 \in \mathcal{Q}(H)$ such that $Q_0^k = A_0^k = 0$, $\pi(Q_0) = \tilde{Q}$, $\pi(A_0) = \tilde{A}$. By Proposition 8.5, $Q_0 \underset{\text{sim}}{\rightarrow} A_0$. A fortiori, $\tilde{Q} \underset{s\text{-sim}}{\rightarrow} \tilde{A}$.

(ii) Assume that $\tilde{Q} \in \tilde{N}^+_{k,h}(A(H))$, $0 \le h \le k-1$. By Lemma 8.10,

$$\{\tilde{W}\tilde{Q}\tilde{W}^{-1}: \tilde{W} \in G(A(H)), \text{ ind } W = 0\} = \tilde{N}^+_{k,h}(A(H))$$

and

$$\{\tilde{V}\tilde{Q}\tilde{V}^{-1}: \tilde{V} \in G(A(H))\} = S(\tilde{Q}) = \tilde{N}^+_k(A(H)).$$

On the other hand, if $\tilde{A} \in \tilde{N}^+_{k,r}(A(H))$, $0 \le r \le k-1$, and $r \ne h$, then it readily follows from Lemma 8.10 (or Corollary 8.11(ii)) that $\|\tilde{W}\tilde{Q}\tilde{W}^{-1} - \tilde{A}\|$ is bounded away from zero for all $\tilde{W} \in G(A(H))$ such that ind $W = 0$.

Hence, $\tilde{Q} \underset{s\text{-sim}}{\not\rightarrow} \tilde{A}$.

It is clear that

$$\tilde{Q} \underset{s\text{-sim}}{\rightarrow} \tilde{A} \Rightarrow \tilde{Q} \underset{w\text{-sim}}{\rightarrow} \tilde{A} \Rightarrow \tilde{A} \in N_k(A(H))$$

and, by our previous observations and Theorem 7.2, it readily follows that $\tilde{Q} \underset{s-\vec{sim}}{} \tilde{A}$ for each $\tilde{A} \in N_k(A(H)) \backslash \tilde{N}_k^+(A(H))$.

(iii) It is completely apparent (from Theorem 5.34) that

$$\tilde{Q}_{\ s-\vec{sim}}\ \tilde{A} \Rightarrow \tilde{Q}_{\ w-\vec{sim}}\ \tilde{A} \Rightarrow \tilde{A} \in N(A(H))^-.$$

Conversely, if $\tilde{A} \in N(A(H))^-$ then (by Proposition 3.45) there exists Q_0, $A_0 \in L(H)$ such that $\pi(Q_0) = \tilde{Q}$, $\pi(A_0) = \tilde{A}$, Q_0 is a quasinilpotent and $\sigma(A_0) = \sigma_e(A_0) = \sigma(\tilde{A})$. It is easily seen that $\tilde{Q}_0{}^k \neq 0$ for all $k = 1,2,\ldots$ and (by Theorem 5.1) that $A_0 \in N(H)^-$. By Proposition 8.3, $Q_0 \ \vec{sim}\ A_0$. A fortiori, $\tilde{Q}_{\ s-\vec{sim}}\ \tilde{A}$. □

COROLLARY 8.13. *(i) For each* $k \geq 2$, $N_k(A(H))$ *is the disjoint union of the sets* $N_k(A(H)) \backslash \tilde{N}_k^+(A(H))$, $\tilde{N}_{k,0}^+(A(H))$, $\tilde{N}_{k,1}^+(A(H)), \ldots, \tilde{N}_{k,k-1}^+(A(H))$.

(ii) $\tilde{N}_{k,h}^+(A(H)) = \{\tilde{W}\tilde{T}_h\tilde{W}^{-1}: \ \tilde{W} \in G(A(H)),\ \text{ind } W = 0\}$, *for each* $h = 0,1,2,\ldots,k-1$.

(iii) $\tilde{N}_k^+(A(H)) = S(\tilde{T}_h)$ *for all* $h = 0,1,2,\ldots,k-1$.

(iv) $\tilde{N}_k^+(A(H))^- \backslash \tilde{N}_k^+(A(H)) = \tilde{N}_{k,h}^+(A(H))^- \backslash \tilde{N}_{k,h}^+(A(H)) = N_k(A(H)) \backslash \tilde{N}_k^+(A(H))$ *for each* h, $h = 0,1,2,\ldots,k-1$.

As in Chapter II, let $[\tilde{T}]_s = \{\tilde{R} \in A(H):\ \tilde{T}_{\ s-\vec{sim}}\ \tilde{R}$ and $\tilde{R}_{\ s-\vec{sim}}\ \tilde{T}\}$ and $[\tilde{T}]_w = \{\tilde{R} \in A(H):\ S(\tilde{R})^- = S(\tilde{T})^-\}$; $[\tilde{A}]_s < [\tilde{T}]_s$ means that $\tilde{T}_{\ s-\vec{sim}}\ \tilde{A}$ and $[\tilde{A}]_w < [\tilde{T}]_w$ means that $\tilde{T}_{\ w-\vec{sim}}\ \tilde{A}$, respectively. These two relations are partial orders in the corresponding quotient sets obtained from $A(H)$ by using the equivalence relation $\#_s$ and $\#_w$, respectively. If $Q(A(H))$ denotes the set of all quasinilpotent elements of $A(H)$, then $(Q(A(H))/\#_s, <)$ is a lattice and $(Q(A(H))/\#_w, <)$ is a chain. Indeed, it is not difficult to see (by using Proposition 8.12) that their structures are the ones depicted in Diagram (8.7) (below), where Q_u denotes an arbitrary universal quasinilpotent operator.

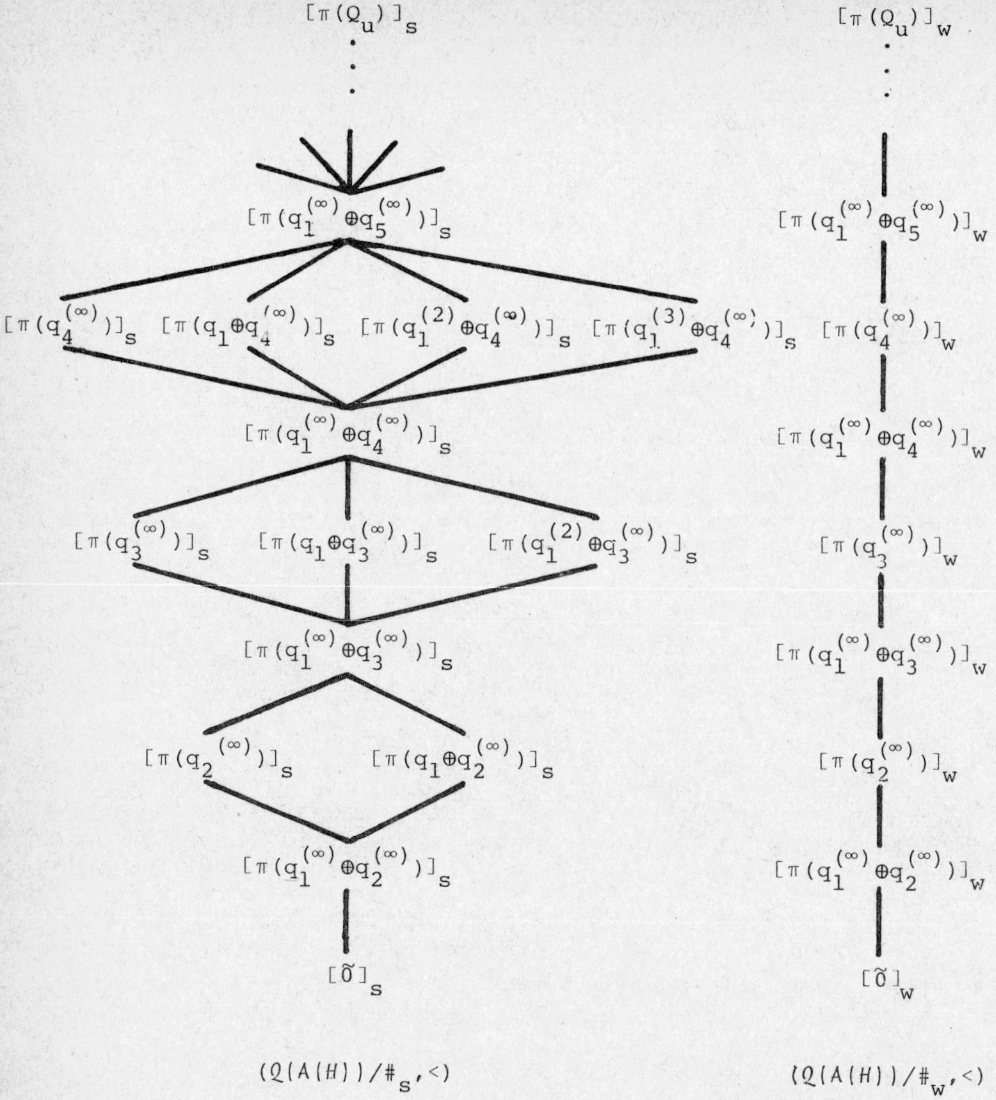

DIAGRAM (8.7)

8.3.2 Nice elements of the Calkin algebra

In the case when T is similar to a compact perturbation of a nice Jordan operator with more than one point in the essential spectrum, the analysis of $S(\tilde{T})$ present some subtlety that cannot be detected with the sole help of Lemma 7.1. The phenomenon will be better explained with an analog of Lemma 8.8 for this case. The proofs of Lemmas 8.14,

8.15 and 8.16 below follow the same lines as those of Lemmas 8.8, 8.9 and 8.10, respectively, and will be omitted.

LEMMA 8.14. *Let* τ *and* τ_1 *be two faithful unital *-homomorphisms of* $L(\mathbb{C}^{k_1})\oplus L(\mathbb{C}^{k_2})\oplus\ldots\oplus L(\mathbb{C}^{k_m})$ *into* $A(H)$; *then*

(i) There exists a unitary element $\tilde{U}\in A(H)$ *such that* $\tilde{U}\tau(X) = \tau_1(X)\tilde{U}$ *for all* $X\in\oplus_{j=1}^{m} L(\mathbb{C}^{k_j})$.

(ii) There exists an integer h, $0\le h\le k_0-1$, *where*

$$k_0 = \mathrm{GCD}\{k_1,k_2,\ldots,k_m\},$$

such that if $\lambda_1,\lambda_2,\ldots,\lambda_m$ *are* m *distinct complex numbers, then*

$$\{\text{ind } W:\ \tilde{W}\tau(\oplus_{j=1}^{m}[\lambda_j+q_{k_j}]) = \tau_1(\oplus_{j=1}^{m}[\lambda_j+q_{k_j}])\tilde{W},\ \tilde{W}\in G(A(H))\} = h+k_0\mathbb{Z}.$$

(iii) There exists an integer h, $0\le h\le k_0-1$, *such that*

$$\{\text{ind } W:\ \tilde{W}\tau(X) = \tau_1(X)\tilde{W},\ \textit{for all } X\in\oplus_{j=1}^{m}L(\mathbb{C}^{k_j}),\ \tilde{W}\in G(A(H))\} = h+k_0\mathbb{Z}.$$

LEMMA 8.15. *Let* C *be a C*-algebra with identity* 1 *and let* $\tau:\oplus_{j=1}^{m} L(\mathbb{C}^{k_j})\to C$ *be a faithful unital *-homomorphism. Assume that* $\lambda_1,\lambda_2,\ldots,\lambda_m$ *are* m *distinct complex numbers and let* $x\in C$ *with minimal polynomial* p, $p(\lambda) = \Pi_{j=1}^{m}(\lambda-\lambda_j)^{k_j}$. *Then there exists* $s = s(x)\in C$ *such that* $\tau(\oplus_{j=1}^{m}[\lambda_j+q_{k_j}])s = sx$ *and*

$$||s-1||\le C_p||x-\tau(\oplus_{j=1}^{m}[\lambda_j+q_{k_j}])||(1+||x||)^{\sum_{j=1}^{m}k_j},$$

where C_p *is a constant depending only on* p.

LEMMA 8.16. *Let* $\lambda_1,\lambda_2,\ldots,\lambda_m$ *be* m *distinct complex numbers and let*

$$T_h = q_1^{(h)}\oplus[\oplus_{j=1}^{m}(\lambda_j+q_{k_j}^{(\infty)})]\in L(H);$$

where $0\le h\le k_0-1$, $k_0 = \mathrm{GCD}\{k_1,k_2,\ldots,k_m\}$.

Then $S(\tilde{T}_0)$ *has exactly* k_0 *components:*

$$S(\tilde{T}_h) = \{\tilde{W}\tilde{T}_0\tilde{W}^{-1}:\ \tilde{W}\in G(A(H)),\ \text{ind } W\equiv h \pmod{k}\},$$

$h = 0,1,2,\ldots,k-1$.

8.4 Compact perturbations of nice Jordan operators

From our observations (at the beginning of Section 8.3.1) and Proposition 8.12(ii), we obtain the following

COROLLARY 8.17. *(i) If* $Q\in L(H)$ *is similar to a compact perturbation of* $q_1^{(h)}\oplus q_k^{(\infty)}$, *where* $k\ge 2$ *and* $0\le h\le k-1$, *then* $S(\dot{Q})^-$ *is the disjoint union of* $S(q_1^{(\infty)}\oplus Q)^-$ *and* $S^+(Q)$ *(defined by* (8.5)*).*

(ii) $S^+(Q)$ *is an open dense subset of* $S(Q)^-$.

(iii) $S^+(Q) \subset N^+_{k,h}(H)$ *and* $S(Q)^- \cap N^+_{k,j}(H) = \emptyset$ *for all* $j \neq h$, $0 \leq j \leq k-1$.

(iv) $N^+_{k,h}(H)$, $0 \leq h \leq k-1$, *and* $N^+_k(H) = \cup_{h=0}^{k-1} N^+_{k,h}(H)$ *intersect* $N_k(H)$ *in open sets; furthermore,* $N^+_k(H) \cap N_k(H)$ *is dense in* $N_k(H)$.

Let Q be as above. If Q is quasinilpotent, then $S(q_1^{(\infty)} \oplus Q)^-$ is completely described by Proposition 8.5. If $\sigma(Q) = \{0\} \cup \{\lambda_1, \lambda_2, \ldots, \lambda_m\}$ is a finite set, then $Q \sim Q_0 \oplus F$, where Q_0 is a quasinilpotent and F acts on a finite dimensional space; in this case $S(q_1^{(\infty)} \oplus Q)^-$ can be easily described by using Lemma 7.1, Proposition 8.5 and Theorem 2.1. If $\sigma(Q) = \{0\} \cup \{\lambda_n\}_{n=1}^{\infty}$, where $\lambda_n \neq 0$ for all n, $\lambda_n \neq \lambda_m$ if $n \neq m$, and $\lambda_n \to 0$ $(n \to \infty)$, then $S(q_1^{(\infty)} \oplus Q)^-$ also admits a simple description, given by Proposition 8.6.

On the other hand, it follows from Corollary 8.17 that $S^+(Q) = S(Q)^- \cap N^+_k(H) = S(Q)^- \cap N^+_{k,h}(H)$. Since every element of $N^+_{k,h}(H)$ is *similar* to some operator of the form

$$q_1^{(h)} \oplus q_k^{(\infty)} + C, \quad C \in K(H), \tag{8.8}$$

we can directly assume that

$$Q = q_1^{(h)} \oplus q_k^{(\infty)} + K, \quad k \geq 2, \quad 0 \leq h \leq k-1, \quad K \in K(H). \tag{8.9}$$

Now it is clear that the structure of $S(Q)^-$ will be completely determined if we can specify which operators of the form (8.8) belong to $S(Q)^-$. Furthermore, by our previous observations, we only have to analyze the case when Q is quasinilpotent and the case when 0 is not isolated in $\sigma(Q)$. There are only partial answers to these problems. However, these partial results are sufficient to exhibit the complexity of the general problem. The case when 0 is not an isolated point of $\sigma(Q)$ will be considered in Section 8.45. Sections 8.41 through 8.44 will be devoted to the case when Q is a quasinilpotent. A conjecture will be presented concerning the classification of the closures of similarity orbits of these operators.

8.4.1 Nice Jordan nilpotents

LEMMA 8.18. *Let* $\{Q_n\}_{n=1}^{\infty}$ *be a sequence of nice nilpotent operators,* $Q_n \sim q_k^{(\infty)} \oplus F_n$ *and let* $Q \sim q_k^{(\infty)} \oplus F$, *where* $F_n \in L(\mathbb{C}^{r_n})$, $F \in L(\mathbb{C}^r)$ *are Jordan nilpotents. Then the following are equivalent.*

(i) There exists operators W_n *in* $G(H)$ *such that*

$$||W_n Q_n W_n^{-1} - Q|| \to 0 \quad (n \to \infty);$$

(ii) There exists a natural number p *such that* $Q_n \underset{\text{sim}}{\to} Q$ *for all* $n \geq p$;

(iii) There exist a natural number p *and non-negative integers* α_n *and* β_n $(n \geq p)$ *such that*

$$k\beta_n + r = k\alpha_n + r_n \text{ and } q_k^{(\alpha_n)} \oplus F_n \underset{\text{sim}}{\to} q_k^{(\beta_n)} \oplus F.$$

PROOF. (iii) => (ii) => (i) are trivial implications.

(i) => (iii) If $Q_n \simeq q_k^{(\infty)} \oplus F_n$, let $H_1^{(n)}$ and $H_2^{(n)}$ denote the subspaces corresponding to $q_k^{(\infty)}$ and F_n, respectively, and let $P_j^{(n)}$ denote the orthogonal projection of H onto $H_j^{(n)}$, $j = 1,2$. Similarly, let $H = H_1 \oplus H_2$, where H_1 is the space of $q_k^{(\infty)}$ and H_2 is the space of F, and let P_j be the orthogonal projection of H onto H_j, $j = 1,2$. Let $Q_n' = W_n Q_n W_n^{-1}$, $X = H_1 \ominus QH_1$ and $X_n = \{x \in X\colon\ W_n P_2^{(n)} W_n^{-1} x = 0\}$.

Define $Y_n = X_n \vee QX_n \vee \ldots \vee Q^{k-1}X_n$ and $Z_n = X_n \vee Q_n' X_n \vee \ldots \vee Q_n'^{k-1} X_n$.

Since the spaces $X_n, QX_n, \ldots, Q^{k-1}X_n$ are pairwise orthogonal and $Q^j|X_n$ is isometric (for $j = 0,1,2,\ldots,k-1$), the algebraic sum $X_n + Q_n' X_n + \ldots + Q_n'^{k-1} X_n$ will be closed and direct for all n large enough; moreover, $\|P_{Y_n} - P_{Z_n}\| \to 0$ $(n \to \infty)$.

Since $W_n P_2^{(n)} W_n^{-1}(X_n) = \{0\}$, it follows that $W_n^{-1}(X_n) \subset H_1^{(n)}$; so that $Q_n'^k X_n = \{0\}$ and therefore Z_n is invariant under Q_n' and, for n large enough, $Q_n'|Z_n \sim q_k^{(\infty)}$. Furthermore, since Y_n has finite codimension, Z_n also has finite codimension. Since Q_n' and Q_n are similar via W_n and $W_n^{-1}(X_n) \subset H_1^{(n)}$, we can use Lemmas 7.9 and 7.8 in order to find an invariant subspace $R_n \subset H_1^{(n)}$ of Q_n such that $H_1^{(n)} = R_n \dot{+} W_n^{-1}(X_n)$ $(\dim R_n < \infty)$.

Since $q_k^{(\infty)} \oplus (Q_n|R_n) \sim q_k^{(\infty)}$, it is easily seen that there exists a non-negative integer α_n such that $Q_n|R_n \sim q_k^{(\alpha_n)}$. On the other hand, Y_n reduces Q and therefore there exists a non-negative integer β_n such that

$$Q|H \ominus Y_n = P_{H\ominus Y_n} Q|H \ominus Y_n \simeq q_k^{(\beta_n)} \oplus F.$$

Since $\|P_{Z_n} - P_{Y_n}\| \to 0$ $(n \to \infty)$, for all n large enough we can find unitary operators $U_n \in L(H)$ such that $\|U_n - 1\| \to 0$ and $U_n^* P_{Z_n} U_n = P_{Y_n}$. Hence, we have:

$$\lim(n \to \infty)\ \|P_{H\ominus Y_n} Q P_{H\ominus Y_n} - U_n^* P_{H\ominus Z_n} Q_n' P_{H\ominus Z_n} U_n\| = 0. \tag{8.10}$$

But

$$P_{H\ominus Z_n} Q_n'|H \ominus Z_n \sim Q_n|R_n \oplus H_2^{(n)} \sim q_k^{(\alpha_n)} \oplus F_n.$$

Hence, (8.10) implies the existence (for n large enough) of invertible operators

$$S_n\colon \mathbb{C}^{n\alpha_n + r_n} \to \mathbb{C}^{n\beta_n + r}$$

such that

$$\lim(n \to \infty) \ \|S_n(q_k^{(\alpha_n)} \oplus F_n)S_n^{-1} - q_k^{(\alpha_n)} \oplus F\| = 0.$$

Now it is easy to derive from this relation that, given $j \geq 1$, then

$$\mathrm{rank}(q_k^{(\beta_n)} \oplus F_n)^j \leq \mathrm{rank}(q_k^{(\alpha_n)} \oplus F)^j \qquad (8.11)$$

for all $n \geq n_o(j)$.

Since $(q_k^{(\beta_n)} \oplus F_n)^m = 0$ for some $m \geq k$, it follows that (8.11) holds for all $j = 1,2,\ldots,$ and for all $n \geq p$.

If $n \geq p$, it follows from Lemma 2.5 that $q_k^{(\alpha_n)} \oplus F_n \underset{\text{sim}}{\to} q_k^{(\beta_n)} \oplus F$. □

PROPOSITION 8.19. *Let* $T \in N_k^+(H)$ *be similar to a nice Jordan nilpotent:*

$$T = W\{q_k^{(\infty)} \oplus [\oplus_{j=1, j\neq k}^m q_j^{(\tau_j)}]\}W^{-1}, \ m \geq k, \ \sum_{j=1, j\neq k}^m j\tau_j = d_T < \infty.$$

Then $S^+(T)$ *is the set of all* $A \in N_k^+(H)$ *such that*

$$A = V\{q_k^{(\infty)} \oplus [\oplus_{j=1, j\neq k}^m q_j^{(\alpha_j)}]\}V^{-1}, \ \sum_{j=1, j\neq k}^m j\alpha_j = d_A < \infty \qquad (8.12)$$

and there exist non-negative integers τ_k, α_k *such that*

$$\begin{cases} \min\{\tau_k, \alpha_k\} = 0, \ k\tau_k + d_T = k\alpha_k + d_A, \ \text{and} \\ \mathrm{rank}[\oplus_{j=1}^m q_j^{(\alpha_j)}]^r \leq \mathrm{rank}[\oplus_{j=1}^m q_j^{(\tau_j)}]^r, \end{cases} \qquad (8.13)$$

for all $r = 1,2,\ldots,m-1$.

Furthermore, there exists a constant $\beta_{k,m} > 0$ *such that if* $B \in S(T)^-$ *and*

$$\|B - q_k^{(\infty)} \oplus [\oplus_{j=1, j\neq k}^m q_j^{(\tau_j)}]\| < \beta_{k,m},$$

then $B \sim q_k^{(\infty)} \oplus [\oplus_{j=1, j\neq k}^m q_j^{(\tau_j)}] \sim T$. *In particular,* $S(T)$ *is an open dense subset of* $S(T)^-$.

PROOF. It readily follows from Lemma 2.5 that if A satisfies (8.12) and there exist τ_k, $\alpha_k \geq 0$ such that (8.13) hold, then

$$\oplus_{j=1}^m q_j^{(\tau_j)} \underset{\text{sim}}{\to} \oplus_{j=1}^m q_j^{(\alpha_j)} \qquad (8.14)$$

and, a fortiori, $T \underset{\text{sim}}{\to} A$.

Conversely, if $\|W_n T W_n^{-1} - A\| \to 0$ $(n \to \infty)$ for a suitable sequence $\{W_n\}_{n=1}^\infty$ in $G(H)$, then it follows from Lemma 8.15 (with $Q_n \equiv T$ and $Q = A$) that there exist non-negative integers τ_k, α_k such that $k\tau_k + d_T = k\alpha_k + d_A$ and (8.14) holds. Since, by Lemma 2.5, this is equivalent to the validity of the inequalities

$$\mathrm{rank}[\oplus_{j=1}^m q_j^{(\alpha_j)}]^r \leq \mathrm{rank}[\oplus_{j=1}^m q_j^{(\tau_j)}]^r, \ r = 1,2,\ldots,m-1,$$

it readily follows that τ_k, α_k can be chosen so that $\min\{\tau_k, \alpha_k\} = 0$.

Assume that $\{B_n\}_{n=1}^{\infty} \subset S(T)^-$ and $\|T - B_n\| \to 0$ $(n \to \infty)$. Since $S^+(T)$ is open and dense in $S(T)^-$ (Corollary 8.17(ii)), B_n must be similar to a compact perturbation of T for all n large enough. Furthermore, since rank $B_n^k \le$ rank T^k for all n and rank T^k is finite, it follows from Corollary 7.24 that B_n is similar to a nice Jordan nilpotent:

$$B_n = V_n\{q_k^{(\infty)} \oplus [\oplus_{j=1,j\neq k}^{m} q_j^{(\beta_{nj})}]\}V_n^{-1}.$$

By Lemma 8.18, there exist a natural number p and non-negative integers β_{nk}, τ_{nk} $(n \ge p)$ such that $\sum_{j=1}^{m} j\beta_{nj} = k\tau_{nk} + \sum_{j=1,j\neq k}^{m} j\tau_j$ and

$$\operatorname{rank}\{q_k^{(\beta_{nk})} \oplus [\oplus_{j=1,j\neq k}^{m} q_j^{(\tau_j)}]\}^r \le \operatorname{rank}\{\oplus_{j=1}^{m} q_j^{(\beta_{nj})}\}^r \qquad (8.15)$$

for all $r = 1,2,\ldots,m-1$, and for all $n \ge p$.

But, on the other hand, the first part of the proof indicates that the reverse inequalities also hold in (8.15), i.e., these inequalities are actually *equalities*. It readily follows that if $n \ge p$, then $\beta_{nj} = \tau_j$ for all $j = 1,2,\ldots,k-1,k+1,k+2,\ldots,m$. I.e., $B_n \sim T$ for all $n \ge p$.

From this result and the proof of Lemma 8.18 we deduce the existence of constants $\beta_{k,m} > 0$, depending only on k and m $(m \ge k)$, such that the intersection of the open ball (in $L(H)$) of radius $\beta_{k,m}$ about $q_k^{(\infty)} \oplus [\oplus_{j=1,j\neq k}^{m} q_j^{(\tau_j)}]$ with $S(T)^-$ only contains points of $S(T)$. Now it is easily seen that if $R = S\{q_k^{(\infty)} \oplus [\oplus_{j=1,j\neq k}^{m} q_j^{(\tau_j)}]\}S^{-1} \sim T$, then the intersection of the open unit ball of radius $\beta_{k,m}(\|S\|.\|S^{-1}\|)^{-1}$ about R with $S(T)^-$ is also included in $S(T)$. Hence, $S(T)$ is an open (and obviously dense) subset of $S(T)^-$. □

From Proposition 8.19 and Theorem 2.7, we obtain the following

COROLLARY 8.20. *For each* $k \ge 2$ *and* h, $0 \le h \le k-1$.

$$(\{T \in N_{k,h}^{+}(H):\ T \text{ is a nice Jordan nilpotent}\}/\#,<)$$

is a lattice.

PROOF. If $T \sim q_k^{(\infty)} \oplus [\oplus_{j=1,j\neq k}^{m} q_j^{(\tau_j)}]$ and $A \sim q_k^{(\infty)} \oplus [\oplus_{j=1,j\neq k}^{m} q_j^{(\alpha_j)}] \in N_{k,h}^{+}(H)$ are nice Jordan nilpotents, then there exist non-negative integers τ_k, α_k such that $\sum_{j=1}^{m} j\tau_j = \sum_{j=1}^{m} j\alpha_j = d$.

If $B = \oplus_{j=1}^{m} q_j^{(\beta_j)}$, $C = \oplus_{j=1}^{m} q_j^{(\gamma_j)} \in N(\mathbb{C}^d)$ are the Jordan nilpotents determined by Theorem 2.7 such that

$$[B] = [\oplus_{j=1}^{m} q_j^{(\tau_j)}] \vee [\oplus_{j=1}^{m} q_j^{(\alpha_j)}],\ [C] = [\oplus_{j=1}^{m} q_j^{(\tau_j)}] \wedge [\oplus_{j=1}^{m} q_j^{(\alpha_j)}],$$

then it readily follows from Proposition 8.17 that

$[T] \vee [A] = [q_k^{(\infty)} \oplus B]$ and $[T] \wedge [A] = [q_k^{(\infty)} \oplus C]$. □

We shall see later that this lattice is not closed under "sup" or "inf" (for denumerable subsets). From this result, Propositions 8.3, 8.4 and 8.5, and Corollary 8.17, we obtain

COROLLARY 8.21.
$([Q(H)\setminus\cup_{k=2}^{\infty}N_k^+(H)]\cup(\cup_{k=2}^{\infty}\{T\in N_k^+(H):$ *T is a nice Jordan nilpotent*$\}/\#,<)$ *is a lattice.*

This lattice contains the infimum of each of its nonempty subsets, but it is not closed under "sup".

CONJECTURE 8.22. $(Q(H)/\#,<)$ is a lattice.

We shall say that a similarity orbit $S(T)$ is *locally closed* if (1) there exists $\varepsilon > 0$ such that the intersection of the *closed* ball $B(T,\varepsilon)$ (in $L(H)$) of radius ε about T with $S(T)$ is closed in $L(H)$.

This notion admits several equivalent formulations; namely: (2) Given $A = WTW^{-1} \in S(T)$ there exists $\varepsilon(A) > 0$ such that $B(A,\varepsilon(A))\cap S(T)$ is closed in $L(H)$ (Take $\varepsilon(A) = \varepsilon(||W||.||W^{-1}||)^{-1}$.); (3) There exists $\varepsilon > 0$ such that $B(T,\varepsilon)\cap S(T)^- \subset S(T)$; (4) Given $A \in S(T)$ there exists $\varepsilon(A) > 0$ such that $B(A,\varepsilon(A))\cap S(T)^- \subset S(T)$; (5) $S(T)$ is an open dense subset of its closure. (Take $0 < \varepsilon < \beta_{k,m}$ to show that (5) => (1). The converse implication is obvious.)

COROLLARY 8.23. *(i) If* T *is a nice Jordan operator, then* $S(T)$ *is locally closed.*

(ii) If T *is a nice Jordan operator, then* $S(\tilde{T})$ *is locally closed in* $A(H)$.

PROOF. (i) By using the same arguments as in the proof of Lemma 7.1, we can reduce the problem to the case when $\sigma(T) = \{0\}$. If rank T is finite (T acting on a finite or an infinite dimensional space), then the result follows from Lemma 2.5, Corollary 2.8 and Proposition 1.12 (i). If T is a nice Jordan nilpotent and rank $T = \infty$, then it follows from Proposition 8.19.

(ii) This follows from Corollaries 8.11 and 8.13. □

PROPOSITION 8.24. *(i) If* $S(T)$ *is locally closed, then* $[T] = \{A \in L(H): A \# T\}$ *coincides with* $S(T)$.

(ii) If $S(\tilde{T})$ *is locally closed, then* $[\tilde{T}] = S(\tilde{T})$.

PROOF. (i) It is completely apparent that $S(T) \subset [T] \subset S(T)^-$ for all $T \in L(H)$.

If $A \# T$, then there exists a sequence $\{A_n\}_{n=1}^{\infty}$ in $S(A)$ such that $\|A_n - T\| \to 0$ $(n \to \infty)$ and $A_n \in S(T)^-$ for all $n = 1,2,\ldots$. If $S(T)$ is locally closed, it readily follows that $A_n \sim T$ for all n large enough. A fortiori, $A \sim A_n \sim T$ (for all n large enough). Therefore, $A \in S(T)$.

The second statement follows by the same argument. □

8.4.2 Nilpotents of order 2

If $T \in N_2(H)$, then either T is nice and similar to $q_1^{(j)} \oplus q_2^{(\infty)}$ for some j, $0 \le j < \infty$, or $T \# q_1^{(\infty)} \oplus q_2^{(\infty)}$, or T is compact. The structure of the poset $(N_2(H)/\#,<)$ is the one given by the following diagram:

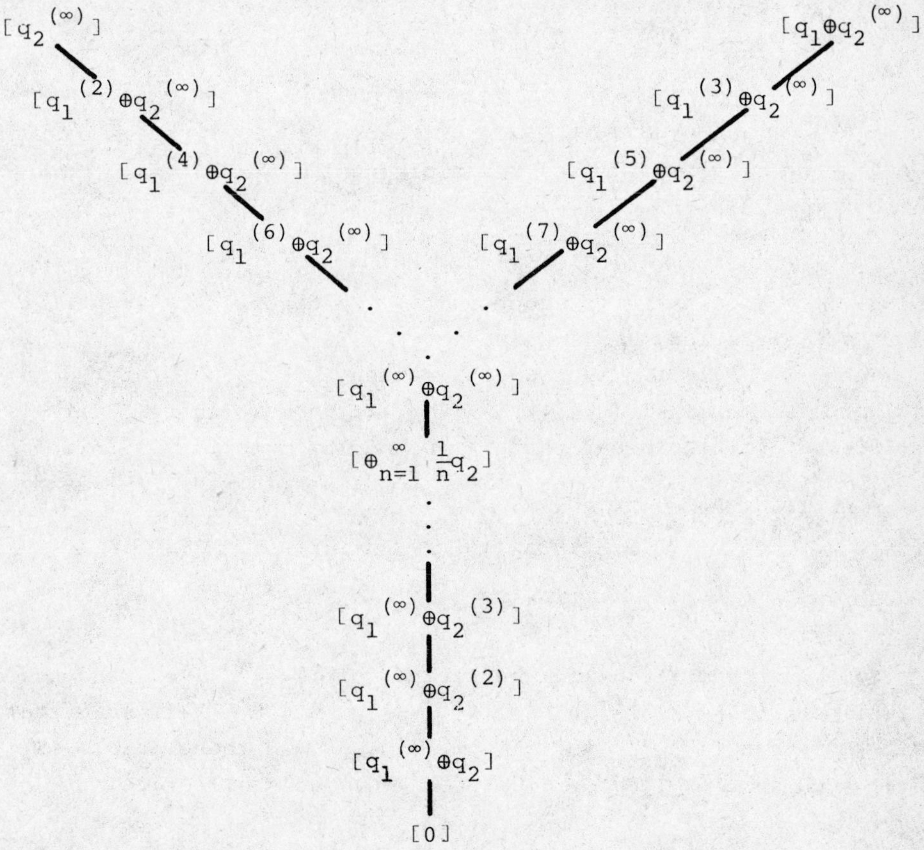

DIAGRAM (8.16)

8.4.3 Quasinilpotent perturbations

The next lemma provides a device to construct quasinilpotent operators T such that $\tilde{T}^k = 0$ and $\tilde{T}^{k-1}+\tilde{T}^*$ is invertible (for some $k \geq 2$).

LEMMA 8.25. *Let* $T_n \in N_k^+(H)$ $(k \geq 2;\ n = 1,2,\dots)$ *be quasinilpotent operators such that* $T_{n+1} \underset{\text{sim}}{\rightarrow} T_n$ *for all* $n = 1,2,\dots$. *Then there exists* $W_n \in G(H)$ *and a quasinilpotent* $T \in N_k^+(H)$ *such that* $\|W_nT_nW_n^{-1} - T\| \to 0$ $(n \to \infty)$ *and* $\{T_n\}_{n=1}^{\infty} \subset S(T)^-$.

PROOF. Let $\varepsilon > 0$ be such that $\tilde{A}^{k-1}+\tilde{A}^*$ is invertible for any $A \in L(H)$ with $\|T_1 - A\| \leq \varepsilon$. We shall recurrently construct $\{S_n, \{R_j^{(n)}\}_{j=1}^{n-1}, \varepsilon_n\}_{n=1}^{\infty}$, where S_n, $R_j^{(n)}$ are invertible operators and $\varepsilon_n > 0$, such that $S_1 = 1$, $\varepsilon_1 = \varepsilon/2$ and for all $n > 1$, we have:

1) $\varepsilon_n < \varepsilon_{n-1}/2$;

2) $\|S_nT_nS_n^{-1} - S_{n-1}T_{n-1}S_{n-1}^{-1}\| < \varepsilon_{n-1}/2$;

3) $\|X - S_nT_nS_n^{-1}\| < \varepsilon_n \Rightarrow \|R_j^{(n)}XR_j^{(n)-1} - T_j\| < 1/n$ $(j = 1,2,\dots,n-1)$.

This will be done as follows. Since $T_n \underset{\text{sim}}{\rightarrow} T_{n-1}$, it is easily seen that we can find $S_n \in G(H)$ satisfying 2) $(n = 2,3,\dots)$. Next we choose $R_j^{(n)}$ such that

$$\|R_j^{(n)}S_nT_nS_n^{-1}R_j^{(n)-1} - T_j\| < 1/n,$$

$j = 1,2,\dots,n-1$. Clearly (by continuity), we can find ε_n such that 1) and 3) are satisfied.

Then $\{S_nT_nS_n^{-1}\}_{n=1}^{\infty}$ is a Cauchy sequence in $N_k^+(H)$; let T denote its limit. Since $\|T - T_1\| < \sum_{n=1}^{\infty} \varepsilon_n \leq \varepsilon$, it follows that $\tilde{T}^{k-1}+\tilde{T}^*$ is invertible. It is also clear that $\tilde{T}^k = 0$ and (by Corollary 1.2(ii)) $\sigma(T) = \{0\}$; moreover, $\|T - S_nT_nS_n^{-1}\| < \sum_{j=n}^{\infty} \varepsilon_j/2 < \varepsilon_n$ so that (by using 3) for $j < n$), we have

$$\|R_j^{(n)}TR_j^{(n)-1} - T_j\| < 1/n.$$

Hence $T \underset{\text{sim}}{\rightarrow} T_j$ for all $j = 1,2,\dots$. □

Some extra notation will be needed. If $\{T_n\}_{n=1}^{\infty}$ is a sequence of quasinilpotent operators in $N_{k,h}^+(H)$ $(k \geq 2,\ 0 \leq h \leq k-1)$ such that $T_{n+1} \underset{\text{sim}}{\rightarrow} T_n$ and $T \underset{\text{sim}}{\rightarrow} T_n$ for all $n = 1,2,\dots$, and there exists $\{W_n\}_{n=1}^{\infty} \subset G(H)$ such that $\|W_nT_nW_n^{-1} - T\| \to 0$ $(n \to \infty)$, we shall write:

$$T_n \underset{\text{sim}}{\uparrow} T$$

The following result summarizes the most immediate properties of "$\underset{\text{sim}}{\uparrow}$".

PROPOSITION 8.26. *(i) Given a quasinilpotent* R *in* $N_{k,h}^+(H)$, *there exist a sequence of nice Jordan nilpotents* $\{T_n\}_{n=1}^{\infty}$ *in* $N_{k,h}^{+}(H)$ *and a sequence* $\{W_n\}_{n=1}^{\infty}$ *in* $G(H)$ *such that* $\|W_nT_nW_n^{-1} - R\| \to 0$ $(n \to \infty)$ *and* $T_{n+1} \underset{sim}{\to} T_n$ *for all* $n = 1,2,\ldots$.

(ii) For every sequence

$$\cdots \underset{sim}{\to} T_{n+1} \underset{sim}{\to} T_n \underset{sim}{\to} \cdots \underset{sim}{\to} T_2 \underset{sim}{\to} T_1$$

of quasinilpotents in $N_{k,h}^{+}(H)$, *there exists a quasinilpotent* $T \in N_{k,h}^{+}(H)$ *such that* $T_n \underset{sim}{\uparrow} T$.

(iii) If $T_n \underset{sim}{\uparrow} T$, *then*

$$S(T)^{-} = [\cup_{n=1}^{\infty} S(T_n)]^{-} . \tag{8.17}$$

(iv) If $T_n \underset{sim}{\uparrow} T$ *for a sequence* $\{T_n\}_{n=1}^{\infty}$ *of quasinilpotents in* $N_{k,h}^{+}(H)$ *and* $T_n \underset{sim}{\not\to} T_{n+1}$ *for infinitely many* n*'s, then* $S(T)$ *is not locally closed. This holds, in particular, if* T_n *is similar to a nice Jordan nilpotent for all* n, *but* T *is not similar to a nice Jordan nilpotent.*

(v) If $T_n \underset{sim}{\uparrow} T$ *and* $A \underset{sim}{\to} T_n$ *for all* $n = 1,2,\ldots$, *then* $A \underset{sim}{\to} T$.

PROOF. (ii) is an immediate consequence of Lemma 8.25 and (iii), (v) and (vi) follow trivially from the definition of " $\underset{sim}{\uparrow}$ ".

(i) From Corollary 7.6, Proposition 7.22 and Corollary 7.24, there exists a sequence $\{R_n\}_{n=1}^{\infty} \subset N_{k,h}^{+}(H)$ of operators similar to nice Jordan nilpotents such that $\|R - R_n\| \to 0$ $(n \to \infty)$.

Define $T_1 = R_1$. By Corollary 8.20, there exists a nice Jordan nilpotent T_n such that $[T_n] = [R_1] \vee [R_2] \vee \ldots \vee [R_n]$ $(n = 2,3,\ldots)$. It is easily seen that $T_{n+1} \underset{sim}{\to} T_n \underset{sim}{\to} R_n$ $(n = 1,2,\ldots)$, whence the result follows.

(iv) Observe that if $T_n \underset{sim}{\not\to} T_{n+1}$ for infinitely many n's, then T cannot be similar to any of the T_n's. This is true, in particular, if T_n is similar to a nice Jordan nilpotent, but T does not have that form. Clearly, $S(T)$ cannot be locally closed. □

REMARK 8.27. Observe that Proposition 8.26(i) and (ii) guarantee that, if $R \in N_{k,h}^{+}(H)$ is a quasinilpotent, then there exists a sequence $\{T_n\}_{n=1}^{\infty}$ of nice Jordan operators in $N_{k,h}^{+}(H)$ satisfying $T_n \underset{sim}{\uparrow} T$ for some quasinilpotent T in $N_{k,h}^{+}(H)$ such that $T \underset{sim}{\to} R$. However, this proposition does not guarantee that the T_n's can be chosen so that $T \# R$, i.e., $T_n \underset{sim}{\uparrow} R$. This is, precisely, the main gap to be filled in order to completely determine the closures of these similarity orbits.

If $J \sim q_k^{(\infty)} \oplus [\oplus_{j=1, j\neq k}^{m} q_j^{(\alpha_j)}] \in N_{k,h}^{+}(H)$ is similar to a nice Jordan nilpotent, then $F = \oplus_{j=1, j\neq k}^{m} q_j^{(\alpha_j)}$ acts on a space of finite dimension equal to $\sum_{j=1, j\neq k}^{m} j\alpha_j = rk+h$ for some $r \geq 0$. Define the number $Stk^m(T)$, the *m-th stable kernel* of T, by

$$Stk^m(T) = \text{nul } F^m - \text{nul}(q_k^{(r)} \oplus q_1^{(h)})^m \quad (m = 1,2,\ldots). \tag{8.18}$$

The following assertions are easy consequences of Lemmas 2.5 and 8.1 and Corollary 7.24:

a) $Stk^m(T)$ does not depend on the particular operator $q_k^{(\infty)} \oplus F$ similar to T. Hence, it is a well-defined similarity invariant;

b) If $T_1, T_2 \in N_{k,h}^+(H)$ are similar to nice Jordan operators, then $T_1 \underset{sim}{\to} T_2$ if and only if $Stk^m(T_1) \leq Stk^m(T_2)$ for all $m = 1,2,\ldots$;

c) If $T, T_n \in N_{k,h}^+(H)$ are similar to nice Jordan nilpotents and $\|T - T_n\| \to 0$ $(n \to \infty)$, then

$\limsup(n \to \infty)\ Stk^m(T_n) \leq Stk^m(T)$, for all $m = 1,2,\ldots$;

Now we shall extend the definition of m-th stable kernel for an arbitrary quasinilpotent T in $N_{k,h}^+(H)$.

$$Stk^m(T) = \inf_{\varepsilon > 0} (\sup\{Stk^m(T') : T' \in N_{k,h}^+(H) \text{ is similar to a nice Jordan nilpotent}, \|T - T'\| < \varepsilon\}). \tag{8.19}$$

By c), it is easily seen that if T is actually similar to a nice Jordan nilpotent, then the new definition (8.19) coincides with the previous one (8.18). It is also clear that this extended m-th stable kernel is a similarity invariant.

PROPOSITION 8.28. *Let* $T, R, T_n, R_n \in N_{k,h}^+(H)$ *be quasinilpotent operators and assume that* T_n, R_n *are similar to nice Jordan nilpotents for all* $n = 1,2,\ldots$; *then*

(i) If $\|T - T_n\| \to 0$ $(n \to \infty)$, *then*

$$\limsup(n \to \infty)\ Stk^m(T_n) \leq Stk^m(T), \text{ for all } m = 1,2,\ldots .$$

(ii) If $T \underset{sim}{\to} R$, *then* $Stk^m(T) \leq Stk^m(R)$ *for all* $m = 1,2,\ldots$.

(iii) If $T_n \underset{sim}{\uparrow} T$, *then* $Stk^m(T) = \inf_n Stk^m(T_n)$, *for all* $m = 1,2,\ldots$.

(iv) If $T_n \underset{sim}{\uparrow} T$ *and* $R_n \underset{sim}{\uparrow} R$, *then* $T \underset{sim}{\to} R$ *if and only if* $Stk^m(T) \leq Stk^m(R)$ *for all* $m = 1,2,\ldots$.

PROOF. (i) follows immediately from c) and (ii) is a consequence of (i).

(iii) Since $T \underset{sim}{\to} T_n$ for all n, it follows that $Stk^m(T) \leq Stk^m(T_n)$ $(n = 1,2,\ldots$; use (ii)). Since T is the limit of a sequence of similarities of the T_n's, we have

$$\limsup(n \to \infty)\ Stk^m(T_n) \leq Stk^m(T) \quad (m = 1,2,\ldots)$$

and $Stk^m(T_n)$ is non-increasing because $T_{n+1} \underset{sim}{\to} T_n$ for each $n = 1,2,\ldots$.

Hence, $\mathrm{Stk}^m(T) = \inf_n \mathrm{Stk}^m(T_n)$ $(m = 1,2,\ldots)$.

(iv) If $T \underset{\text{sim}}{\rightarrow} R$, it follows from (ii) that $\mathrm{Stk}^m(T) \le \mathrm{Stk}^m(R)$ $(m = 1,2,\ldots)$.

Conversely, if $\mathrm{Stk}^m(T) \le \mathrm{Stk}^m(R)$ for all $m = 1,2,\ldots$, and R_n is a nilpotent of order p_n, then we can use (iii) to see that $\mathrm{Stk}^m(R) \le \mathrm{Stk}^m(R_n)$ $(m = 1,2,\ldots)$ and to show that there exists s_n such that $\mathrm{Stk}^m(T_{s_n}) \le \mathrm{Stk}^m(R_n)$ for $1 \le m \le p_n$. Since R_n is a nilpotent of order p_n, this guarantees that $T_{s_n} \underset{\text{sim}}{\rightarrow} R_n$ and therefore $T \underset{\text{sim}}{\rightarrow} R_n$ (for all $n = 1,2,\ldots$). Since $R_n \underset{\text{sim}}{\uparrow} R$, it readily follows that $T \underset{\text{sim}}{\rightarrow} R$. □

It follows from Proposition 8.28 that the conditions $\mathrm{Stk}^m(T) \le \mathrm{Stk}^m(R)$ $(m = 1,2,\ldots)$ are necessary to have $T \underset{\text{sim}}{\rightarrow} R$ and, in the case when $T_n \underset{\text{sim}}{\uparrow} T$ and $R_n \underset{\text{sim}}{\uparrow} R$ for suitable sequences $\{T_n\}_{n=1}^{\infty}$ and $\{R_n\}_{n=1}^{\infty}$ of nice Jordan nilpotents, these conditions are also sufficient. So it is natural to raise the following

CONJECTURE 8.29. If $T, R \in \mathcal{N}_{k,h}^{+}(\mathcal{H})$ are quasinilpotent operators and $\mathrm{Stk}^m(T) \le \mathrm{Stk}^m(R)$ for all $m = 1,2,\ldots$, then $T \underset{\text{sim}}{\rightarrow} R$. Moreover, given T as above there exists a sequence $\{T_n\}_{n=1}^{\infty} \subset \mathcal{N}_{k,h}^{+}(\mathcal{H})$ of nice Jordan operators such that $T_n \underset{\text{sim}}{\uparrow} T$.

An affirmative answer (together with the previous results) could be regarded as a complete solution to the problem of characterizing the closures of the similarity orbits of quasinilpotent operators. By Proposition 8.26, an affirmative answer to this conjecture would also provide an affirmative answer to Conjecture 8.22: Indeed, if $\{A_m\}_{m=1}^{\infty}$ is a denumerable family of quasinilpotents in $\mathcal{N}_{k,h}^{+}(\mathcal{H})$ and $T_{mn} \underset{\text{sim}}{\uparrow} A_m$, $m = 1,2,\ldots$, then we can define T_n be $[T_n] = \vee[T_{ij}]_{1\le i,j\le n}$; then $T_{n+1} \underset{\text{sim}}{\rightarrow} T_n$ for all $n = 1,2,\ldots$, and $T_n \underset{\text{sim}}{\uparrow} T \in \mathcal{N}_{k,h}^{+}(\mathcal{H})$, where $[T] = \vee_m[A_m]$.

Another consequence would be that $\mathcal{S}(T)$ (T as above) cannot be locally closed unless T is similar to a nice Jordan nilpotent.

8.4.4 Universal operators in $\mathcal{N}_{k,h}^{+}(\mathcal{H})$

Let $T \in \mathcal{N}_{k,h}^{+}(\mathcal{H})$ be a quasinilpotent operator. If T is not a nilpotent, then it is easily seen that rank $T^j = \infty-$ for all $j \ge k$. If $T^t = 0$, $T^{t-1} \neq 0$, rank $T^j = \infty-$ for $j = k,k+1,\ldots,r-1$ and rank T^j is finite for $r \le j \le t-1$, then it follows from Lemmas 7.9 and 7.8 that $T \sim R \oplus [\oplus_{j=r+1}^{t}$

$q_j^{(\tau_j)}]$, where $d = \sum_{j=r+1}^{t} j\tau_j < \infty$, $R^r = 0$ and rank $R^j = \infty-$ for $j = k,k+1,k+2,\dots,r-1$; moreover, by (2.2),

$$\tau_j = \text{rank } T^{j-1} - 2\text{rank } T^j + \text{rank } T^{j+1} \ge 0, \text{ for } j = r+1, r+2, \dots, t.$$

(A finite sequence of finite ranks of operators will be called *compatible* if the above conditions are fulfilled.)

It is also clear that $R \in N^+_{k,h'}(H_o)$, where H_o is a subspace of codimension d and h', $0 \le h' \le k-1$, satisfies $d+h' \equiv h \pmod k$.

The following results provide some insight about the structure of the poset $(Q(H) \cap N^+_{k,h}(H)/\#, <)$.

PROPOSITION 8.30. *(i) Given* $k \ge 2$, h $(0 \le h \le k-1)$, $r \ge k+1$ *and a compatible finite sequence* $\{m_j\}_{j=r}^{t-1}$ *of ranks, there exists* $T_u \in N^+_{k,h}(H)$, $T_u^t = 0$ *such that*

$$\text{rank } T_u^j = \begin{cases} \infty, & \text{for } 1 \le j \le k-1 \\ \infty-, & \text{for } k \le j \le r-1 \\ m_j, & \text{for } r \le j \le t-1 \end{cases} \qquad (8.20)$$

and T_u *is a universal operator for the class* $N^+_{k,h}(H;r;\{m_j\}_{j=r}^{t-1}) = \{A \in N_t(H) \cap N^+_{k,h}(H):\ \{\text{rank } A^j\}_{j=1}^{t-1}$ *satisfies* (8.20)$\}$.

(ii) Besides the maximal element $[T_u]$, $(N^+_{k,h}(H;r;\{m_j\}_{j=r}^{t-1})/\#, <)$ *contains a chain* $\{[T_{n,1}]\}_{n \in \mathbb{Z}}$ *order-isomorphic to* $\mathbb{Z}$, *whose supremum is equal to* $[T_u]$.

(iii) If $k > 2$ *and* $r \ge k+2$, *then* $(N^+_{k,h}(H;r;\{m_j\}_{j=r}^{t-1})/\#, <)$ *contains at least two "parallel" chains* $\{[T_{n,1}]\}_{n \in \mathbb{Z}}$ *and* $\{[T_{n,2}]\}_{n \in \mathbb{Z}}$ *order-isomorphic to* $\mathbb{Z}$. *The supremum of each chain is* $[T_u]$, $T_{m,1} \overset{\not\to}{\text{sim}} T_{n,2}$ *for any pair* (m,n) *and* $T_{m,2} \overset{\to}{\text{sim}} T_{n,1}$ *if and only if* $m \ge n$.

(This partial structure of $(N^+_{k,h}(H;r;\{m_j\}_{j=r}^{t-1})/\#, <)$, for the case when $k > 2$ and $r \ge k+2$ is depicted in Diagram (8.21).)

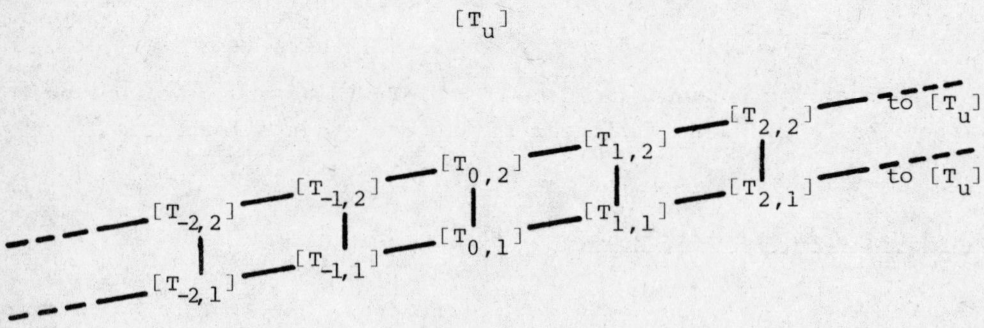

DIAGRAM (8.21)

We shall need the following auxiliary result.

LEMMA 8.31. *(i) If* K *is compact,* $k \geq 2$, $1 \leq p \leq k$, *and*

$$Q_k(K;p) = \begin{pmatrix} 0 & 1 & & & & & & & & & & \\ 0 & 0 & 1 & & & & & & & & & \\ 0 & 0 & 0 & . & & & & & & & & \\ . & . & . & . & . & & & 0 & & & & \\ . & . & . & & . & . & & & & & & \\ 0 & 0 & 0 & . & . & 0 & 1 & & & & & \\ K & 0 & 0 & . & . & 0 & 0 & 1 & & & & \\ 0 & 0 & 0 & . & . & 0 & 0 & 0 & 1 & & & \\ . & . & . & & & . & . & . & . & . & & \\ . & . & . & & & . & . & . & . & . & . & \\ 0 & 0 & 0 & . & . & 0 & 0 & 0 & . & . & 0 & 1 \\ 0 & 0 & 0 & . & . & 0 & 0 & 0 & . & . & 0 & 0 \end{pmatrix} \begin{matrix} H_1 \\ H_2 \\ H_3 \\ . \\ . \\ H_{p-1} \\ H_p \\ H_{p+1} \\ . \\ . \\ H_{k-1} \\ H_k \end{matrix}, \tag{8.22}$$

then $Q_k(K;p) \underset{\text{sim}}{\rightarrow} Q_k(C;p)$ *for each* C *such that* $K \underset{\text{sim}}{\rightarrow} C$.

(ii) If $p < k$, $0 \leq j \leq k-p-1$ *and* $Q_k(K;p,j)$ *is the operator defined by* (8.22) *with* K *in the* (p,1)*-entry replaced by* 0 *and the* 0 *in the* (p+j,j+1)*-entry replaced by* K, *and* K *is quasinilpotent, then* $Q_k(K;p,j) \# Q_k(K;p)$.

(iii) If K_s *is a compact nilpotent of order* $s \geq 2$ *such that* rank $K_s^{\,j} = \infty$ *for* $j = 1,2,\ldots,s-1$, *(namely,* $K_s = \oplus_{n=1}^{\infty} \frac{1}{n} q_s$*), then*

$$q_k^{(\infty)} \oplus [q_{k+p(s-1)} \oplus q_{k-p}^{(s-1)}]^{(n)} \underset{\text{sim}}{\uparrow} Q_k(K_s;p)$$

and

$$\mathrm{Stk}^m(Q_k(K_s;p)) = \begin{cases} 0, & \text{for } m \geq k+p(s-1) \\ -\infty, & \text{for } k-p < m < k+p(s-1) \\ 0, & \text{for } 1 \leq m \leq k-p. \end{cases} \tag{8.23}$$

(In particular, $\mathrm{Stk}^m(Q_k(K_s;k)) = -\infty$ *for* $1 \leq m < ks$.*)*

(iv) If K_u *is a universal compact quasinilpotent (namely,* $K_u = \oplus_{n=1}^{\infty} \frac{1}{n} q_n$*), then*

$$q_k^{(\infty)} \oplus q_{k+p(n-1)} \oplus q_{k-p}^{(n-1)} \underset{\text{sim}}{\uparrow} Q_k(K_u;p)$$

and

$$\mathrm{Stk}^m(Q_k(K_u;p)) = \begin{cases} -\infty, & \text{for } m > k-p \\ 0, & \text{for } 1 \leq m \leq k-p. \end{cases} \tag{8.24}$$

(In particular, $\mathrm{Stk}^m(Q_k(K_u;k)) = -\infty$ *for all* $m \geq 1$.*)*

PROOF. (i) If $W \in G(H_1)$, then $W^{(k)} Q_k(K;p) W^{(k)^{-1}} = Q_k(WKW^{-1};p)$. Now the result follows by taking limits as W runs over suitable sequences in $G(H_1)$.

(ii) Similarly, $K \underset{\text{sim}}{\rightarrow} C$ implies $Q_k(K;p,j) \underset{\text{sim}}{\rightarrow} Q_k(C;p,j)$. Since K is a compact quasinilpotent, it follows from Proposition 8.5 that

$S(K)^-$ is uniquely determined by the finite rank nilpotents that it contains.

If $F = 0\oplus[\oplus_{i=2}^m q_i^{(\alpha_i)}]$ is a finite rank nilpotent, then $Q_k(F;p,j) \simeq \oplus_{i=2}^m Q_k(0\oplus q_i;p,j)^{(\alpha_i)}$ and a straightforward analysis of these operators shows that

$$Q_k(0\oplus q_i;p,j) \sim q_k^{(\infty)}\oplus q_{k+p(i-1)}\oplus q_{k-p}^{(i-1)}, \text{ for } 0 \le j \le k-p-1. \tag{8.25}$$

Hence, $Q_k(F;p,j) \sim Q_k(F;p)$ for $0 \le j \le k-p-1$, whence the result follows.

(iii) The formula $q_k^{(\infty)}\oplus[q_{k+p(s-1)}\oplus q_{k-p}^{(s-1)}]^{(n)} \underset{\text{sim}}{\uparrow} Q_k(K_s;p)$ is an obvious consequence of (i) and (8.25). Now (8.23) follows from Proposition 8.28(iii) and some straightforward computations.

(iv) This follows by the same arguments as in (iii). □

PROOF OF PROPOSITION 8.30. (i) By our previous observations, all the operators in $N_{k,h}^+(H;r;\{m_j\}_{j=r}^{t-1})$ have the form $T \sim R\oplus F$, where $F = \oplus_{j=r+1}^t q_j^{(\tau_j)}$, $\tau_j = m_{j-1}-2m_j+m_{j+1}$ $(j = r+1,r+2,\ldots,t;$ m_t is defined equal to 0) and $R \in N_{k,h'}^+(H';r;\emptyset)$.

By Lemma 2.5 (see also the proof of Corollary 8.2(i), there exists a sequence $\{C_n\}_{n=1}^\infty$ in $L(\mathbb{C}^{kr})$ such that $C_n \sim q_r^{(k)}$ and $\|C_n-q_k^{(r)}\| \to 0$ $(n \to \infty)$.

Define $\mathbf{T}_u = (\oplus_{n=1}^\infty C_n)\oplus q_1^{(h')}\oplus F$.

It is completely apparent from this definition (and Lemma 2.5) that $q_k^{(\infty)}\oplus q_r^{(kn)}\oplus q_1^{(h')}\oplus F \underset{\text{sim}}{\uparrow} T_u$ and therefore, by Proposition 8.26(iii),

$$S(\mathbf{T}_u)^- = [\cup_{n=1}^\infty S(q_k^{(\infty)}\oplus q_r^{(kn)}\oplus q_1^{(h')}\oplus F)]^-. \tag{8.26}$$

Let A be an arbitrary operator in $N_{k,h}^+(H;r;\{m_j\}_{j=r}^{t-1})$. By Proposition 8.26(i) there exists a sequence $\{A_i\}_{i=1}^\infty$ of operators similar to nice Jordan nilpotents such that $\|A - A_i\| \to 0$ $(i \to \infty)$, $A_{i+1} \underset{\text{sim}}{\to} A_i$ $(i = 1,2,\ldots)$ and $\operatorname{rank} A_i^j \le \operatorname{rank} A^j$ for all $j = 1,2,\ldots,t-1$, and for all $i = 1,2,\ldots$.

Clearly, $A_i \sim q_k^{(\infty)}\oplus[\oplus_{j=1}^t q_j^{(\alpha_{ij})}]$, where $\sum_{j=1}^t j\alpha_{ij} \equiv h \pmod k$, $\alpha_{ij} = \tau_j$ for $j = r+1,r+2,\ldots,t$ (for all $i \ge i_o$) and $\alpha_{ik} = 0$ for all $i = 1,2,\ldots$. Fix $i \ge i_o$. Since $\operatorname{rank}[q_r^{(k)}]^p - \operatorname{rank}[q_k^{(r)}]^p \ge 1$ for all $p = 1,2,\ldots,r-1$, it is not difficult to see that there exist $n_i \ge 1$ and $\alpha_i \ge 0$ such that

$$kr\alpha_i+\sum_{j=1}^t j\alpha_{ij} = krn_i+\sum_{j=r+1}^t j\tau_j+h'$$

and

$$\operatorname{rank}(q_k^{(r\alpha_i)}\oplus[\oplus_{j=1}^t q_j^{(\alpha_{ij})}])^p \le \operatorname{rank}(q_r^{(kn_i)}\oplus q_1^{(h')}\oplus[\oplus_{j=r+1}^t q_j^{(\tau_j)}])^p$$

for all $p = 1,2,\ldots,t-1$. (Observe that $n_i - \alpha_i = (kr)^{-1} \sum_{j=1}^{k-1} j\alpha_{ij}+\sum_{j=r+1}^t$

$j(\alpha_{ij}-\tau_j) - h']$.)

By Lemma 2.5, it follows that $q_k^{(\infty)} \oplus q_r^{(kn_i)} \oplus q_1^{(h')} \oplus F \underset{\text{sim}}{\vec{\ }} A_i$.

By (8.26), we conclude that $T_u \underset{\text{sim}}{\vec{\ }} A_i$ (for all $i = 1,2,\ldots$) and, a fortiori, that $T_u \underset{\text{sim}}{\vec{\ }} A$.

Hence T_u is universal for $N^+_{k,h}(H;r;\{m_j\}_{j=r}^{t-1})$.

(ii) Define

$$T_{0,1} = q_1^{(h')} \oplus F \oplus Q_k(K_{r-k+1};1),$$

where $Q_k(K_{r-k+1};1)$ is the operator given by (8.22) with K_{r-k+1} as in (iii) of Lemma 8.31 and

$$T_{n,1} = \begin{cases} T_{0,1} \oplus q_r^{(nk)}, & \text{if } n > 0, \\ T_{0,1} \oplus q_1^{(-nk)}, & \text{if } n < 0. \end{cases}$$

By Lemma 8.31(iii), $T_i = q_k^{(\infty)} \oplus q_r^{(i)} \oplus q_{k-1}^{(i(r-k))} \oplus q_1^{(h')} \oplus F \underset{\text{sim}}{\uparrow} T_{0,1}$, $T_i \oplus q_r^{(nk)} \underset{\text{sim}}{\uparrow} T_{n,1}$ $(n > 0)$ and $T_i \oplus q_1^{(-nk)} \underset{\text{sim}}{\uparrow} T_{n,1}$ $(n < 0)$. By (i) and Lemma 2.5 $T_u \underset{\text{sim}}{\vec{\ }} T_{n+1,1} \underset{\text{sim}}{\vec{\ }} T_{n,1}$ for all $n \in \mathbb{Z}$. It only remains to show that $T_{n,1} \underset{\text{sim}}{\not\vec{\ }} T_{n+1,1}$ for any n.

Formula (8.23) implies that

$$\text{Stk}^{k-1}(T_{n,1}) = c_{k-1}(F) - \begin{cases} n(r-k)(k-1), & \text{if } n \geq 0, \\ nk, & \text{if } n < 0 \end{cases} \tag{8.27}$$

(where $c_{k-1}(F)$ is a constant that only depends on F).

Now the result follows from Proposition 8.28(iv).

(iii) Assume that $k > 2$ and $r - k \geq 2$. Define $T_{n,1}$ as in (ii) and

$$T_{n,2} = T_{n,1} \oplus Q_k(K_2;2).$$

By Lemma 8.31(iii),

$$T_i \oplus q_r^{(nk)} \oplus (q_{k+2} \oplus q_{k-2})^{(i)} \underset{\text{sim}}{\uparrow} T_{n,2} \quad (n \geq 0)$$

and

$$T_i \oplus q_1^{(-nk)} \oplus (q_{k+2} \oplus q_{k-2})^{(i)} \underset{\text{sim}}{\uparrow} T_{n,2} \quad (n < 0).$$

Since $q_{k+2} \oplus q_{k-2} \underset{\text{sim}}{\vec{\ }} q_{k+1} \oplus q_{k-1} \underset{\text{sim}}{\vec{\ }} q_k^{(2)}$ (Lemma 2.5), we can see that $T_u \underset{\text{sim}}{\vec{\ }} T_{n+1,2} \underset{\text{sim}}{\vec{\ }} T_{n,2} \underset{\text{sim}}{\vec{\ }} T_{n,1}$ for each $n \in \mathbb{Z}$.

Since (use (8.23)) $\text{Stk}^{k-1}(T_{n,2}) = -\infty$ for all $n \in \mathbb{Z}$, it follows from (8.27) and Proposition 8.28(ii) that $T_{m,1} \underset{\text{sim}}{\not\vec{\ }} T_{n,2}$ for any pair of integers m, n.

By (8.23)

$$\text{Stk}^{k-2}(T_{n,1}) = \text{Stk}^{k-2}(T_{n,2}) = c_{k-2}(F) - \begin{cases} n(r-k)(k-2), & \text{if } n \geq 0, \\ nk, & \text{if } n < 0 \end{cases}$$

(where $c_{k-2}(F)$ is a constant that only depends on F). By Proposition 8.28(ii),

$$T_{n-1,2} \not\vec{\text{sim}}\ T_{n,1} \quad \text{for any } n \in \mathbb{Z}.$$

On the other hand, since $\text{Stk}^{k-2}(T_{n-1,2}) - \text{Stk}^{k-2}(T_{n,2}) > 0$ for all n, Proposition 8.28(ii) also implies that

$$T_{n-1,2} \not\vec{\text{sim}}\ T_{n,2} \quad \text{for any } n \in \mathbb{Z}. \qquad \square$$

REMARKS 8.32. (i) Let k, h, r and $\{m_j\}_{j=r}^{t-1}$ be as in Proposition 8.30. The same argument of the above proof can be used to show that $(N_{k,h}^{+}(H;r;\{m_j\}_{j=r}^{t-1})/\#,<)$ actually contains s "parallel chains" $\{[T_{n,1}]\}$, $\{[T_{n,2}]\},\ldots,\{[T_{n,s}]\}$ $(n \in \mathbb{Z})$, order-isomorphic to $\mathbb{Z}$, where $s = \max\{r-k,k-1\}$. The supremum of each chain is equal to $[T_u]$ and $T_{n,s}\ \vec{\text{sim}}\ T_{n',s'}$ if and only if $n \geq n'$ and $s \geq s'$.

(ii) The chains of Proposition 8.30(ii) and (iii) are not maximal (except for the case (ii) when $k=2$): If $k \geq 3$ and $n \geq 2$, then $q_{kn}\oplus T_{0,1}\ \vec{\text{sim}}\ q_{kn-1}\oplus q_1\oplus T_{0,1}\ \vec{\text{sim}}\ q_{kn-2}\oplus q_2\oplus T_{0,1}\ \vec{\text{sim}}\ q_{kn-3}\oplus q_3\oplus T_{0,1}\ \vec{\text{sim}} \cdots$ $\cdots\ \vec{\text{sim}}\ q_{kn-(k-2)}\oplus q_{k-2}\oplus T_{0,1}\ \vec{\text{sim}}\ q_{k(n-1)}\oplus T_{0,1}$, and none of these arrows can be reversed. However, $q_{kn-(k-1)}\oplus q_{k+1}\oplus T_{0,1}\ \#\ q_{k(n-1)}\oplus T_{0,1}$.

(iii) Besides these chains, $(N_{k,h}^{+}(H;r;\{m_j\}_{j=r}^{t-1})/\#,<)$ contains many other elements; namely,

$$[q_4\oplus q_1^{(2)}\oplus Q_5(K_2;1)] \text{ and } [q_3^{(2)}\oplus Q_5(K_2;1)]$$

are two *incomparable* elements of $(N_{5,1}^{+}(H;1;\emptyset)/\#,<)$.

The same phenomena appear in the case when we consider quasinilpotent non-nilpotent operators. The proof of the following proposition follows exactly the same lines as the one given for Proposition 8.30 and will be omitted.

PROPOSITION 8.33. *Given* $k \geq 2$ *and* h, $0 \leq h \leq k-1$, *the operator* $T_u = Q_k(K_u;k)\oplus q_1^{(h)}$ *is universal for the class*

$$N_{k,h}^{+}(H;\infty) = \{A \in \mathcal{Q}(H)\cap N_{k,h}^{+}(H):\ \text{rank } A^j = \infty\ \textit{for all}\ j \geq k\}.$$

and satisfies

$$\text{Stk}^m(T_u) = -\infty \ \textit{for all}\ m = 1,2,\ldots .$$

If $T_{0,p} = Q_k(K_u;p)\oplus q_1^{(h)}$, $T_{n,p} = T_{0,p}\oplus q_{nk}$ $(n > 0)$ *and* $T_{n,p} = T_{0,p}\oplus q_1^{(-nk)}$ $(n < 0)$, $p = 1,2,\ldots,k-1$, *then*

$$\text{Stk}^m(T_{n,p}) = \begin{cases} -\infty, & \textit{if } m \geq k-p \\ -m(n+1), & \textit{if } 1 \leq m < k-p \textit{ and } n \geq 0 \\ -n(k-m), & \textit{if } 1 \leq m < k-p \textit{ and } n < 0 \end{cases}$$

and $T_{n',p'}\ \vec{\text{sim}}\ T_{n,p}$ *if and only if* $n' \geq n$ *and* $p' \geq p$.

(This partial structure of $(N^{+}_{k,j}(H;\infty)/\#,<)$ is depicted in Diagram 8.28.)

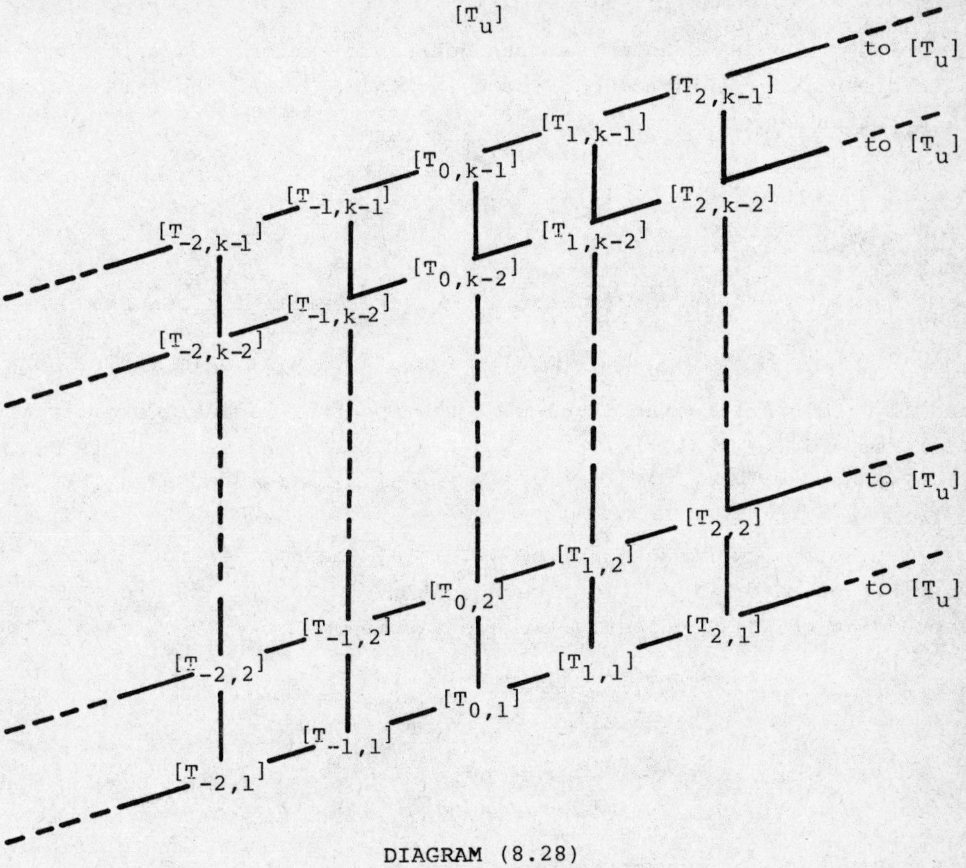

DIAGRAM (8.28)

8.4.5 A general criterion for universality

All universal elements (for the above studied quasinilpotent classes) satisfy $\mathrm{Stk}^{m}(T_u) = -\infty$, for $1 \le m \le k$, and this seems to be the "abstract" characterization of universality. It will be shown that, if $T \in N^{+}_{k,h}(H)$, then T is a universal operator for its own class if and only if $T \underset{\text{sim}}{\to} T \oplus q_{k+1}{}^{(k)}$. In order to make it more precise, we shall establish the result (and proof) as three separate propositions. In the case when T is a quasinilpotent, this result provides some support to the above conjecture about universality.

We shall need two auxiliary results.

LEMMA 8.34. *Let* $T = q_1^{(h)} \oplus q_k^{(\infty)} + K$ *(*$k \geq 2$, $0 \leq h \leq k-1$, $K \in K(H)$*). If* $sp(T) < \varepsilon$ *(for some* $\varepsilon > 0$*), then there exist* $W_\varepsilon \in G(H)$ *and* $m = m(\varepsilon)$ *such that* $\|W_\varepsilon T W_\varepsilon^{-1} - q_1^{(h+km(\varepsilon))} \oplus q_k^{(\infty)}\| < 2\varepsilon$*; moreover,* $\|W_\varepsilon\|\|W_\varepsilon^{-1}\|$ *only depends on* $\phi(r) = \max\{\|(\lambda - T)^{-1}\|: |\lambda| = r\}$ *(*$r > \varepsilon$*).*

PROOF. Let P_m denote the orthogonal projection of H onto the finite dimensional subspace corresponding to $q_1^{(h)}$ and the first m copies of q_k; then

$$T = \begin{pmatrix} A_{11}^{(m)} & A_{12}^{(m)} \\ A_{21}^{(m)} & q_k^{(\infty)} + A_{22}^{(m)} \end{pmatrix} \begin{matrix} H_m \\ H \ominus H_m \end{matrix},$$

where $H_m = \text{ran } P_m$.

Since $P_m \to 1$ (strongly, as $m \to \infty$), it readily follows that

$$\|A_{12}^{(m)}\| + \|A_{21}^{(m)}\| + \|A_{22}^{(m)}\| = \eta_m \to 0 \ (m \to \infty)$$

and (by the upper semicontinuity of the spectral radius; Corollary 1.2 (ii)) $sp(A_{11}^{(m)}) < (3/2)\varepsilon$ for all $m > m_0(\varepsilon)$. Furthermore, it is also clear that $\|(\lambda - A_{11}^{(m)})^{-1}\| < 2\phi(r)$ for all $r > (3/2)\varepsilon$ and all $m > m_1(\varepsilon) > m_0(\varepsilon)$.

By Rota's construction (see (3.5) in Remark 3.34) there exist invertible operators $V_m \in G(H_m)$ such that $\|V_m\|.\|V_m^{-1}\| < M$, where M only depends on $\phi((3/2)\varepsilon)$ (Clearly, we can assume that $\|V_m\|,\|V_m^{-1}\| < M^{1/2}$), and

$$\|V_m A_{11}^{(m)} V_m^{-1}\| < (3/2)\varepsilon.$$

Define $W_m \in G(H)$ by $W_m|H_m = V_m$, $W_m|H_m^{\perp} = 1$; then

$$W_m T W_m^{-1} = \begin{pmatrix} V_m A_{11}^{(m)} V_m^{-1} & V_m A_{12}^{(m)} \\ A_{21}^{(m)} V_m^{-1} & q_k^{(\infty)} + A_{22}^{(m)} \end{pmatrix} \simeq q_1^{(h+km)} \oplus q_k^{(\infty)} + E(m;\varepsilon),$$

where $E(m;\varepsilon)$ is a compact operator such that $\|E(m;\varepsilon)\| < (3/2)\varepsilon + M^{1/2}\eta_m$ ($m > m_1(\varepsilon)$). Since $\eta_m \to 0$ ($m \to \infty$), we can find $m(\varepsilon) > m_1(\varepsilon)$ such that $M^{1/2}\eta_m < \varepsilon/2$ for all $m \geq m(\varepsilon)$.

Define $W_\varepsilon = W_{m(\varepsilon)}$. It is completely apparent that W_ε and $m = m(\varepsilon)$ satisfy all our requirements. □

LEMMA 8.35. *Let* $L = q_1^{(h)} \oplus Q(\oplus_{n=1}^{\infty} R_n; 1)$ *be the operator defined by* (8.22) *with* K *replaced by* $\oplus_{n=1}^{\infty} R_n$, *where* $R_n = \lambda_n + |\lambda_n| J_n$, J_n *is a Jordan nilpotent acting on a space of (finite) dimension* d_n *(*$n = 1,2,\ldots$*) and* $\{\lambda_n\}_{n=1}^{\infty}$ *is a sequence of distinct non-zero complex numbers converging to* 0. *It* T_u *is a universal operator for the class* $N_{k,0}^{+}(H;k+1;\emptyset)$, *then* $L \oplus T_u \underset{sim}{\to} L \oplus Q_u$, *where* Q_u *is a universal operator for the class* $N_{k,0}^{+}(H;\infty)$.

PROOF. By Proposition 8.33 and Lemma 8.31, it is enough to show that

$$L \oplus T_u \underset{\text{sim}}{\to} L \oplus q_k^{(\infty)} \oplus q_{km}$$

for all $m = 1, 2, \ldots$ Clearly, for all practical purposes we can (and shall!) assume that $h = 0$, i.e., that the direct summand $q_1^{(h)}$ is absent. Define

$$\hat{R}_n = \begin{pmatrix} R_n & 1_n & & & & \\ & 0 & 1_n & & 0 & \\ & & 0 & \cdot & & \\ & & & \cdot\;\cdot & & \\ & & & & \cdot\;\cdot & \\ & 0 & & & \cdot\;\cdot & \\ & & & & 0 & 1_n \\ & & & & & 0 \end{pmatrix}$$

($k \times k$ operator matrix; 1_n denotes the identity on a space of dimension d_n) and let $D_j = \{n\colon\ d_n \equiv j (\mathrm{mod}\ k)\}$, $j = 0, 1, \ldots, k-1$. Clearly, it is possible to write $\oplus_{n=1}^{\infty} R_n = R_o \oplus [\oplus_{n=1}^{\infty} S_n]$, where $R_o = \oplus\{R_n\colon\ n \in D_j$ for some j such that D_j is a finite set$\}$ acts on a finite dimensional space, S_n acts on a space of finite dimension kr_n, and $\sigma(R_o) \cap \sigma(S_n) = \sigma(S_{n'}) \cap \sigma(S_{n''}) = \emptyset$ for all n, n', n'', $n' \neq n''$ (each S_n is the sum of k R_m's such that the corresponding dimensions d_m are all congruent modulus k).

It is completely apparent that L can be written as $L = \hat{R}_o \oplus L_o$, where $L_o = \oplus_{n=1}^{\infty} \hat{S}_n$ and $\hat{R}_o$ and $\hat{S}_n$ are defined in the same way as the operator matrix $\hat{R}_n$.

This reduces our task to showing that

$$(\oplus_{n=1}^{\infty} \hat{S}_n) \oplus T_u = L_o \oplus T_u \underset{\text{sim}}{\to} L_o \oplus q_k^{(\infty)} \oplus q_{km} \quad (m = 1, 2, \ldots).$$

Let e_n be a unit vector in the space of S_n such that $S_n e_n = \mu_n e_n$, where μ_n is the eigenvalue of S_n of largest modulus (by definition of S_n, $\mu_r \neq 0$ for all $r = 1, 2, \ldots$, and $\mu_{r'} \neq \mu_{r''}$ if $r' \neq r''$). By Lemma 3.36,

$$\hat{S}_n = \begin{pmatrix} \mu_n & * \\ 0 & S_n' \end{pmatrix} \begin{matrix} \bigvee\{e_n\} \\ \{e_n\}^{\perp} \end{matrix} \underset{\text{sim}}{\to} \mu_n \oplus S_n'$$

and it is straightforward to check that there exists $S_n'' \simeq q_{k-1} \oplus q_k^{(c_n)}$ ($c_n = k^2 r_n - k$) such that $\|S_n' - S_n''\| \leq 2|\mu_n|$.

Fix m and let $p > 1$; then (by Proposition 4.21) $L_o \simeq_a q_k^{(\infty)} \oplus L_o$ and

$$q_k^{(\infty)} \oplus L_o = q_k^{(\infty)} \oplus (\oplus_{n=1}^{\infty} \hat{S}_n) = q_k^{(\infty)} \oplus (\oplus_{n=1}^{p} \hat{S}_n) \oplus (\oplus_{n=p+1}^{p+km} \hat{S}_n) \oplus (\oplus_{n=p+km+1}^{\infty} \hat{S}_n) \underset{\text{sim}}{\to} q_k^{(\infty)} \oplus (\oplus_{n=1}^{p} \hat{S}_n) \oplus [(\mathrm{diag}\{\mu_{p+1}, \mu_{p+2}, \ldots, \mu_{p+km}\}) \oplus (\oplus_{n=p+1}^{p+km} S_n')]$$
$$\oplus (\oplus_{n=p+km+1}^{\infty} \hat{S}_n) \simeq [(\oplus_{n=p+1}^{p+km} S_n') \oplus q_k^{(\infty)}] \oplus [(\oplus_{n=1}^{p} \hat{S}_n) \oplus q_k^{(k^2 \sum_{n=p+1}^{p+km} r_n)}$$

$$\oplus(\oplus_{n=p+km+1}^{\infty}\hat{S}_n)]\oplus(\operatorname{diag}\{\mu_{p+1},\mu_{p+2},\dots,\mu_{p+km}\}) \sim [(\oplus_{n=p+1}^{p+km} S''_n)\oplus q_k^{(\infty)}$$

$$-(\oplus_{n=p+1}^{p+km}\{S''_n - S'_n\})\oplus 0]\oplus[(\oplus_{n=1}^{p}\hat{S}_n)\oplus(\oplus_{n=p+1}^{p+km}\{\hat{S}_n - R_n\oplus 0\oplus 0\oplus\dots\oplus 0\})$$

$$\oplus(\oplus_{n=p+km+1}^{\infty}\hat{S}_n)]\oplus(\operatorname{diag}\{\mu_{p+1},\mu_{p+2},\dots,\mu_{p+km}\})$$

$$\sim [(\oplus_{n=p+1}^{p+km} S''_n)\oplus q_k^{(\infty)} - (\oplus_{n=p+1}^{p+km}\{S''_n - S'_n\})\oplus 0]\oplus[(\oplus_{n=1}^{\infty}\hat{S}_n) - (\oplus_{n=1}^{p} 0)$$

$$\oplus(\oplus_{n=p+1}^{p+km} R_n\oplus 0\oplus 0\oplus\dots\oplus 0)\oplus(\oplus_{n=p+km+1}^{\infty} 0)]$$

$$\oplus\begin{pmatrix} \mu_{p+1} & 1 & & & & & \\ & \mu_{p+2} & 1 & & & 0 & \\ & & \mu_{p+3} & \cdot & & & \\ & & & \cdot & \cdot & & \\ & & & & \cdot & \cdot & \\ & 0 & & & & \cdot & \cdot \\ & & & & & \mu_{p+km-1} & 1 \\ & & & & & & \mu_{p+km} \end{pmatrix}$$

$$\simeq q_{k-1}^{(km)}\oplus q_k^{(\infty)}\oplus L_0\oplus q_{km} - E(p;m),$$

where $E(p;m) \simeq [(\oplus_{n=p+1}^{p+km}\{S''_n - S'_n\})\oplus 0]\oplus[0\oplus(\oplus_{n=p+1}^{p+km} R_n\oplus 0\oplus 0\oplus\dots\oplus 0)\oplus 0]\oplus(\operatorname{diag}\{\mu_{p+1},\mu_{p+2},\dots,\mu_{p+km}\})$ is a finite rank operator such that

$$\|E(p;m)\| \le 2\max\{sp(S_n):\ n > p\} \to 0\ (p\to\infty)$$

Hence,

$$q_k^{(\infty)}\oplus L \underset{\text{sim}}{\to} [q_{km}\oplus q_k^{(\infty)}]\oplus L\oplus q_{k-1}^{(km)}$$

for all $m = 1,2,\dots$. Since T_u is a universal operator for the class $N_{k,0}^{+}(H;k+1;\emptyset)$, it readily follows that

$$L\oplus T_u \simeq_a L\oplus q_k^{(\infty)}\oplus T_u \underset{\text{sim}}{\to} [q_{km}\oplus q_k^{(\infty)}]\oplus L\oplus q_{k-1}^{(km)}\oplus q_{k+1}^{(km)}\oplus q_k^{(\infty)}$$

$$\underset{\text{sim}}{\to} [q_{km}\oplus q_k^{(\infty)}]\oplus L\oplus q_k^{(2)}\oplus q_k^{(\infty)} \simeq [q_{km}\oplus q_k^{(\infty)}]\oplus L \quad (m=1,2,\dots).$$

The proof is complete now. □

PROPOSITION 8.36. *Let* $T \in N_{k,h}^{+}(H)$ *and assume that* $\sigma(T)$ *is the union of* $\{0\}$ *and a sequence* $\{\lambda_n\}_{n=1}^{\infty}$ *of infinitely many distinct non-zero complex numbers converging to* 0. *Then the following are equivalent*

(i) $S^{+}(T) = \{A \in N_{k,h}^{+}(H):\ \sigma(A) = \sigma(T)$ *and* $\dim H(\lambda_n;A) = \dim H(\lambda_n;T)$ *and* $T|H(\lambda_n;T) \underset{\text{sim}}{\to} A|H(\lambda_n;A)$ *for all* $n = 1,2,\dots\}$ *(i.e.,* T *is universal for its class).*

(ii) $T \# T\oplus Q_u$, *where* Q_u *is a universal operator for the class* $N_{k,0}^{+}(H;\infty)$.

(iii) $T \# T\oplus T_u$, *where* T_u *is a universal operator for the class*

$N^{+}_{k,0}(H;k+1;\emptyset)$.

(iv) $T \underset{\text{sim}}{\to} T\oplus Q_u$ *(Q_u as in (ii)).*

(v) $T \underset{\text{sim}}{\to} T\oplus T_u$ *(T_u as in (iii)).*

(vi) $T \underset{\text{sim}}{\to} q_{k+1}^{(k)}\oplus T$.

PROOF. (i) => (ii) <=> (iv) => (iii) <=> (v) => (vi) are trivial implications. On the other hand, it is clear that if $T \underset{\text{sim}}{\to} T\oplus q_{k+1}^{(k)}$ then (use Proposition 4.21(ii)) we also have $T \underset{\text{sim}}{\to} T\oplus q_k^{(\infty)}\oplus q_{k+1}^{(k)} \underset{\text{sim}}{\to} T\oplus q_k^{(\infty)}\oplus q_{k+1}^{(2k)} \underset{\text{sim}}{\to} \cdots \underset{\text{sim}}{\to} T\oplus q_k^{(\infty)}\oplus q_{k+1}^{(km)}$ $(m=1,2,\ldots)$. By using Lemma 8.31 and Proposition 8.30, we see that $T \underset{\text{sim}}{\to} T\oplus T_u$.

Since (by Lemma 8.35) $T\oplus T_u \underset{\text{sim}}{\to} T\oplus Q_u$, we conclude that (ii) - (vi) are equivalent statements. Thus, in order to complete the proof it only remains to show that $T\oplus Q_u$ is a universal operator for the class of T. This will be done in two steps.

1) Clearly, there exists a unique (up to unitary equivalence) operator $L = q_1^{(h)}\oplus Q(\oplus_{n=1}^{\infty} R_n;1) \in N^{+}_{k,h}(H)$ such that $\sigma(L) = \sigma(T)$ and $L|H(\lambda_n;L) \sim T|H(\lambda_n;T)$ for all $n=1,2,\ldots$. It will be shown that $T\oplus Q_u \underset{\text{sim}}{\to} L\oplus Q_u$. Observe that $T\oplus Q_u \underset{\text{sim}}{\to} T\oplus q_k^{(\infty)}\oplus T_u\oplus Q_u$ and (with the notation of Lemma 8.35)

$$T\oplus T_u\oplus q_k^{(\infty)} \simeq (T\oplus q_k^{(d_o+1)})\oplus(q_k^{(\infty)}\oplus T_u) \underset{\text{sim}}{\to} T_o\oplus R_o\oplus q_{k-1}^{(d_o)}\oplus q_1^{(d_o+k)}$$

$$\oplus(q_k^{(\infty)}\oplus T_u) \sim q_1^{(h)}\oplus\hat{R}_o\oplus[T_o\oplus q_1^{(d_o+k-h)}]\oplus[q_k^{(\infty)}\oplus T_u],$$

where $T_o = T|H(\sigma(T)\setminus\sigma(R_o);T)$.

It is completely apparent that $A = T_o\oplus q_1^{(d_o+k-h)} \in N^{+}_{k,0}(H)$, $\sigma(A) = \sigma(T)\setminus\sigma(R_o)$, $A|H(\lambda_n;A) \sim T|H(\lambda_n;T)$ for all $\lambda_n \in \sigma(T)\setminus\sigma(R_o)$ and that our problem has been reduced to showing that

$$A\oplus T_u\oplus q_k^{(\infty)} \underset{\text{sim}}{\to} L_o = \oplus_{n=1}^{\infty} \hat{S}_n.$$

For each m $(m=1,2,\ldots)$, $A \sim (\oplus_{n=1}^{m} S_n)\oplus A_m$, where $\sigma(A_m) = \sigma(A)\setminus\sigma(\oplus_{n=1}^{m} S_m)$. Since S_n acts on a space of dimension kr_n, it is not difficult to see that A_m is similar to $q_k^{(\infty)}+K_m$ (for some compact K_m) and $\text{sp}(A_m) \to 0$ $(m\to\infty)$. Proceeding as above (see also the proof of Lemma 8.35), we see that

$$A\oplus T_u\oplus q_k^{(\infty)} \sim (\oplus_{n=1}^{m} S_n)\oplus q_k^{(k\sum_{n=1}^{m} r_n)}\oplus A_m\oplus T_u\oplus q_k^{(\infty)}$$

$$\underset{\text{sim}}{\to} (\oplus_{n=1}^{m} \hat{S}_n)\oplus[q_1^{(k\sum_{n=1}^{m} r_n)}\oplus A_m]\oplus T_u\oplus q_k^{(\infty)}.$$

By Lemma 8.34, we can find a subspace H_m of the space of $q_1^{(k\sum_{n=1}^{m} r_n)}\oplus A_m$ whose dimension is a (finite) multiple of k and W_m invertible such that

$$W_m[q_1^{(k\sum_{n=1}^m r_n)}\oplus A_m]W_m^{-1} = \begin{pmatrix} A_{11}^{(m)} & A_{12}^{(m)} \\ A_{21}^{(m)} & q_k^{(\infty)}+A_{22}^{(m)} \end{pmatrix} \begin{matrix} H_m \\ H_m^{\perp} \end{matrix}$$

and $\|(A_{ij}^{(m)})_{i,j=1}^2\| \le 2\ sp(A_m)$.

Let $kc_m = \dim H_m$; then $T_u \underset{sim}{\to} T_u \oplus q_{k+1}^{(k(k-1)c_m)}$ and, proceeding as in the proof of Lemma 8.34, we conclude that

$$\operatorname{dist}[q_k^{(\infty)}, S(q_1^{(k\sum_{n=1}^m r_n)}\oplus A_m \oplus q_{k+1}^{(k(k-1)c_m)}] = \eta_m \to 0\ (m\to\infty).$$

A fortiori,

$$\operatorname{dist}[(\oplus_{n=1}^m \hat{S}_n)\oplus q_k^{(\infty)}\oplus T_u, S(A\oplus T_u \oplus q_k^{(\infty)})] = \eta_m \to 0\ (m\to\infty).$$

Since $\|(\oplus_{n=1}^m \hat{S}_n)\oplus q_k^{(\infty)} - L_o\| \le 2\ sp(A_m) \to 0\ (m\to\infty)$, it readily follows from the universality of T_u and Proposition 4.21 that

$$A\oplus T_u \oplus q_k^{(\infty)} \underset{sim}{\to} L_o \oplus T_u.$$

Hence, $T\oplus Q_u \underset{sim}{\to} L\oplus Q_u = q_1^{(h)}\oplus Q(\oplus_{n=1}^\infty R_n;1)\oplus Q_u = q_1^{(h)}\oplus(\oplus_{n=1}^\infty \hat{R}_n)\oplus Q_u$.

2) Let $B = q_1^{(h)}\oplus q_k^{(\infty)}+C$, $C\in K(H)$, be any operator such that $\sigma(B)=\sigma(T)$, $\dim H(\lambda_n;B) = \dim H(\lambda_n;T)$ for all $\lambda_n\in\sigma(T)$ and $T|H(\lambda_n;T) \underset{sim}{\to} B|H(\lambda_n;B)$ for each n, $n=1,2,\dots$. By Proposition 4.21, $B \simeq_a B\oplus q_k^{(\infty)}$. Given $\varepsilon>0$, choose m large enough to guarantee that $d_n \notin \cup\{D_j: D_j$ is a finite set$\}$ and $|\lambda_n| < \varepsilon/4$ for all $n>m$. Then

$$B = \begin{pmatrix} A_\varepsilon & C_\varepsilon \\ 0 & B_\varepsilon \end{pmatrix} \begin{matrix} H(\{\lambda_n\}_{n=1}^m;B) \\ H(\{\lambda_n\}_{n=1}^m;B)^{\perp} \end{matrix}$$

and $sp(B_\varepsilon) < \varepsilon/4$. As in Proposition 3.45, we can find a compact normal operator K_ε, $\|K_\varepsilon\| < \varepsilon/4$, such that $M_\varepsilon = B_\varepsilon - K_\varepsilon$ is quasinilpotent. By Corollary 3.22,

$$\begin{pmatrix} A_\varepsilon & C_\varepsilon \\ 0 & M_\varepsilon \end{pmatrix} \sim A_\varepsilon \oplus M_\varepsilon, \text{ and } \left\| B - \begin{pmatrix} A_\varepsilon & C_\varepsilon \\ 0 & M_\varepsilon \end{pmatrix} \right\| < \varepsilon/4,$$

so that

$$\begin{pmatrix} A_\varepsilon & C_\varepsilon \\ 0 & M_\varepsilon \end{pmatrix} \oplus q_k^{(\infty)} \sim A_\varepsilon \oplus q_k^{(\infty)} \oplus M_\varepsilon.$$

Let $d_\varepsilon = kr_\varepsilon + h_o$ be the dimension of $H(\{\lambda_n\}_{n=1}^m;B)$. Then $3k-h_o+h > k$, $3k+h_o-h > k$ and $A_\varepsilon\oplus q_k^{(\infty)}\oplus q_{3k-h_o+h}\oplus q_{3k+h_o-h}\oplus M_\varepsilon \underset{sim}{\to} A_\varepsilon \oplus q_k^{(\infty)}\oplus q_{3k}^{(2)}\oplus M_\varepsilon$ $\underset{sim}{\to} A_\varepsilon\oplus q_k^{(\infty)}\oplus q_k^{(6)}\oplus M_\varepsilon = A_\varepsilon \oplus q_k^{(\infty)}\oplus M_\varepsilon$ (use Lemma 2.5); the above construction and the universality of Q_u indicate that

$$L\oplus Q_u \underset{sim}{\to} A_\varepsilon \oplus (\oplus_{n=m+1}^\infty \hat{R}_n)\oplus q_{3k-h_o+h}\oplus q_{3k+h_o-h}\oplus M_\varepsilon$$

$$\underset{sim}{\to} A_\varepsilon \oplus(\oplus_{n=m+1}^\infty \hat{R}_n)\oplus q_k^{(6)}\oplus M_\varepsilon.$$

Since $\|q_k^{(\infty)} - \oplus_{n=m+1}^\infty \hat{R}_n\| \le 2\max\{|\lambda_n|: n>m\} < \varepsilon/2$, we see that

$$\operatorname{dist}\left[\begin{pmatrix} A_\varepsilon & C_\varepsilon \\ 0 & M_\varepsilon \end{pmatrix} \oplus q_k^{(\infty)}, S^+(L\oplus \underset{\sim}{Q}_u)\right] < \varepsilon/2.$$

Therefore,

$$\operatorname{dist}[B, S^+(L\oplus \underset{\sim}{Q}_u)] < \varepsilon.$$

Since ε can be chosen arbitrary small, it readily follows that $L\oplus Q_u \underset{\text{sim}}{\to} B$.

The proof is now complete. □

CONJECTURE 8.37. If T is similar to a compact perturbation of $q_1^{(h)} \oplus q_k^{(\infty)}$ (for some $k \geq 2$ and $0 \leq h \leq k-1$) and $\sigma(T)$ contains infinitely many points, then $T \# L\oplus \underset{\sim}{Q}$, where L is the operator defined in the proof of Lemma 8.35 and $\underset{\sim}{Q}$ is a suitable quasinilpotent compact perturbation of $q_k^{(\infty)}$.

PROPOSITION 8.38. *If* $T \in N^+_{k,h}(H;r;\{m_j\}_{j=r}^{t-1})$, *then the following are equivalent*

(i) T *is a universal operator for the class* $N^+_{k,h}(H;r;\{m_j\}_{j=r}^{t-1})$.

(ii) $T \# T\oplus T_u$, *where* T_u *is a universal operator for the class* $N^+_{k,0}(H;k+1;\emptyset)$.

(iii) $T \underset{\text{sim}}{\to} T\oplus T_u$, *where* T_u *has the form of (ii).*

(iv) $T \underset{\text{sim}}{\to} T\oplus q_{k+1}^{(k)}$.

PROOF. (i) => (ii) and the equivalence between (ii), (iii) and (iv) follow exactly as in the first part of the proof of the previous proposition.

(iii) => (i) Combining Lemmas 7.8 and 7.9 we see that, for each m ($m = 1,2,\ldots$), $T \sim F\oplus q_r^{(km)} \oplus T_m$, where $F = \oplus_{j=r+1}^{t} q_j^{(\tau_j)}$ is a nilpotent acting on a finite dimensional space such that $\operatorname{rank} F^j = \operatorname{rank} T^j$ for all $j \geq r$ and $T_m \in N^+_{k,h'}(N;r;\emptyset)$ with $0 \leq h' \leq k-1$, $h' \equiv h - \sum_{j=r+1}^{t} j\tau_j$ (mod k).

Now we can proceed as in the proof of Lemma 8.34 and the first part of the proof of Proposition 8.36 in order to show that

$$T \underset{\text{sim}}{\to} T\oplus T_u \underset{\text{sim}}{\to} F\oplus q_r^{(km)} \oplus T_m \oplus T_u \underset{\text{sim}}{\to} A_m = F\oplus q_r^{(km)} \oplus q_k^{(\infty)}\ (m = 1,2,\ldots).$$

Since, by Proposition 8.30 and Lemma 8.31, $A_m \underset{\text{sim}}{\uparrow} R_u$, where R_u is a universal operator for the class $N^+_{k,h}(H;r;\{m_j\}_{j=r}^{t-1})$, we are done. □

The argument for the proof corresponding to the third case is closely related to the one used in the proof of Lemma 8.1.

PROPOSITION 8.39. *If* $T \in N^+_{k,h}(H;\infty)$, *then the following are equiv-*

alent

(i) T *is a universal operator for the class* $N^+_{k,h}(H;\infty)$.

(ii) $T \# T\oplus Q_u$, *where* Q_u *is a universal operator for the class* $N^+_{k,0}(H;\infty)$.

(iii) $T \# T\oplus T_u$, *where* T_u *is a universal operator for the class* $N^+_{k,0}(H;k+1;\emptyset)$.

(iv) $T \underset{\text{sim}}{\to} T\oplus Q_u$, *where* Q_u *has the form of (ii).*

(v) $T \underset{\text{sim}}{\to} T\oplus T_u$, *where* T_u *has the form of (iii).*

(vi) $T \underset{\text{sim}}{\to} T\oplus q_{k+1}^{(k)}$.

PROOF. Once again, (i) => (ii) and the equivalence between any two of the statements (ii) - (vi) follow exactly in the same way as in the proof of Proposition 8.36.

(iv) => (i) By hypothesis, we can directly assume that T is a quasinilpotent compact perturbation of $q_1^{(h)}\oplus q_k^{(\infty)}$ such that $T^j \neq 0$ for all $j = 1,2,\ldots$.

Let $m \geq 2$ and let e_1 be a unit vector such that $T^j e_1 \neq 0$ for all $j = 1,2,\ldots$ (see the introductory paragraph of Section 8.1.1). Let P_s denote the orthogonal projection of H onto the finite dimensional subspace corresponding to the direct sum of $q_1^{(h)}$ and the first s copies of q_k $(s = 1,2,\ldots)$; then $P_s \to 1$ (strongly, as $s \to \infty$).

Let $\{e_n\}_{n=1}^{\infty}$ be the Gram-Schmidt orthonormalization of the sequence $\{T^{n-1}e_1\}_{n=1}^{\infty}$. It is not difficult to check that T admits a matrix representation of the form

$$T = \left(\begin{array}{cccccccc|c} t_{11} & t_{12} & t_{13} & \cdot & \cdot & \cdot & t_{1,p-1} & t_{1p} & \\ t_{21} & t_{22} & t_{23} & \cdot & \cdot & \cdot & t_{2,p-1} & t_{2p} & \\ & t_{32} & t_{33} & \cdot & \cdot & \cdot & t_{3,p-1} & t_{3p} & \\ & & \cdot & & & & \cdot & \cdot & * \\ & & & \cdot & & & \cdot & \cdot & \\ & 0 & & & \cdot & & \cdot & \cdot & \\ & & & & & \cdot & \cdot & \cdot & \\ & & & & & & t_{p,p-1} & t_{pp} & \\ \hline 0 & 0 & 0 & \cdot & \cdot & \cdot & 0 & t_{p+1,p} & \\ & & & & & & & 0 & \\ & & & & & & & 0 & \\ & & & 0 & & & & \cdot & * \\ & & & & & & & \cdot & \\ & & & & & & & \cdot & \end{array}\right) \begin{array}{l} e_1 \\ e_2 \\ e_3 \\ \cdot \\ \cdot \\ \cdot \\ \cdot \\ e_p \\ \\ \\ \\ N \\ \\ \\ \end{array}, \qquad (8.29)$$

where $p = km+h$ and N is the orthogonal complement of $M = \bigvee\{e_j\}_{j=1}^{p}$.

Since M is finite dimensional, it is not difficult to check that $\|P_M(1-P_s)\| + \|(1-P_s)P_M\| = \varepsilon_s \to 0$ $(s \to \infty)$. This means that M is "almost" orthogonal to ker P_s for all s large enough. Fix $r > m$ so that $\varepsilon_r < 1$; then there exists an invertible operator $W = 1+G$ (for some finite rank operator G) such that $WM \perp \ker P_r$. Thus, for practical purposes (replacing, if necessary, T by WTW^{-1}) we can directly assume that $M \perp \ker P_s$ for all $s > r$.

Given $\varepsilon > 0$, we can proceed as in Lemma 8.34 and Step 1) of the proof of Proposition 8.36 in order to show that

$$T \sim \begin{pmatrix} A & B & C \\ D & E & F \\ 0 & H & q_k^{(\infty)}+J \end{pmatrix} \begin{matrix} M \\ \operatorname{ran} P_{s(\varepsilon)} \ominus M \\ \ker P_{s(\varepsilon)} \end{matrix}$$

(for some $s(\varepsilon) > r$ depending on ε), where

$$A = \begin{pmatrix} a_{11} & a_{12} & a_{13} & \cdot & \cdot & \cdot & a_{1,p-1} & a_{1p} \\ a_{21} & a_{22} & a_{23} & \cdot & \cdot & \cdot & a_{2,p-1} & a_{2p} \\ & a_{32} & a_{33} & \cdot & \cdot & \cdot & a_{3,p-1} & a_{3p} \\ & & \cdot & & & & \cdot & \cdot \\ & & & \cdot & & & \cdot & \cdot \\ & & & & \cdot & & \cdot & \cdot \\ & 0 & & & & \cdot & \cdot & \cdot \\ & & & & & & a_{p,p-1} & a_{pp} \end{pmatrix} \begin{matrix} e_1 \\ e_2 \\ e_3 \\ \cdot \\ \cdot \\ \cdot \\ \cdot \\ e_p \end{matrix}$$

(i.e., the matrix of A looks like the (1,1) block of (8.29)),

$$D = \begin{pmatrix} 0 & 0 & 0 & \cdot & \cdot & \cdot & 0 & a_{p,p+1} \\ & & & & & & & 0 \\ & & & & & & & 0 \\ & & & 0 & & & & \cdot \\ & & & & & & & \cdot \\ & & & & & & & \cdot \end{pmatrix}$$

(i.e., the matrix of D looks like the (2,1) block of (8.29)) and

$$\left\| \begin{pmatrix} A & B & C \\ D & E & F \\ 0 & H & J \end{pmatrix} \right\| < \varepsilon/(8p).$$

Now we can proceed exactly as in the proof of Lemma 8.1 in order to show that

$$T \sim q_{km+h} \oplus q_1^{(k[s(\varepsilon)-m])} \oplus q_k^{(\infty)} + E(m;\varepsilon),$$

where $E(m;\varepsilon)$ is a compact operator such that $\|E(m;\varepsilon)\| < \varepsilon$.

On the other hand (by Corollary 8.2(ii)), given $\eta > 0$ we can find $W_\eta \in G(\mathbb{C}^{k^2})$ such that

$$\|W_\eta [q_1 \oplus q_{k+1}^{(k-1)}] W_\eta^{-1} - q_k^{(k)}\| < \eta.$$

The definitions of Q_u, T_u and the above observation indicate that

$$T \underset{\text{sim}}{\to} T\oplus Q_u \underset{\text{sim}}{\to} T\oplus T_u \underset{\text{sim}}{\to} \{[q_{km+h}\oplus q_1{}^{(k[s(\varepsilon)-m])}\oplus q_k{}^{(\infty)}]+E(m;\varepsilon)\}$$

$$\oplus q_{k+1}{}^{((k-1)k[s(\varepsilon)-m])}\oplus q_k{}^{(\infty)}$$

$$\simeq T_{m,\varepsilon} = \{q_{km+h}\oplus q_k{}^{(\infty)}\oplus[q_1\oplus q_{k+1}{}^{(k-1)}]^{(k[s(\varepsilon)-m])}\}+E'(m;\varepsilon),$$

where $E'(m;\varepsilon) \simeq E(m;\varepsilon)\oplus 0$.

Clearly, we have

$$(1\oplus 1\oplus W_\eta{}^{(k[s(\varepsilon)-m])})T_{m,\varepsilon}(1\oplus 1\oplus W_\eta{}^{(k[s(\varepsilon)-m])})^{-1}$$
$$= q_{km+h}\oplus q_k{}^{(\infty)}\oplus[q_k{}^{(k)}]^{(k[s(\varepsilon)-m])}+E''(m;\varepsilon),$$

where $E''(m;\varepsilon)$ is a compact operator such that

$$\|E''(m;\varepsilon)\| < \eta+\varepsilon(\|W_\eta\|.\|W_\eta{}^{-1}\|).$$

Since ε can be chosen arbitrarily small, we can assume that $\varepsilon(\|W_\eta\|.\|W_\eta{}^{-1}\|) < \eta$, whence it readily follows that

$$\text{dist}[q_{km+h}\oplus q_k{}^{(\infty)}, S^+(T)] < 2\eta.$$

Since η can be chosen arbitrarily small, we conclude that $T \underset{\text{sim}}{\to} T_m = q_{km+h}\oplus q_k{}^{(\infty)}$ $(m=2,3,\ldots)$.

By Proposition 8.30 and Lemma 8.31, $T_m \underset{\text{sim}}{\uparrow} Q_u$, whence the result follows. □

COROLLARY 8.40. *If* $T \in N^+_{k,h}(H)$, *then given* $\varepsilon > 0$ *there exists* $K \in K(H)$, $\|K\| < \varepsilon$, *such that* $T+K$ *is a universal operator for the class of* T.

PROOF. We can directly assume that $T = q_1{}^{(h)}\oplus q_k{}^{(\infty)}+C$ $(k \geq 2,\ 0 \leq h \leq k-1$, C compact). By Proposition 4.21, there exists $K_1 \in K(H)$, $\|K_1\| < \varepsilon/2$, such that $T+K_1 \simeq T\oplus q_k{}^{(\infty)}$. Now it is easy to find a second compact operator K_2, $\|K_2\| < \varepsilon/2$, whose action only modifies the direct summand $q_k{}^{(\infty)}$, such that $T+K \simeq T\oplus T_u$ for some universal operator T_u for the class $N^+_{k,0}(H;k+1;\emptyset)$, where $K = K_1+K_2$.

Now the result follows from Propositions 8.36, 8.38 and 8.39.□

8.5 Separation of isolated points of the essential spectrum affiliated with nilpotents

Let $T \in L(H)$ and assume that μ is an isolated point of $\sigma_e(T)$. If $r > 0$ is small enough to guarantee that $D(\mu;r)^- \cap \sigma_e(T) = \{\mu\}$ and $\gamma = \partial D(\mu;r)$ (positively oriented), then

$$\tilde{E}_\mu = \frac{1}{2\pi i}\int_\gamma (\lambda-\tilde{T})^{-1}\,d\lambda$$

is a non-zero idempotent of $A(H)$ and can be lifted to an idempotent E_μ

$\in L(H)$. Let $W \in G(H)$ be such that $P_\mu = WE_\mu W^{-1}$ is an orthogonal projection and let $H = H_\mu \oplus H_\mu^\perp$, where $H_\mu = \text{ran } P_\mu$ and $H_\mu^\perp = \ker P_\mu$. It is easily seen that

$$WTW^{-1} = \begin{pmatrix} A & B \\ C & D \end{pmatrix} \begin{matrix} H_\mu \\ H_\mu^\perp \end{matrix},$$

where B and C are compact operators, $\sigma_e(A) = \{\mu\}$ and $\mu \notin \sigma_e(D)$. Hence $\tilde{T} \sim (\mu + \tilde{Q}_\mu) \oplus \tilde{D}$, where $Q_\mu = A - \mu$ is an essential quasinilpotent.

It will be convenient to introduce the following definition: Let $T \in L(H)$ and $\lambda \in \mathbb{C}$

$$k(\lambda;\tilde{T}) = \begin{cases} 0, & \text{if } \lambda \notin \sigma_e(T), \\ n, & \text{if } \lambda \text{ is an isolated point of } \sigma_e(T) \text{ and } \tilde{Q}_\lambda \text{ (as defined above) is a nilpotent of order } n, \\ \infty, & \text{otherwise}. \end{cases} \tag{8.30}$$

If $2 \le k(\lambda;\tilde{T}) < \infty$ and $\tilde{Q}_\lambda$ is similar to $\pi(q_n^{(\infty)})$ we shall say that λ is an isolated point of $\sigma_e(T)$ affiliated with $q_n^{(\infty)}$. If either $k(\lambda;\tilde{T}) = 1$ or $2 \le k(\lambda;\tilde{T}) < \infty$ and $\tilde{Q}_\lambda$ is not similar to $\pi(q_n^{(\infty)})$, we shall say that λ is affiliated with $q_1^{(\infty)} \oplus q_n^{(\infty)}$. (It is completely apparent that this definition is independent of the particular invertible operator W above chosen.)

From Proposition 6.19, we obtain the following

COROLLARY 8.41. *Let* $T \in L(H)$, *let* $\Lambda = \{\lambda_1, \lambda_2, \dots, \lambda_m\}$ *be a finite set of isolated points of* $\sigma_e(T)$, *let* Ω_j *be the component of* $\sigma_e(T)$ *such that* $\lambda_j \in \Omega_j^-$ *(*Ω_j *either coincides with or is disjoint from* Ω_h, *if* $j \neq h$*) and assume that* $\text{ind}(\lambda - T) \neq 0$ *for all* $\lambda \in \Omega = \cup_{j=1}^m \Omega_j$. *Given* $\varepsilon > 0$, *there exists* $K \in K(H)$, $\|K\| < \varepsilon$, *such that*

$$T - K \sim A \oplus [\oplus_{j=1}^m (\lambda_j + Q_j)]$$

where $\sigma_e(A) \cap \Lambda = \emptyset$, $\text{min.ind}(\lambda - A) = 0$ *for every* $\lambda \in \Omega \cap \Lambda$ *and* $Q_1, Q_2, \dots, Q_m$ *are quasinilpotent operators.*

Furthermore, if $1 \le k(\lambda;T) = n_j < \infty$, *then* Q_j *can be chosen nilpotent of order* n_j, *but* $Q_j^{n_j - 1}$ *cannot be compact for any possible choice of* Q_j, $j = 1, 2, \dots, m$.

PROOF. Assume that $\text{ind}(\lambda - A_1) > 0$ for all $\lambda \in \Omega_1$.

By Proposition 6.19 there exists a compact operator K_1, $\|K_1\| < \varepsilon/3m$, such that

$$T - K_1 = \begin{pmatrix} A_1 & * \\ 0 & R_1 \end{pmatrix} \sim A_1 \oplus R_1, \tag{8.31}$$

where $\sigma(R_1) = \{\lambda_1\}$, $\lambda_1 \notin \sigma_e(A_1)$ and $\text{nul}(\lambda - A_1) = \text{ind}(\lambda - T)$ and $\text{nul}(\lambda - A_1)^* = 0$ for all $\lambda \in \Omega_1$. (If $\text{ind}(\lambda - T) < 0$ for all $\lambda \in \Omega_1$, then we can apply

Proposition 6.19 to T^* in order to obtain the same decomposition with $\sigma(R_1) = \{\lambda_1\}$, $\lambda_1 \notin \sigma_e(A_1)$ and $\mathrm{nul}(\lambda - A_1) = 0$ and $\mathrm{nul}(\lambda - A_1)^* = -\mathrm{ind}(\lambda - T)$ for all $\lambda \in \Omega_1$.)

By Theorem 7.2, $R_1 = R_1' + K_1'$, where $(R_1')^{n_1} = 0$ and K_1' is compact. Now we can repeat the argument of the proof of Proposition 6.19: First replace K_1' by a suitable finite rank operator F_1' so that $\|K_1' - F_1'\| < \varepsilon/3m$ and

$$R_1' + F_1' = \begin{pmatrix} F_1'' & * \\ 0 & R_1'' \end{pmatrix}, \tag{8.32}$$

where F_1'' acts on a finite dimensional space, $\sigma(F_1'') \subset \Omega_1$ and R_1'' is a nilpotent of order n_1. Thus the operator

$$\begin{pmatrix} A_1 & * & * \\ 0 & F_1'' & * \\ 0 & 0 & R_1'' \end{pmatrix} \tag{8.33}$$

differs from T by a compact operator whose norm is less than $2\varepsilon/3m$.

By Theorem 3.49, there exists a compact operator K_3', $\|K_3'\| < \varepsilon/3m$ such that

$$A_1'' = \begin{pmatrix} A_1 & * \\ 0 & F_1'' \end{pmatrix} - K_3'$$

satisfies the properties: (1) $\lambda_1 \notin \sigma_e(A_1'')$, and (2) $\mathrm{nul}(\lambda - A_1'')^* = 0$ for all $\lambda \in \Omega_1$.

It follows from Corollary 3.22 that, if $C_1 = K_1 + (0 \oplus (K_1' - F_1'))_1 + (K_3 \oplus 0)_2$ (where $(\ldots)_1$ corresponds to the decomposition (8.31) and $(\ldots)_2$ corresponds to the decomposition (8.33)), then $C_1 \in K(H)$, $\|C_1\| < \varepsilon/m$ and

$$T - C_1 = \begin{pmatrix} A_1'' & * \\ 0 & R_1'' \end{pmatrix} \sim A_1'' \oplus R_1''.$$

It is completely apparent that $(R_1'')^{n_1 - 1}$ cannot be compact. Now the result follows by an obvious inductive argument. □

PROPOSITION 8.42. *Let* $T \in L(H)$, *let* $\Lambda = \{\lambda_1, \lambda_2, \ldots, \lambda_m\}$ *be a finite set of isolated points of* $\sigma_e(T)$ *such that* $\Lambda \subset$ interior $\sigma(T)$ *and assume that* $1 \le k(\lambda_j; T) = n_j < \infty$ *for each* $j = 1, 2, \ldots, m$. *Given* $\varepsilon > 0$ *there exists* $K \in K(H)$, $\|K\| < \varepsilon$, *such that*

$$T - K \sim T_1 = A \oplus [\oplus_{j=1}^{m} (\lambda_j + Q_j)],$$

where $\sigma_e(A) \cap \Lambda = \emptyset$, $\mathrm{min.ind}(\lambda - T_1)^k = \mathrm{min.ind}(\lambda - T)^k$ *for all* $\lambda \in \rho_{s-F}(T)$ *and all* $k = 1, 2, \ldots$, $\tilde{Q}_j$ *is a nilpotent of order* n_j *for each* $j = 1, 2, \ldots,$ m, *and, moreover*

(i) If $\lambda_j \notin [\rho_{s-F}^{s}(T)]^-$, *then* Q_j *is quasinilpotent;*

(ii) If $\lambda_j \in [\rho^s_{s-F}(T)]^-$, *then*

$\sigma(\lambda_j+Q_j) = \{\lambda_j\}\cup\{\lambda \in \rho^s_{s-F}(T): |\lambda-\lambda_j| \le \tfrac{1}{4}\mathrm{dist}[\lambda_j,\sigma_e(T)\setminus\{\lambda_j\}]\}$;

(iii) If $\lambda_j \notin [\rho^s_{s-F}(T)]^-$ *and* λ_j *is affiliated with* $q_{n_j}^{(\infty)}$, *then* Q_j *is a universal operator for* $N^+_{n_j,0}(H_j;\infty)$;

(iv) If $\lambda_j \in [\rho^s_{s-F}(T)]^-$, λ_j *is affiliated with* $q_{n_j}^{(\infty)}$ *and*

$\sigma(\lambda_j+Q_j) = \{\lambda_j\}\cup\{\lambda_j+\mu_{j,n}\}_{n=1}^{\infty}$ *(*$\{\lambda_j+\mu_{j,n}\}_{n=1}^{\infty} \subset \rho^s_{s-F}(T)$ *only accumulates at* λ_j*), then* $Q_j \# Q_{n_j}(\oplus_{n=1}^{\infty}[\mu_{j,n}+|\mu_{j,n}|J_{j,n}];1)\oplus Q_u$ *(defined as in Propositition 8.36(ii)) for suitably Jordan nilpotents* $J_{j,n}$, $n \ge 1$.

Furthermore, there also exists $R \in L(H)$, $\|R\| < \varepsilon$, *such that*

$$T - K - R \sim T_2 = A\oplus[\oplus_{j=1}^{m} (\lambda_j+R_j)],$$

where

(v) If λ_j *is affiliated with* $q_{n_j}^{(\infty)}$, *then* $R_j = Q_j$;

(vi) If $\lambda_j \notin [\rho^s_{s-F}(T)]^-$ *and* λ_j *is affiliated with* $q_1^{(\infty)}\oplus q_{n_j}^{(\infty)}$, *then* $R_j \sim q_1^{(\infty)}\oplus q_{n_j}^{(\infty)}$;

(vii) If $\lambda_j \in [\rho^s_{s-F}(T)]^-$, λ_j *is affiliated with* $q_1^{(\infty)}\oplus q_{n_j}^{(\infty)}$ *and* $\sigma(\lambda_j+Q_j) = \{\lambda_j\}\cup\{\lambda_j+\mu_{n,j}\}_{n=1}^{\infty}$ *(*$\{\lambda_j+\mu_{n,j}\}_{n=1}^{\infty} \subset \rho^s_{s-F}(T)$ *only accumulates at* λ_j*), then*

$$R_j \sim \{\oplus_{n=1}^{\infty}[\mu_{j,n}+|\mu_{j,n}|J_{j,n}]\}\oplus q_{n_j}^{(\infty)}$$

for suitably chosen Jordan nilpotents $J_{j,n}$, $n = 1,2,\ldots$.

PROOF. As in Corollary 8.41, we can confine ourselves to the case when $\Lambda = \{\lambda_1\}$ (The general case follows by an inductive argument). Furthermore, we can also translate the operator and assume that $\lambda_1 = 0$ is an isolated point of $\sigma_e(T)$ such that $1 \le k(0;\tilde{T}) = k < \infty$.

Let

$$T = \begin{pmatrix} T_r & T_{12} & T_{13} \\ 0 & T_o & T_{23} \\ 0 & 0 & T_\ell \end{pmatrix} \begin{matrix} H_r(T) \\ H_o(T) \\ H_\ell(T) \end{matrix}$$

be the triangular representation (3.8) of T. Applying Corollary 8.41 to T_r and T_ℓ, we can find compact operators K_r, K_ℓ, $\|K_r\| < \varepsilon/3$, $\|K_\ell\| < \varepsilon/3$, such that

$$T_r - K_r = \begin{pmatrix} A_r & * \\ 0 & B_r \end{pmatrix} \quad \text{and} \quad T_\ell - K_\ell = \begin{pmatrix} B_\ell & * \\ 0 & A_\ell \end{pmatrix},$$

where B_r and B_ℓ are nilpotent operators, B_r and $\tilde{B}_r$ have the same order of nilpotency, B_ℓ and $\tilde{B}_\ell$ have the same order of nilpotency, $0 \notin \sigma_e(A_r)\cup\sigma_e(A_\ell)$, $T_r - K_r \sim A_r\oplus B_r$ and $T_\ell - K_\ell \sim B_\ell\oplus A_\ell$. Thus, if $K_1 = K_r\oplus 0\oplus K_\ell$, then K_1 is compact, $\|K_1\| < \varepsilon/3$ and (by Corollary 3.22)

$$T-K_1=\begin{pmatrix} A_r & & & & \\ & B_r & & * & \\ & & T_o & & \\ & 0 & & B_\ell & \\ & & & & A_\ell \end{pmatrix} \sim A_r\oplus\begin{pmatrix} B_r & * & * \\ 0 & T_o & * \\ 0 & 0 & B_\ell \end{pmatrix}\oplus A_\ell, \qquad (8.34)$$

where 0 is an isolated point of the essential spectrum of the second direct summand, not contained in the interior of the spectrum of this operator.

(iii) If 0 is an isolated point of the spectrum of the second direct summand of (8.34), then it follows from the Riesz' decomposition theorem that $T-K_1 \sim A_r\oplus T_1\oplus Q\oplus A_\ell$, where $0 \notin \sigma(T_1)$ and Q is a quasinilpotent such that $\tilde{Q}^k=0$, $\tilde{Q}^{k-1}\neq 0$. If 0 is an isolated point of $\sigma_e(A)$ affiliated with $q_k^{(\infty)}$, then it is easily seen that $\tilde{Q}^{k-1}+\tilde{Q}^*$ is invertible and Q is similar to a compact perturbation of $q_1^{(h)}\oplus q_k^{(\infty)}$ for some h, $0\le h\le k-1$. Clearly, we can directly assume that $Q=q_1^{(h)}\oplus q_k^{(\infty)}+C$ (where C is compact).

Assume that $0\in\sigma(A_r)$ and let P_n be the orthogonal projection in the space of Q onto the span of the subspace corresponding to $q_1^{(h)}$ and the space corresponding to the first n copies of q_k; then

$$\|Q-(P_nQP_n)\oplus q_k^{(\infty)}\| = \|C-(P_nQP_n)\| \to 0 \ (n\to\infty)$$

and P_nQP_n acts on a finite dimensional space.

By Theorem 3.49, there exists a compact operator K_n of arbitrarily small **prescribed** norm such that $A_r\oplus P_nQP_n - K_n$ is smooth. It readily follows that we can find a compact operator K_2, $\|K_2\|<\varepsilon/3$, such that

$$T-K_1-K_2 \sim A_r'\oplus T_1\oplus q_k^{(\infty)}\oplus A_\ell,$$

where A_r' is smooth.

By Corollary 8.40, we can find a compact operator K_3, $\|K_3\|<\varepsilon/3$, such that

$$T-K \sim A\oplus Q_u,$$

where $K=K_1+K_2+K_3\in K(H)$, $\|K\|<\varepsilon/3+\varepsilon/3+\varepsilon/3=\varepsilon$, $A=A_r'\oplus T_1\oplus A_\ell$ and Q_u is a universal operator for $N_{k,0}^{+}(H;\infty)$. (If $0\notin\sigma(A_r)$, then $0\in\sigma(A_\ell)$ and we can apply the above argument to $Q\oplus A_\ell$.)

(i) and (vi) If 0 is isolated in the spectrum of the second direct summand of (8.34), but it is also an isolated point of $\sigma_e(T)$ affiliated with $q_1^{(\infty)}\oplus q_k^{(\infty)}$, then we can use Theorem 7.2 in order to obtain the decomposition $Q=Q_1+C_1$, where $Q_1^k=0$ and C_1 is compact. By Corollary 7.6 we can find a compact operator K_2, $\|K_2\|<\varepsilon/3$, such that $T-K_1-K_2 \sim A_r\oplus T_1\oplus Q_2\oplus F_2\oplus A_\ell$, where $\tilde{Q}_2^k=0$, $\tilde{Q}_2^{k-1}\neq 0$, $\tilde{Q}_2^{k-1}+\tilde{Q}_2^*$ is not invertible, $\sigma(F_2)=\{0\}\subset\sigma(A_r)\cup\sigma(A_\ell)$ and F_2 acts on a finite dimensional space. By Theorem 3.48, there exists a compact operator K_3, $\|K_3\|<\varepsilon/3$, such that $T-K \sim A\oplus Q_2$, where $K=K_1+K_2+K_3 \in K(H)$, $\|K\|<\varepsilon/3+\varepsilon/3+\varepsilon/3=\varepsilon$ and $A=A_r'\oplus T_1\oplus A_\ell$ (where A_r' is a smooth compact perturbation of $A_r\oplus F_2$, in the case when $0\in\sigma(A_r)$), or $A=A_r\oplus T_1\oplus A_\ell'$ (where A_ℓ' is a smooth compact perturbation of $F_2\oplus A_\ell$ in the case when $0\notin\sigma(A_r)$ and $0\in\sigma(A_\ell)$).

Since $\tilde{Q}_2^{k-1}+\tilde{Q}_2^*$ is not invertible, it follows from Proposition 8.5 that there exists $Q_3 \sim q_1^{(\infty)}\oplus q_k^{(\infty)}$ such that $\|Q_2-Q_3\|<\varepsilon$, whence we obtain (vi).

(iv) If 0 is not an isolated point of the spectrum of

$$B=\begin{pmatrix} B_r & * & * \\ 0 & T_o & * \\ 0 & 0 & B_\ell \end{pmatrix},$$

let $\{\mu_n\}_{n=1}^{\infty}$ be an enumeration of the normal eigenvalues of B contained in $D(0;r)^-$, where $r=\frac{1}{4}\text{dist}[0,\sigma_e(T)\setminus\{0\}]$ and let J_n be the canonical Jordan form of the nilpotent $B-\mu_n|H(\mu_n;B)$. By Riesz' decomposition theorem, $B\sim B_o\oplus B_1$, where $\sigma(B_o)=\sigma(B)\cap D(0;r)^-$, $\sigma(B_1)=\sigma(B)\setminus D(0;r)^-$ and $B_o-\mu_n|H(\mu_n;B_o)\sim J_n$ for all $n=1,2,\ldots$.

Clearly, $\tilde{B}_o^{\,k}=0$, $\tilde{B}_o^{\,k-1}\neq 0$; moreover, 0 is an isolated point of $\sigma_e(T)$ affiliated with $q_k^{(\infty)}$ if and only if B_o can be chosen unitarily equivalent to $q_1^{(h)}\oplus q_k^{(\infty)}+C$, where $0\le h\le k-1$ and C is compact. By Proposition 4.21 and Corollary 8.40, there exists a compact operator K_2', $\|K_2'\|<\varepsilon/3$, such that

$$T-K_1-K_2' \sim A_r\oplus(B_o\oplus Q_u\oplus q_k^{(\infty)})\oplus B_1\oplus A_\ell,$$

where $B_o\oplus Q_u \# q_1^{(h)}\oplus Q_k(\oplus_{n=1}^{\infty}[\mu_n+|\mu_n|J_n];1)\oplus Q_u$. It is easily seen that $B_o\oplus Q_u\oplus q_{k-h}\oplus q_k^{(\infty)} \# Q_k(\oplus_{n=1}^{\infty}[\mu_n+|\mu_n|J_n];1)\oplus Q_u$ (See the proof of Proposition 8.36 and that (by using Lemma 2.5) K_2' can be replaced by a compact operator K_2, $\|K_2\|<\varepsilon/3$, such that

$$T-K_1-K_2 \sim A_r\oplus[(B_o\oplus Q_u\oplus q_{k-h}\oplus q_k^{(\infty)})\oplus q_{k+h}]\oplus B_1\oplus A_\ell.$$

Now the proof follows exactly as in case (iii). (If $0\in\sigma(A_r)$, q_{k+h} is "absorbed" in A_r via a small compact perturbation; if $0\notin\sigma(A_r)$, then $0\in\sigma(A_\ell)$ and q_{k+h} is "absorbed" in A_ℓ.)

(vii) Assume that 0 is not isolated in $\sigma(B)$, but it is an isolated point of $\sigma_e(T)$ affiliated with $q_1^{(\infty)}\oplus q_k^{(\infty)}$. Then it follows from Proposition 8.6 that $B_o \# [\oplus_{n=1}^{\infty}(\mu_n+|\mu_n|J_n)]\oplus q_k^{(\infty)}$. Hence, there exists $B_o' \sim [\oplus_{n=1}^{\infty}(\mu_n+|\mu_n|J_n)]\oplus q_k^{(\infty)}$ such that $\|B_o-B_o'\|<\varepsilon$.

It is completely apparent from the above constructions that

$$\text{min.ind}(\lambda-T_1)^k=\text{min.ind}(\lambda-T_2)^k=\text{min.ind}(\lambda-T)^k$$

for all $\lambda\in\rho_{s-F}(T)$ and for all $k=1,2,\ldots$.

The proof of Proposition 8.42 is complete. □

Let T, Λ, ε be as in Proposition 8.42 and let

$$T-K \sim T_1 = A\oplus[\oplus_{j=1}^{m}(\lambda_j+Q_j)]$$

be the decomposition provided by that proposition with $K \in K(H)$, $\|K\| < \varepsilon/3$, where the Q_j's satisfy the conditions (i) - (iv).

If Q_j is similar to a compact perturbation of $q_{n_j}^{(\infty)}$ ($n_j \geq 2$), then by a formal repetition of the arguments of the proof we can find a compact operator C_j of arbitrarily small prescribed norm such that $Q_j - C_j \sim Q_j'\oplus F_j$, where $Q_j' \simeq q_{n_j}^{(\infty)}$, F_j acts on a space whose dimension is a finite multiple of n_j and $\sigma(F_j) \subset \sigma(A)$. If Q_j does not have that form, then we can find C_j as above such that $Q_j - C_j \sim Q_j'\oplus F_j$, where F_j acts on a finite dimensional space, $\sigma(F_j) \subset \sigma(A)$ and $Q_j' \# q_1^{(\infty)}\oplus q_{n_j}^{(\infty)}$.

Hence, there exists $K' \in K(H)$, $\|K'\| < \varepsilon/3$, such that

$$T-K-K' \sim [A\oplus(\oplus_{j=1}^{m} F_j)]\oplus[\oplus_{j=1}^{m}(\lambda_j+Q_j')].$$

Finally, by Theorem 3.48, there exists $K'' \in K(H)$, $\|K''\| < \varepsilon/3$, such that

$$T-C \sim A'\oplus[\oplus_{j=1}^{m}(\lambda_j+Q_j')],$$

where $C = K+K'+K'' \in K(H)$, $\|C\| < \varepsilon$ and A' is a smooth compact perturbation of A. Thus, we have the following

COROLLARY 8.43. *Let* $T \in L(H)$, *let* $\Lambda = \{\lambda_1, \lambda_2, \ldots, \lambda_m\}$ *be a finite set of isolated points of* $\sigma_e(T)$ *such that* $\Lambda \subset$ interior $\sigma(T)$ *and assume that* $1 \leq k(\lambda;\tilde{T}) = n_j < \infty$ *for each* $j = 1,2,\ldots,m$. *Given* $\varepsilon > 0$ *there exists* $C \in K(H)$, $\|C\| < \varepsilon$, *such that*

$$T-C \sim A'\oplus[\oplus_{j=1}^{m}(\lambda_j+Q_j')],$$

where A' *is smooth and either* $n_j \geq 2$ *and* $\tilde{Q}_j'^{\,n_j-1}+\tilde{Q}_j'^*$ *is invertible and* $Q_j' \simeq q_{n_j}^{(\infty)}$, *or* $\tilde{Q}_j'^{\,n_j-1}+\tilde{Q}_j'^*$ *is not invertible and* $Q_j' \# q_1^{(\infty)}\oplus q_{n_j}^{(\infty)}$.

8.6 Notes and remarks

Proposition 8.3 has been independently (and almost simultaneously) proved by C. Apostol [13] and by D. A. Herrero [138]. In fact, the results of [137] and [138] provide a complete characterization of the set of all universal quasinilpotents for $L(H)$ and for each closed bilateral ideal of $L(H)$ strictly larger than $K(H)$, for the case when H is not necessarily separable (provided the ideal admits some universal quasinilpotent.) On the other hand, the characterization of the set of all universal *compact* quasinilpotents (Proposition 8.4) was obtained by C. Apostol in [13]. The proofs given here combine C. Apostol's Lemma 8.1, the technical Corollary 8.2 (essentially contained in [44]) and an argument of [138].

The results contained in Sections 8.2, 8.3 and 8.4 were not published before. In [44], J. Barría and D. A. Herrero extended their previous results about finite rank operators [43] to arbitrary nilpotents and nilpotent elements of the Calkin algebra. Unfortunately, the article contained several errors and the main result did not cover all possibilities as claimed there. Correct proofs of the same results (Indeed, slightly better ones) were independently obtained by C. Apostol and D. Voiculescu in [34], by using a different approach.

The proofs of Propositions 8.5 and 8.6 are essentially the same as the ones given in [44], corrected with the help of the structural results (due to L. A. Fialkow and D. A. Herrero) given in Sections 7.5 and 7.6 of Chapter VII.

The proofs of Propositions 8.12 and 8.17 given here have been taken from [34]. They are more illuminating and "coördinate free" than the corresponding (correct) proofs given in [44]. C. Apostol and D. Voiculescu's approach is heavily based on their Lemma 7.20 and Lemma 8.8, which can also be derived from [182, Lemma 3.3].

Lemma 8.8 can be regarded as the link between the present problem and Brown-Douglas-Fillmore theory. In order to make it more explicit, it will be necessary to introduce some definitions: Let C be a separable C*-algebra with identity. By an extension of $K(H)$ by C we shall mean a pair (E,ϕ), where E is a C*-subalgebra of $L(H)$ containing $K(H)$ and the identity operator and $\phi: E \to C$ is a *-homomorphism with kernel $K(H)$. Such a pair yields the short exact sequence

$$0 \to K(H) \xrightarrow{\iota} E \xrightarrow{\phi} C \to 0,$$

where ι is the inclusion map. Two such extensions (E_1,ϕ_1), (E_2,ϕ_2) are called *strongly equivalent* if there exists a unitary operator $V \in L(H)$ such that the following diagram commutes:

$$\begin{array}{ccc} E_1 & \xrightarrow{\phi_1} & \\ \alpha_V \downarrow & & C \\ E_2 & \xrightarrow{\phi_2} & \end{array} \qquad (\alpha_V(T) = VTV^*). \qquad (8.35)$$

$\mathrm{Ext}^s(C)$ will denote the set of all strongly equivalent classes of extensions of $K(H)$ by C.

(E_1,ϕ_1) and (E_2,ϕ_2) are *weakly equivalent* if there exists a *Fredholm* partial isometry $V \in L(H)$ such that Diagram (8.35) commutes. Let $\mathrm{Ext}(C)$ denote the set of all weakly equivalent classes of extensions of $K(H)$ by C.

If (E,ϕ) is an extension of $K(H)$ by C, then $\pi(E)$ is *-isomorphic to C and this defines a faithful unital *-homomorphism $\tau:C \to A(H)$. Conversely, if $\tau:C \to A(H)$ is a faithful unital *-homomorphism, then $(\pi^{-1}[\tau(C)],\pi)$ defines a extension of C. Thus, we can identify extensions with *-monomorphisms from C into $A(H)$. Two *-monomorphisms τ_1 and τ_2 are weakly equivalent (strongly equivalent, resp.) if and only if there exists a unitary element $\tilde{U} \in A(H)$ (a unitary operator $U \in L(H)$, resp.) such that $\tau_1 = \tilde{U}\tau_2\tilde{U}^*$.

For two *-monomorphisms τ_1 and τ_2 from C into $A(H)$ we define $\tau_1 + \tau_2$ to be the *-monomorphism from C into $A(H\oplus H)$ such that

$$(\tau_1+\tau_2)(x) = \tau_1(x)\oplus\tau_2(x),\ x \in C.$$

Since $H \simeq H\oplus H$, this defines a structure of abelian semigroup in $\mathrm{Ext}(C)$ ($\mathrm{Ext}^s(C)$) by:

$$[\tau_1]+[\tau_2]=[\tau_1+\tau_2] \quad ([\tau_1]_s+[\tau_2]_s=[\tau_1+\tau_2]_s,\ \text{resp.}).$$

In fact, $\mathrm{Ext}(C)$ and $\mathrm{Ext}^s(C)$ are abelian semigroups with identity and, in many cases, they are actually groups. It is also apparent that $[\tau]_s \to [\tau]$ defines an epimorphism of semigroups (of groups, in case they are actually groups) from $\mathrm{Ext}^s(C)$ onto $\mathrm{Ext}(C)$. The reader is referred to [60], [61], [182] and, especially, to [40] and the references given in the different articles contained there.

In particular, if $C = L(\mathbb{C}^k)$ (for some $k \geq 2$), then $\mathrm{Ext}[L(\mathbb{C}^k)] = \{0\}$ and $\mathrm{Ext}^s[L(\mathbb{C}^k)] = \mathbb{Z}_k = \mathbb{Z}/k\mathbb{Z}$, so that the natural epimorphism

$$\mathrm{Ext}^s[L(\mathbb{C}^k)] \to \mathrm{Ext}[L(\mathbb{C}^k)]$$

is a trivial mapping [182,Lemma 3.3].

This interesting connection with the Brown-Douglas-Fillmore theory is not apparent from the proofs of Propositions 8.12 and 8.19 given in [44].

Corollary 8.23 and Proposition 8.24 are due to L. A. Fialkow and D. A. Herrero. These two results are related with the local analysis of the similarity orbits (see [16] and [94]).

The results of Section 8.4.3 belong to C. Apostol and D. Voiculescu [34] and partially fill the gap in the main theorem of [44]. Sections 8.4.4 and 8.4.5 contain recent results of D. A. Herrero. (Actually, the peculiar behavior of the operators $q_{kn}\oplus Q_k(K_u;1)$, where K_u is a universal compact quasinilpotent has been partially analyzed in [44, Section 7].)

Finally, Corollary 8.41 is Corollary 1 of [42] and Proposition 8.42 can be regarded as a nontrivial improvement of Theorem 3 of that reference.

References

1. G. R. Allan On one-sided inverses in Banach algebras of holomorphic vector-valued functions. J. London Math. Soc. 42 (1967) 463-470.
2. G. R. Allan Holomorphic vector-valued functions on a domain of holomorphy. J. London Math. Soc. 42 (1967) 509-513.
3. J. H. Anderson, On compact perturbations of operators. Can. J. Math. 26 (1974) 247-250.
4. C. Apostol Quasitriangularity in Hilbert space. Indiana Univ. Math. J. 22 (1973) 817-825.
5. C. Apostol On the norm-closure of nilpotents. Rev. Roum. Math. Pures et Appl. 19 (1974) 277-282.
6. C. Apostol Matrix models for operators. Duke Math. J. 42 (1975) 779-785.
7. C. Apostol Quasitriangularity in Banach space. I. Rev. Roum. Math. Pures et Appl. 20 (1975) 131-170.
8. C. Apostol Quasitriangularity in Banach space. III. Rev. Roum. Math. Pures et Appl. 20 (1975) 389-410.
9. C. Apostol Hypercommutativity and invariant subspaces. Rev. Roum. Math. Pures et Appl. 17 (1972) 335-339.
10. C. Apostol The correction by compact perturbations of the singular behavior of operators. Rev. Roum. Math. Pures et Appl. 21 (1976) 155-175.
11. C. Apostol On the norm-closure of nilpotents. III. Rev. Roum. Math. Pures et Appl. 21 (1976) 143-153.
12. C. Apostol Inner derivations with closed range. Rev. Roum. Math. Pures et Appl. 21 (1976) 249-265.
13. C. Apostol Universal quasinilpotent operators. Rev. Roum. Math. Pures et Appl. 25 (1980) 135-138.
14. C. Apostol and K. Clancey, Generalized inverses and spectral theory. Trans. Amer. Math. Soc. 215 (1976) 293-300.
15. C. Apostol and K. Clancey, On generalized resolvents. Proc. Amer. Math. Soc. 58 (1976) 163-168.
16. C. Apostol, L. A. Fialkow, D. A. Herrero and D. Voiculescu, Approximation of Hilbert space operators. II. (Boston-London-Toronto: Pitman Publ. Inc., to appear).
17. C. Apostol and C. Foiaş, On the distance to biquasitriangular operators. Rev. Roum. Math. Pures et Appl. 20 (1975) 261-265.
18. C. Apostol, C. Foiaş and C. M. Pearcy, That quasinilpotent operators are norm-limits of nilpotent operators revisited. Proc. Amer. Math. Soc. 73 (1979) 61-64.
19. C. Apostol, C. Foiaş and L. Szidó, Some results on nonquasitriangular operators. Indiana Univ. Math. J. 22 (1973) 1151-1161.
20. C. Apostol, C. Foiaş and D. Voiculescu, Structure spectrale des opérateurs nonquasitriangulaïres. C. R. Acad. Sci. Paris (Sér. A) 276 (1973) 49-52.
21. C. Apostol, C. Foiaş and D. Voiculescu, Some results on nonquasitriangular operators. II. Rev. Roum. Math. Pures et Appl. 18 (1973) 159-181.
22. C. Apostol, C. Foiaş and D. Voiculescu, Some results on nonquasitriangular operators. III. Rev. Roum. Math. Pures et Appl. 18 (1973) 309-324.
23. C. Apostol, C. Foiaş and D. Voiculescu, Some results on nonquasi-

triangular operators. IV. Rev. Roum. Math. Pures et Appl. 18 (1973) 487-514.

24. C. Apostol, C. Foiaş and D. Voiculescu, Some results on nonquasitriangular operators. V. Rev. Roum. Math. Pures et Appl. 18 (1973) 1133-1140.

25. C. Apostol, C. Foiaş and D. Voiculescu, Some results on nonquasitriangular operators. VI. Rev. Roum. Math. Pures et Appl. 18 (1973) 1473-1494.

26. C. Apostol, C. Foiaş and D. Voiculescu, On the norm-closure of nilpotents, II. Rev. Roum. Math. Pures et Appl. 19 (1974) 549-577.

27. C. Apostol and B. B. Morrel, On uniform approximation of operators by simple models. Indiana Univ. Math. J. 26 (1977) 427-442.

28. C. Apostol, C. M. Pearcy and N. Salinas, Spectra of compact perturbations of operators. Indiana Univ. Math. J. 26 (1977) 345-350.

29. C. Apostol and N. Salinas, Nilpotent approximations and quasinilpotent operators. Pac. J. Math. 61 (1975) 327-337.

30. C. Apostol and J. G. Stampfli, On derivation ranges. Indiana Univ. Math. J. 25 (1976) 857-869.

31. C. Apostol and L. Szidó, Ideals in W*-algebras and the function η of A. Brown and C. Pearcy. Rev. Roum. Math. Pures et Appl. 18 (1973) 1151-1170.

32. C. Apostol and D. Voiculescu, On a problem of Halmos. Rev. Roum. Math. Pures et Appl. 19 (1974) 283-284.

33. C. Apostol and D. Voiculescu, Quasitriangularity in Banach space. II. Rev. Roum. Math. Pures et Appl. 20 (1975) 171-179.

34. C. Apostol and D. Voiculescu, Closure of similarity orbits of nilpotent and quasinilpotent Hilbert space operators. (Preprint, not for publication, 1977).

35. N. Aronszajn, Theory of reproducing kernels. Trans. Amer. Math. Soc. 68 (1950) 337-404.

36. N. Aronszajn and K. T. Smith, Invariant subspaces of completely continuous operators. Ann. of Math. 60 (1954) 345-350.

37. W. B. Arveson, An invitation to C*-algebras. Grad. Texts in Math. Vol. 39. (New York-Heidelberg-Berlin: Springer-Verlag, 1976).

38. W. B. Arveson, Notes on extensions of C*-algebras. Duke Math. J. 44 (1977) 329-355.

39. W. B. Arveson and J. Feldman, A note on invariant subspaces. Mich. Math. J. 15 (1968) 61-64.

40. M. F. Atiyah et al., K-theory and operator algebras. Athens, Georgia 1975. Lect. Notes in Math. Vol. 575. (Berlin-Heidelberg-New York: Springer-Verlag, 1977).

41. J. A. Ball Rota's theorem for general functional Hilbert spaces. Proc. Amer. Math. Soc. 64 (1977) 55-61.

42. J. Barría and D. A. Herrero, Closure of similarity orbits of Hilbert space operators. IV: Normal operators. J. London Math. Soc. (2) 17 (1978) 525-536.

43. J. Barría and D. A. Herrero, Closure of similarity orbits of nilpotent operators. I. Finite rank operators. J. Operator Theory 1 (1979) 177-185.

44. J. Barría and D. A. Herrero, Closure of similarity orbits of nilpotent operators. II. (Preprint, not for publication, 1977).

45. S. K. Berberian, Approximate proper vectors. Proc. Amer. Math. Soc. 13 (1962) 111-114.

46. I. D. Berg Index theory for perturbations of direct sums of normal operators and weighted shifts. Can. J. Math. 30

(1978) 1152-1165.
47. I. D. Berg An extension of the Weyl-von Neumann theorem to normal operators. Trans. Amer. Math. Soc. 160 (1971) 365-371.
48. I. D. Berg On approximation of normal operators by weighted shifts. Mich. Math. J. 21 (1974) 377-383.
49. C. A. Berger and B. I. Shaw, Self-commutators of multicyclic hyponormal operators are always trace class. Bull. Amer. Math. Soc. 79 (1973) 1193-1199.
50. C. A. Berger and B. I. Shaw, Intertwining, analytic structure, and the trace norm estimate, Proceedings of a conference on operator theory, Halifax, Nova Scotia 1973. Lect. Notes in Math. Vol. 345. (Berlin-Heidelberg-New York: Springer-Verlag, 1973).
51. A. R. Bernstein and A. Robinson, Solution of an invariant subspace problem of K. T. Smith and P. R. Halmos. Pac. J. Math. 16 (1966) 421-431.
52. R. Bouldin The triangular representation of C. Apostol. Proc. Amer. Math. Soc. 57 (1976) 256-260.
53. R. Bouldin Essential minimum modulus. Indiana Univ. Math. J. (To appear).
54. A. Brown, C.-K. Fong and D. W. Hadwin, Parts of operators on Hilbert space. Illinois J. Math. 22 (1978) 306-314.
55. A. Brown and C. M. Pearcy, Spectra of tensor product of operators. Proc. Amer. Math. Soc. 17 (1966), 162-169.
56. A. Brown and C. M. Pearcy, Compact restrictions of operators. Acta Sci. Math. (Szeged) 32 (1971) 271-282.
57. A. Brown and C. M. Pearcy, Compact restrictions of operators. II. Acta Sci. Math. (Szeged) 33 (1972) 161-164.
58. A. Brown and C. M. Pearcy, Jordan loops and decompositions of operators. Can. J. Math. 29 (1977) 1112-1119.
59. A. Brown and C. M. Pearcy, On the spectra of derivations on norm ideals. (Preprint).
60. L. G. Brown, R. G. Douglas and P. A. Fillmore, Unitary equivalence modulo the compact operators and extensions of C*-algebras, Proceedings of a conference on operator theory, Halifax, Nova Scotia 1973. Lect. Notes in Math. Vol. 345. (Berlin-Heidelberg-New York: Springer-Verlag, 1973) 58-128.
61. L. G. Brown, R. G. Douglas and P. A. Fillmore, Extensions of C*-algebras, operators with compact self-commutator and K-homology. Bull. Amer. Math. Soc. 79 (1973) 973-978.
62. L. G. Brown, R. G. Douglas and P. A. Fillmore, Extensions of C*-algebras and K-homology. Ann. of Math. 105 (1977) 265-324.
63. J. Bunce and N. Salinas, Completely positive maps on C*-algebras and the left matricial spectrum of an operator, Duke Math. J. 43 (1976) 747-774.
64. S. L. Campbell and R. Gellar, On asymptotic properties of several classes of operators. Proc. Amer. Math. Soc. 66 (1977) 79-84.
65. B. Charles Opérateurs linéaires sur un espace de Banach et modules sur un anneau principal. Symposia Math. 23 (1979) 121-143.
66. K. Clancey Seminormal operators. Lect. Notes in Math. Vol. 742. (Berlin-Heidelberg-New York: Springer-Verlag, 1979).
67. I. Colojara and C. Foiaş, Theory of generalized spectral operators. (New York: Gordon and Breach, 1968).
68. R. E. Curto, The spectrum of elementary operators. (Preprint, 1981).
69. C. Davis and P. Rosenthal, Solving linear operator equations.

Can. J. Math. 26 (1974) 1384-1389.
70. D. Deckard, R. G. Douglas and C. M. Pearcy, On the invariant subspaces of quasitriangular operators. Amer. J. Math. 91 (1969) 634-647.
71. J. Dixmier Étude sur les variétés et les opérateurs de Julia avec quelques applications. Bull Soc. Math. France 77 (1949) 11-101.
72. J. Dixmier Les C*-algèbres et leur représentations. (Paris: Gauthier-Villars, 1964).
73. R. G. Douglas, Banach algebras techniques in operator theory. (New York and London: Academic Press, 1972).
74. R. G. Douglas and C. M. Pearcy, Invariant subspaces of nonquasitriangular operators, Proceedings of a conference on operator theory, Halifax, Nova Scotia 1973. Lect. Notes in Math. Vol. 345. (Berlin-Heidelberg-New York: Springer-Verlag, 1973) 13-57.
75 R. G. Douglas and C. M. Pearcy, A note on quasitriangular operators, Duke Math. J. 37 (1970) 177-188.
76. N. Dunford and J. T. Schwartz, Linear operators. Part I: General theory. (New York: Interscience, 1957).
77. N. Dunford and J. T. Schwartz, Linear operators. Part II: Spectral theory. Self-adjoint operators in Hilbert space. (New York-London: Interscience, 1963).
78. G. Edgar, J. Ernest and S.-G. Lee, Weighing operator spectra. Indiana Univ. Math. J. 21 (1971) 61-80.
79. M. B. Embry and M. Rosenblum, Spectra, tensor products, and linear operators. Pac. J. Math. 53 (1974) 95-107.
80. J. Ernest Charting the operator terrain. Memoirs Amer. Math. Soc. Vol. 171. (Providence, Rhode Island: Amer. Math. Soc., 1976).
81. L. A. Fialkow, A note on non-quasitriangular operators. Acta Sci. Math. (Szeged) 36 (1974) 209-214.
82. L. A. Fialkow, A note on non-quasitriangular operators. II. Indiana Univ. Math. J. 23 (1973) 213-220.
83. L. A. Fialkow, The similarity orbit of a normal operator. Trans. Amer. Math. Soc. 210 (1975) 129-137.
84. L. A. Fialkow, A note on direct sums of quasinilpotent operators. Proc. Amer. Math. Soc. 48 (1975) 125-131.
85. L. A. Fialkow, A note on limits of unitarily equivalent operators. Trans. Amer. Math. Soc. 232 (1977) 205-220.
86. L. A. Fialkow, A note on the operator $X \to AX - XB$. Trans. Amer. Math. Soc. 243 (1978) 147-168.
87. L. A. Fialkow, A note on norm ideals and the operator $X \to AX - XB$. Israel J. Math. 32 (1979) 331-348.
88. L. A. Fialkow, Elements of spectral theory for generalized derivations. J. Operator Theory 3 (1980) 89-113.
89. L. A. Fialkow, Elements of spectral theory for generalized derivations. II: The semi-Fredholm domain. Can. J. Math. (To appear).
90. L. A. Fialkow, A note on the range of the operator $X \to AX - XB$. Illinois J. Math. 25 (1981) 112-124.
91. L. A. Fialkow, Essential spectra of elementary operators. (Preprint).
92. L. A. Fialkow, Spectral theory of elementary multiplications. Trans. Amer. Math. Soc. (To appear).
93. L. A. Fialkow and D. A. Herrero, Inner derivations with closed range in the Calkin algebra. (Preprint).
94. L. A. Fialkow and D. A. Herrero, Characterization of Hilbert space operators with similarity cross sections. (Preprint, not for publication, 1978).
95. P. A. Fillmore, J. G. Stampfli and J. P. Williams, On the essential spectrum, the essential numerical range and a

problem of Halmos. Acta Sci. Math. (Szeged) 33 (1972) 179-192.

96. P. A. Fillmore and J. P. Williams, On operator ranges. Adv. in Math. 7 (1971) 254-281.

97. C. Foiaş, C. M. Pearcy and D. Voiculescu, Biquasitriangular operators and quasisimilarity, Linear spaces and approximation. (Proc. Conf. Oberwolfach 1977). (Basel: Birkhäusser-Verlag, 1978) 47-52.

98. C. Foiaş, C. M. Pearcy and D. Voiculescu, The staircase representation of a biquasitriangular operator. Mich. Math. J. 22 (1975) 343-352.

99. C.-K. Fong Some aspects of derivations on $\mathcal{B}(H)$. Seminar Notes. University of Toronto, 1978.

100. C.-K. Fong and A. R. Sourour, On the operator identity $\sum A_k X B_k \equiv 0$. Can. J. Math. 31 (1979) 845-857.

101. M. Fujii, M. Kajiwara, Y. Kato and F. Kubo, Decompositions of operators in Hilbert spaces. Math. Japonicae 21 (1976) 117-120.

102. T. W. Gamelin, Uniform algebras. (Englewood Cliffs, New Jersey: Prentice Hall, 1969).

103. R. Gellar Two sublattices of weighted shifts invariant subspaces. Indiana Univ. Math. J. 23 (1973) 1-10.

104. R. Gellar and L. Page, Limits of unitarily equivalent normal operators. Duke Math. J. 41 (1974) 319-322.

105. F. Gilfeather, Norm conditions on resolvents of similarities of Hilbert space operators and applications to direct sums and integrals of operators. Proc. Amer. Math. Soc. 68 (1978) 44-48.

106. I. C. Gohberg and M. G. Krein, The basic propositions on defect numbers, roots numbers and indices of linear operators. Uspehi Mat. Nauk SSSR 12 (1957) no. 2 (74) 43-118; English transl. Amer. Math. Soc. Transl. (2) 13 (1960) 185-264.

107. I. C. Gohberg and M. G. Krein, Introduction to the theory of linear nonselfadjoint operators. (Moskow: Nauka, 1965); English transl. Transl. of Mathemetical Monographs Vol. 18. (Providence, Rhode Island: Amer. Math. Soc., 1969).

108. S. Grabiner, Nilpotents in Banach algebras. J. London Math. Soc. (2) 14 (1976) 7-12.

109. L. J. Gray Jordan representation for a class of nilpotent operators. Indiana Univ. Math. J. 26 (1977) 57-64.

110. L. J. Gray On bi-quasitriangular operators. Proc. Amer. Math. Soc. 64 (1977) 291-294.

111. D. W. Hadwin, Closures of unitary equivalence classes. Dissertation. (Indiana University, 1975).

112. D. W. Hadwin, An operator-valued spectrum. Indiana Univ. Math. J. 26 (1977) 329-340.

113. D. W. Hadwin, Continuous functions of operators: A functional calculus. Indiana Univ. Math. J. 27 (1978) 113-125.

114. D. W. Hadwin, An asymptotic double commutant theorem for C*-algebras. Trans. Amer. Math. Soc. 244 (1978) 273-297.

115. D. W. Hadwin, Approximating direct integrals of operators by direct sums. Mich. Math. J. 25 (1978) 123-127.

116. D. W. Hadwin, Approximate equivalence and completely positive maps. (Preprint).

117. P. R. Halmos, Introduction to the theory of Hilbert spaces and the theory of spectral multiplicity. (New York: Chelsea Publ. Co., 1951).

118. P. R. Halmos, Invariant subspaces of polynomially compact operators. Pac. J. Math. 16 (1966) 433-437.

119. P. R. Halmos, A Hilbert space problem book. (Princeton, New Jer-

sey: D. Van Nostrand, 1967).
120. P. R. Halmos, Irreducible operators. Mich. Math. J. 15 (1968) 215-223.
121. P. R. Halmos, Quasitriangular operators. Acta Sci. Math. (Szeged) 29 (1968) 283-293.
122. P. R. Halmos, Permutations of sequences and the Schröder-Bernstein theorem. Proc. Amer. Math. Soc. 19 (1968) 509-510.
123. P. R. Halmos, Ten problems in Hilbert space. Bull. Amer. Math. Soc. 76 (1970) 887-933.
124. P. R. Halmos, Capacity in Banach algebras. Indiana Univ. Math. J. 20 (1971) 855-863.
125. P. R. Halmos, Limits of shifts. Acta Sci. Math. (Szeged) 34 (1973) 131-139.
126. P. R. Halmos and L. J. Wallen, Powers of partial isometries. J. Math. Mech. 19 (1970) 657-663.
127. R. E. Harte, Tensor products, multiplication operators and the spectral mapping theorem. Proc. Royal Irish Acad. 73 (1973) 285-302.
128. R. E. Harte, Berberian-Quigley and the ghost of a spectral mapping theorem. Proc. Royal Irish Acad. 78 (1978) 63-68.
129. J. H. Hedlund, Limits of nilpotent and quasinilpotent operators. Mich. Math. J. 19 (1972) 249-255.
130. H. Helson Lectures on invariant subspaces. (New York-London: Academic Press, 1964).
131. D. A. Herrero, Formal Taylor series and complementary invariant subspaces. Proc. Amer. Math. Soc. 45 (1974) 83-87.
132. D. A. Herrero, Normal limits of nilpotent operators. Indiana Univ. Math. J. 23 (1974) 1097-1108.
133. D. A. Herrero, Toward a spectral characterization of the set of norm limits of nilpotent operators. Indiana Univ. Math. J. 24 (1975) 847-864.
134. D. A. Herrero, Erratum: "Toward a spectral characterization of the set of norm limits of nilpotent operators", Indiana Univ. Math. J. 24 (1975) 847-864. Indiana Univ. Math. J. 25 (1976) 593.
135. D. A. Herrero, Norm limits of nilpotent operators and weighted spectra in non-separable Hilbert spaces. Rev. Un. Mat. Argentina 27 (1975) 83-105.
136. D. A. Herrero, Corrigendum: Norm limits of nilpotent operators and weighted spectra in non-separable Hilbert spaces. Rev. Un. Mat. Argentina 27 (1975) 195-196.
137. D. A. Herrero, Universal quasinilpotent operators. Acta Sci. Math. (Szeged) 38 (1976) 291-300.
138. D. A. Herrero, Almost every quasinilpotent Hilbert space operator is a universal quasinilpotent. Proc. Amer. Math. Soc. 71 (1978) 212-216.
139. D. A. Herrero, Clausura de las órbitas de similaridad de operadores en espacios de Hilbert. Rev. Un. Mat. Argentina 27 (1976) 244-260.
140. D. A. Herrero, Closure of similarity orbits of Hilbert space operators. II: Normal operators. J. London Math. Soc. (2) 13 (1976) 299-316.
141. D. A. Herrero, Closure of similarity orbits of Hilbert space operators. III. Math. Ann. 232 (1978) 195-204.
142. D. A. Herrero, A Rota universal model for operators with multiply connected spectrum. Rev. Roum. Math. Pures et Appl. 21 (1976) 15-23.
143. D. A. Herrero, On analytically invariant subpaces and spectra. Trans. Amer. Math. Soc. 233 (1977) 37-44.
144. D. A. Herrero, On the spectra of the restrictions of an operator. Trans. Amer. Math. Soc. 233 (1977) 45-58.
145. D. A. Herrero, The Volterra operator is a compact universal quasi-

nilpotent. Integral Equations and Operator Theory 1 (1978) 580-588.
146. D. A. Herrero, On multicyclic operators. Integral Equations and Operator Theory 1 (1978) 57-102.
147. D. A. Herrero, Quasisimilar operators with different spectra. Acta Sci. Math. (Szeged) 41 (1979) 101-118.
148. D. A. Herrero, Lifting essentially (G_1) operators. Rev. Un. Mat. Argentina 29 (1980) 113-119.
149. D. A. Herrero, Quasidiagonality, similarity and approximation by nilpotent operators. Indiana Univ. Math. J. 30 (1981) 199-233.
150. D. A. Herrero, Unitary orbits of power partial isometries and approximation by block-diagonal nilpotents, Topics in Modern Operator Theory, 5th International Conference on Operator Theory, Timişoara-Herculane (Romania) June 2-11,1980. Operator Theory: Advances and Applications, Vol.2. (Basel: Birkhausser, 1981)171-210.
151. D. A. Herrero, The distance to a similarity-invariant set of operators. Integral Equations and Operator Theory. (To appear).
152. D. A. Herrero and N. Salinas, Operators with disconnected spectra are dense. Bull. Amer. Math. Soc. 78 (1972) 525-536.
153. T. Kato Perturbation theory for linear operators. (New York: Springer-Verlag, 1966).
154. G. R. Luecke, A new proof of a theorem on quasitriangular operators. Proc. Amer. Math. Soc. 36 (1972) 535-536.
155. G. R. Luecke, A note on quasidiagonal and quasitriangular operators. Pac. J. Math. 56 (1975) 179-185.
156. G. R. Luecke, Essentially (G_1) operators and essentially convexoid operators. Illinois J. Math. 19 (1975) 389-399.
157. G. Lumer and M. Rosenblum, Linear operator equations. Proc. Amer. Math. Soc. 10 (1959) 32-41.
158. P. Meyer-Nieberg, Quasitriangulierbare Operatoren und invariante Untervektorraume stetiger linearer Operatoren. Arch. Math. (Basel) 22 (1971) 186-199.
159. C. L. Olsen, A structure theorem for polynomially compact operators. Amer. J. Math. 93 (1971) 686-698.
160. C. M. Pearcy, Some recent developments in operator theory. CBMS 96. (Providence, Rhode Island: Amer. Math. Soc., 1977).
161. C. M. Pearcy and N. Salinas, An invariant subspace theorem. Mich. Math. J. 20 (1973) 21-31.
162. C. M. Pearcy and N. Salinas, Compact perturbations of seminormal operators. Indiana Univ. Math. J. 22 (1973) 789-793.
163. C. M. Pearcy and N. Salinas, Operators with compact self-commutators. Can. J. Math. 26 (1974) 115-120.
164. C. M. Pearcy and N. Salinas, Finite dimensional representations of C*-algebras and the reducing matricial spectra of an operator. Rev. Roum. Math. Pures et Appl. 20 (1975) 567-598.
165. C. M. Pearcy and N. Salinas, The reducing essential matrical spectra of an operator. Duke Math. J. 42 (1975) 423-434.
166. C. M. Pearcy and N. Salinas, Extensions of C*-algebras and the reducing essential matricial spectra of an operator. K-theory and operator algebras, Athens, Georgia 1975. Lect. Notes in Math. 575. (Berlin-Heidelberg-New York: Springer-Verlag, 1977) 96-112.
167. C. R. Putnam, Commutation properties of Hilbert space operators and related topics. Ergebnisse der Math. und ihrer Grenz. 36. (New York: Springer-Verlag, 1967).
168. C. R. Putnam, An inequality for the area of hyponormal spectra. Math. Z. 28 (1971) 473-477.
169. C. R. Putnam, Operators satisfying a G_1-condition. Pac. J. Math.

84 (1979) 413-426.
170. H. Radjavi and P. Rosenthal, The set of irreducible operators is dense. Proc. Amer. Math. Soc. 21 (1969) 256.
171. H. Radjavi and P. Rosenthal, Invariant subspaces. Ergebnisse der Math. und ihrer Grenz. 77. (New York-Heidelberg-Berlin: Springer-Verlag, 1973).
172. C. Rickart General theory of Banach algebras. (Princeton, New Jersey: D. Van Nostrand, 1960).
173. F. Riesz and B. Sz.-Nagy, Functional analysis. (New York: Ungar, 1955).
174. M. Rosenblum, On the operator equation $BX - XA = Q$. Duke Math. J. 23 (1956) 263-269.
175. G.-C. Rota On models for linear operators. Comm. Pure Appl. Math. 13 (1960) 469-472.
176. S. Sakai C*-algebras and W*-algebras. Ergebnisse der Math. und ihrer Grenz. 60. (Berlin-Heidelberg-New York: Springer-Verlag, 1971).
177. N. Salinas Operators with essentially disconnected spectrum. Acta Sci. Math. (Szeged) 33 (1972) 193-205.
178. N. Salinas A characterization of the Browder essential spectrum. Proc. Amer. Math. Soc. 38 (1973) 369-373.
179. N. Salinas Reducing essential eigenvalues. Duke Math. J. 40 (1973) 561-580.
180. N. Salinas Subnormal limits of nilpotent operators. Acta Sci. Math. (Szeged) 37 (1975) 117-124.
181. N. Salinas On the distance to the set of compact perturbations of nilpotent operators. J. Operator Theory 3 (1980) 179-194.
182. N. Salinas Hypoconvexity and essentially n-normal operators. Trans. Amer. Math. Soc. 256 (1979) 325-351.
183. R. Schatten Norm ideals of completely continuous operators. Ergebnisse der Math. und ihrer Grenz. 27, 2nd. ed.. (Berlin-Heidelberg-New York: Springer-Verlag, 1970).
184. A. L. Shields, Weighted shift operators and analytic function theory. Math. Surveys 13. (Providence, Rhode Island: Amer. Math. Soc., 1974) 49-128.
185. M. A. Shubin, On holomorphic families of subspaces of a Banach space. Mat. Issled. (Kishinev) 5, vyp. 4 (18) (1970) 153-165; English tranl., Integral Equations and Operator Theory 2 (197) 407-420.
186. W. Sikonia The von Neumann converse of Weyl's theorem. Indiana Univ. Math. J. 21 (1971) 121-123.
187. M. Smith Spectra of operator equations. Mich. Math. J. 23 (1976) 151-153.
188. R. A. Smucker, Quasidiagonal and quasitriangular operators. Dissertation. (Indiana University, 1973).
189. J. G. Stampfli, The norm of a derivation. Pac. J. Math. 33 (1970) 737-747.
190. J. G. Stampfli, Compact perturbations, normal eigenvalues and a problem of Salinas. J. London Math. Soc. (2) 9 (1974) 165-175.
191. W. Szymanski, Decompositions of operator-valued functions in Hilbert spaces. Stud. Math. 50 (1974) 265-280.
192. D. Voiculescu, Some extensions of quasitriangularity. Rev. Roum. Math. Pures et Appl. 18 (1973) 1303-1320.
193. D. Voiculescu, Some extensions of quasitriangularity. II. Rev. Roum. Math. Pures et Appl. 18 (1973) 1439-1456.
194. D. Voiculescu, Norm-limits of algebraic operators. Rev. Roum. Math. Pures et Appl. 19 (1974) 371-378.
195. D. Voiculescu, A non-commutative Weyl-von Neumann theorem. Rev. Roum. Math. Pures et Appl. 21 (1976) 97-113.
196. D. Voiculescu, A note on quasitriangularity and trace-class self-

commutators. Acta Sci. Math. (Szeged) 42 (1980) 195-199.
197. L. R. Williams, On quasisimilarity of operators on Hilbert space. Dissertation. (University of Michigan, 1976).
198. L. R. Williams, Similarity invariants for a class of nilpotent operators. Acta Sci. Math. (Szeged) 38 (1976) 423-428.
199. F. Wolf On the essential spectrum of partial differential boundary value problems. Comm. Pure Appl. Math. 12 (1959) 211-228.
200 J. Zemánek and B. Aupetit, Local behavior of the spectral radius in Banach algebras, J. London Math. Soc. (2) 23 (1981) 171-178.

LEMMA 8.34. *Let* $T = q_1^{(h)}$

Index

Symbols and notation